T0122965

The Rise of Fishes

The Rise of Fishes

500 Million Years of Evolution

SECOND EDITION

John A. Long

The Johns Hopkins University Press
Baltimore

Printed in China on acid-free paper
9 8 7 6 5 4 3 2 1

The Johns Hopkins University Press
2715 N. Charles Street
Baltimore, Maryland 21218-4363
www.press.jhu.edu

Library of Congress Cataloging-in-Publication Data

Long, John A., 1957–
The rise of fishes : 500 million years of evolution / John A. Long. —2nd ed.
p. cm.
Includes bibliographical references and index.
ISBN-13: 978-0-8018-9695-8 (hardcover : alk. paper)
ISBN-10: 0-8018-9695-9 (hardcover : alk. paper)
1. Fishes—Evolution. I. Title.
QL618.2.L66 2010
597.13'8—dc22 2010009524

A catalog record for this book is available from the British Library.

Special discounts are available for bulk purchases of this book. For more information, please contact Special Sales at 410-516-6936 or specialsales@press.jhu.edu.

The Johns Hopkins University Press uses environmentally friendly book materials, including recycled text paper that is composed of at least 30 percent post-consumer waste, whenever possible. All of our book papers are acid-free, and our jackets and covers are printed on paper with recycled content.

Contents

Foreword

The evolutionary tree of the vertebrates was assembled piecemeal, beginning in the nineteenth century, firstly on the basis of the living species, and, later on, by the insertion of an increasingly large number of extinct fossil species. Fossil vertebrates are relatively rare, compared with mollusks or other invertebrates, but the complexity of their skeletons can be readily compared with that of the living forms and thus interpreted in terms of anatomical relationships, evolutionary transitions, or adaptations. In addition, most vertebrates, notably jawed vertebrates, show a rather good fit between their evolutionary relationships and the fossil record through time.

John Long's new edition of *The Rise of Fishes* provides a wealth of information, with spectacular illustrations, about a segment of vertebrate phylogeny that was once poorly understood; that is, between the emergence of the vertebrates within chordates and the rise of the four-legged vertebrates, or tetrapods. During the twentieth century, most advances in the elucidation of early vertebrate structure and relationships came not only from the exploration of fossil sites worldwide, but also from new preparation and observation techniques. Erik Stensiö, a Swedish paleontologist whom we regard as the father of our modern way of studying fossil fishes, considered that they should no longer be regarded as beautiful collection specimens but rather be prepared, "dissected," or even sliced in the same way as biologists do with extant specimens. He was blamed for the use of such destructive techniques but through these techniques provided the first comparative framework on which our present knowledge of early fish anatomy is based. Fortunately, the new computed tomography technologies, which John Long uses, now allows a nondestructive access to the slightest anatomical features of such millions of years old fossils and sometimes yields precious evidence for soft tissues that are exceptionally mineralized during fossilization.

In the 1950s, German entomologist Willi Hennig developed new concepts and methods for elucidating the relationships between species in an objective and refutable way. Hennig's views, now known as "cladistics" and widely used by comparative biologists, became soon applied in the 1960s to living and fossil fish systematics, thanks to the influence of American and British fish specialists Gareth Nelson and Colin Patterson. *The Rise of Fishes* follows the line of this conceptual framework, which, during the past fifty years, helped in shaping the structure of the vertebrate tree and has allowed us to readily highlight the still unresolved relationships that are due to conflicting or missing data, as in the case of the stem of the jawed vertebrates. Many fish specialists are not familiar with the cladistic principle of classification that imposes the exclusive naming of nested monophyletic and unranked groups, and John Long rightly provides a simplified classification of living and fossil fishes, which agrees with Joe Nelson's widely used *Fishes of the World.*

The Rise of Fishes admirably integrates early vertebrate evolution in time and space on an ever-changing geography and environment. No doubt, it will incite students to plunge into the still obscure, Deep Past of vertebrate history.

Philippe Janvier
Muséum National d'Histoire Naturelle, Paris

Preface to the Second Edition

Almost thirty years ago, when I was a young, green postgraduate student attempting a thesis describing a Devonian ray-finned fish, I was fortunate enough to meet up with three venerable men of fossil fish studies, Brian Gardiner, Peter Forey, and Colin Patterson, for a few pints of ale and vigorous discussions down at the Cranberry Arms hotel, not far from that illustrious temple of paleontological knowledge, the Natural History Museum. I shall always recall Brian's immortal words to me about why every student of zoology should include some time studying fossil fishes. I don't remember the exact words but his advice went something like this: "Once you understand the complexity of the fish skull, and how it came to evolve, the rest of vertebrate anatomy is quite simple!" I've held this view ever since, that once fish left the water and invaded land as early four-legged animals, the rest of vertebrate evolution, from amphibians through to mammals, was just fine-tuning of an existing body plan.

Today many of my colleagues might not agree with this view, or see it as overly simplistic, but it holds true that most of the significant steps in vertebrate evolution did take place within the fishes. From their first appearance as "glorified swimming worms" about half a billion years ago, fishes have since given us the origins of bone, the development of jaws and teeth, and complex skeletal structures that protect the brain and allow powerful muscles to anchor for efficient swimming. Fishes evolved the first strong limbs, advanced sensory systems, inner ear bones, and even copulatory methods of reproduction. Fishes also gave us multichambered hearts, complex brains, lungs, and an ability to breathe air. Many of these vital stages in their evolution were showcased in the first edition of this book, which appeared in print in 1995, and these evolutionary scenarios mostly hold true today.

However, in the 15 years since the first edition of *The Rise of Fishes* appeared, there has been an extraordinary wealth of new publications on the early evolution and diversification of fossil fishes as well as significant major advances in the systematics and evolutionary biology of the most successful of all extant fishes, the teleosteans. The most significant of these new finds include the oldest fossils of the first fishes from China (*Haikouichhtys, Myllokungmingia*), a deeper understanding of many of the early jawless fishes from new studies on osteostracans, galeaspids, thelodonts, and anaspids; the first finds of near complete primitive sharks from the Early Devonian that bridge the gap with other primitive jawed fishes such as 'acanthodians' and placoderms (*Ptomacanthus, Doliodus*); complete early shark braincases (*Pucapumpella*), combined with wonderful descriptions of many new bizarre early sharks (*Akmonistion, Thrinacoides*); placoderm embryos and maternal feeding structures (*Materpiscis*), and the fact that some arthrodires were sexually dimorphic (*Incisoscutum*); the first discoveries of near-complete early bony fishes (osteichthyans) from the Silurian of China (*Guiyu, Psarolepis*); beautiful 3-D skeletons of osteichthyans with intact braincases (*Ligulalepis*); the deep origins of the neopterygians (*Discoserra*); fossils providing data on the critical stages in the early radiation of sarcopterygians (*Psarolepis, Achoania, Styloichthys*) and of the tetrapod-like fishes (*Kenichthys, Goologongia*), as well as new key discoveries that provide evidence for the seamless transition of fishes to early tetrapods (*Panderichthys, Tiktaalik, Ventastega*); and the earliest known land trackways of tetrapods from the Middle Devonian of Poland, revamping our views about the evolutionary split between fishes and amphibians. Furthermore we have uncovered new data on the origins of the world's most successful fishes, the teleosteans and even discovered vital links that show intermediate stages in the evolution of flatfishes, a problem that left Darwin perplexed but is now resolved from new fossil evidence (*Heteronectes*).

These are but a few of the many outstanding fossil fish finds over the past 15 years that have invigorated evolutionary biology. More significantly, these finds have enabled scientists studying them to formulate tighter phylogenetic frameworks and hypotheses of relationships are now emerging that are robust and have stood the test of time with only minor tweaking as new finds have been plugged into them. This informs our development of more substantial classifications, and enables biogeographic patterns back in time to be tested against geological events that have affected life on Earth, such as mass extinction events, major plate tectonic movements, and global shifts in climate.

The knowledge we have gained about early evolution of fishes over the past decade has been of major interest to the wider scientific communities, as evidenced by the high numbers of early fish papers appearing in the prestigious journal *Nature*, where discoveries of new Silurian and Devonian fish that impact on evolutionary hypotheses are almost as numerous as new papers on dinosaurs or mammals. Many of the new fossil discoveries also enable us to more precisely tune the molecular clock divergence times for major evolutionary events. Others shed light on how hox gene expressions observed in living fish may have been drivers of major evolutionary events. Although much attention has been focused on the transition from fishes to tetrapods over the past 15 years or so, the greatest unsolved mystery in vertebrate history would now appear to be solving the enigma of the origin of jaws. We have very few specimens that show any intermediate stages between the jawless radiation of fishes and the first jawed species, any future discoveries that shed light on this great evolutionary step are eagerly anticipated.

With this in mind I have attempted to thoroughly revise the book to include these updates of new finds as well as capturing images of the most important of these new discoveries. It was my intention to keep the tone of this book pitched at a similar level of public accessibility as in the previous edition and to not get too bogged down with detailed scientific discussions. The book presents a relatively simple scientific overview of the origins and early evolution of fishes but carries a detailed bibliography for further reading if the reader wants the full story about certain discoveries and their implications. Cladograms depicted throughout the book are provided to represent current schemes of relationships for major fish groups, but the reader must treat these with caution as new finds will alter positions of certain species.

This book celebrates the joys of scientific discovery, the complexity and beauty of evolution, and the major contribution ancient fishes have given to the world through elucidating some of the biggest steps in vertebrate history. Some people love to collect fossil fishes to marvel at their extraordinary preservation as a thing of intrinsic beauty, while many others research fossil fishes to unlock the mysteries of their and our evolution. No matter what entices you engage with fish fossils, I hope this new edition continues to inspire wonder and curiosity about the nature of early vertebrate evolution and serves to encourage a new generation of budding paleontologists to look farther afield than the terrestrial vertebrate animals for their specialist areas of study.

Acknowledgments

This book has benefited from the input of many of my colleagues who over the years have provided vital dialogue and discussion on the material, reviewing draft chapters or providing images or use of artwork. My sincerest thanks go out to the following: Gavin Young, Ken Campbell, Richard Barwick, Alex Ritchie (all Australian National University, Canberra), Carole Burrow (Queensland Museum, Brisbane), Susan Turner (Brisbane), Philippe Janvier and Daniel Goujet (Museé Nationale de l'Histoire Naturelle, Paris), Zhu Min and Chang Meeman (Institute of Vertebrate Paleontology and Paleoanthropology, Beijing) John Maisey (American Museum of Natural History), Mark Wilson (University of Alberta, Edmonton), Gloria Arratia and Hans-Peter Schultze (University of Kansas, Lawrence), Michael Coates (University of Chicago), Oliver Hampe (Natural History Museum, Berlin), Per Ahlberg (University of Upsalla), Zerina Johanson (Natural History Museum, London), Brian Choo (Museum Victoria), Robert Sansom (University of Leicester), Moya M. Smith (Guys Hospital London), Jenny Clack and Kenneth McNamara (both Cambridge University). I thank my former and recent PhD students, Katherine Trianjstic (Curtin University, Perth), Brian Choo, Tim Holland, and Alice Clement and colleagues David Pickering, Thomas Rich (all Museum Victoria) and Patricia Vickers-Rich (Monash University) for their support and enthusiastic discussion during the writing of this book.

For providing images and use of artworks, I sincerely thank the following (affiliations not repeated if listed above): Gavin Young, Richard Barwick, Zhu Min, Gloria Arratia, Tim Senden (Australian National University, Canberra), Peter Schouten (Marlee, New South Wales), Peter Trusler (Ringwood, Victoria), Oleg Lebedev (Paleontological Institute, Moscow), John Maisey, Michael Coates, Ted Daeschler (Philadelphia Academy of Sciences), Gloria Arratia, Anna Paganoni (Museo Civico di Scienze Naturali, Bergamo), Eileen Grogan, Richard Lund (Carnegie Museum), Fiona Ferguson (Age of Fishes Museum, Canowindra); Ivars Zupins , Erviks Luksevic (both Natural History Museum of Latvia, Riga), Clay Bryce (Western Australian Museum, Perth), Rudy Kuiter, John Broomfield (both Museum Victoria, Melbourne), Martin Brazeau (Museum of Natural History, Berlin), Matt Friedman (Oxford University), David Ward (Orpington, UK), Michal Ginter, Piotr Szrek (both University of Warsaw) and Yoshitaka Yabumoto (Kitakyushu Museum of Natural History and Human History).

The following institutions permitted use of their images in the book under licensing agreements: The Natural History Museum, London; Cambridge University Zoology Museum, Cambridge; The Natural History Museum, Sweden; The Museé Nationale de Histoire Naturélle (Paris); The Natural History Museum of Latvia, Riga; The Cleveland Museum of Natural History, Cleveland; The American Museum of Natural History, New York; The Philadelphia Academy of Sciences, Philadelphia; The Western Australian Museum, Perth; Museum Victoria, Melbourne.

I would also like to thank the editorial team at the Johns Hopkins University Press for their input toward improving the overall quality of this new edition. Finally, I would especially like to thank my wife, Heather Robinson, for her unending support and her much-needed help with typing and editing parts of the manuscript.

The Rise of Fishes

CHAPTER 1

Earth, Rocks, Evolution, and Fish

Background information to understanding fish evolution

Fishes are an integral part of human culture. We eat them, keep them as pets, spend endless hours trying to catch them, fear them when we are in their environment, yet often respect them for their religious symbolism. Fishes have been on this planet for just over half a billion years, and their long history is intricately tied to the origin and developments of Earth's diverse habitats. As continents drifted and collided, so ocean currents shifted and rerouted. As sea levels rose and fell, connections between landmasses opened and closed. The migrations and adaptations of fishes followed such events and eventually ensconced these animals into the various faunal provinces they inhabit today.

To understand the history of fishes is to understand our own origins, as most of the human form first appeared hundreds of million of years ago in ancient fishes. Step by step, as fishes evolved more complex biologies, they became closer to settling on a body plan that would one day empower them to leave the water and invade land. As early land animals, these highly evolved fishes would eventually be capable of laying hard-shelled eggs, later developing fur and upright postures, and one day building cities around the globe. Today the most successful of all backboned animals, fishes comprise some 30,000 living species and account for the greatest biomass of all vertebrates. Their evolution is an epic story set against a dramatic background of changing climates, moving continents, and apocalyptic mass extinction

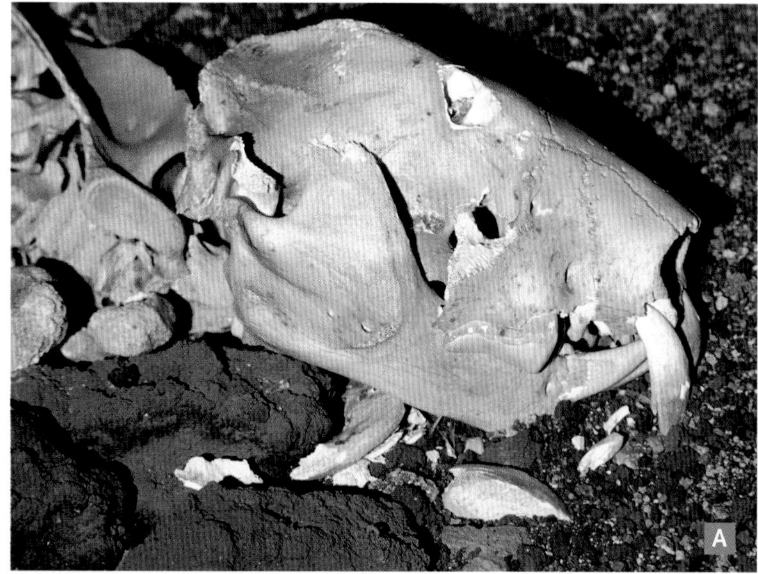

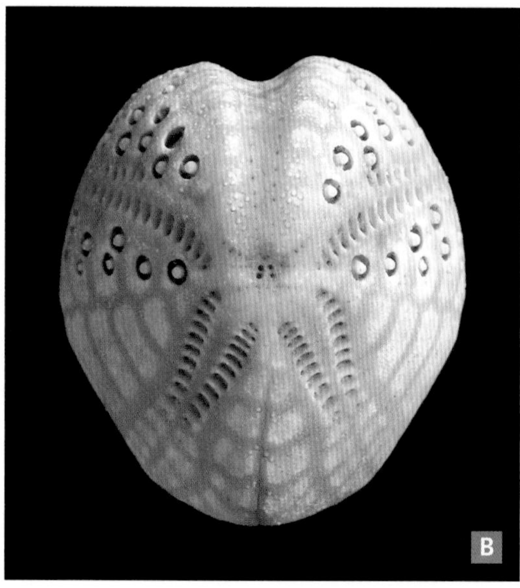

Fossils are the remains of former life on this planet such as bones (*A*, skull of *Thylacoleo*), shells (*B*, sea urchin fossils, *Lovenia*), and leaves or wood (*C*, fossil leaves). They also include impressions where animals once walked (*D*, dinosaur tracks). Fossils are generally preserved by replacement of the original material with new minerals or by alteration or addition of chemicals to original hard parts. Bone may be strengthened with additional calcium minerals, replacing the phosphate component, or be completely replicated by a new mineral (as in opalized bones). In rare cases, soft tissues or skin, fur, or feathers may be preserved (*E*, dinosaur head showing feather impressions).

Fossils and Rocks

There are three types of rock. When molten magma from deep underground rises to the surface through volcanos and chills, it forms volcanic, or igneous, rock. These are typically rocks with very small crystals of minerals due to rapid cooling, such as basalts. Magmas that chill slowly in the Earth's crust form coarse crystalline rocks such as granite, and these plutonic rocks are also classified by geologists as a kind of volcanic rock.

Rocks exposed at the surface of the Earth will eventually erode through weathering into small particles, which are washed by water and wind to accumulate either in river basins or lakes or on the seafloors. Rock formed by such piles of sediment is known as sedimentary. These can be made up of boulders, pebbles, sand, silt, or mud and may be cemented together by chemical solutions of iron or calcium carbonate that harden them into rocks. As thick piles of sediment build up in a low-lying basin, the lowest layers begin to undergo compression from the weight above them.

The third kind of rock forms when sedimentary rocks or volcanic rocks are thrust deep into the Earth from plate tectonic actions and become subjected to intense pressures and high temperatures that restructure their very fabric. These are known as metamorphic rocks. Of these three kinds of rocks, it is primarily sedimentary rocks that can bury life through catastrophic events (e.g., floods) or by gradual accumulation on the seafloors, and the remains of life forms that are preserved by these processes are called fossils.

Fossils are preserved in a variety of ways. In most cases, dead organisms are scavenged by invertebrates or by bacteria and other microorganisms. This process breaks down all the organic components, leaving just the bones. Some scavengers eat the bones to get the organic materials inside them. With exposure to weathering or underwater turbulence, the bones can break down to miniscule fragments and not be preserved. To be preserved as a fossil, the organism must be buried under certain ideal conditions. Rapid burial is best, in basins with high sedimentation rates, as this ensures the animal is buried whole. In such cases, whole body fossils might be preserved. Chemical processes operating in the burial environment then determine whether bone will be preserved or dissolved (as occurs in acidic environments).

Sedimentary rocks form by accumulation of sediments such as sand, silt, and muds that later became cemented together as rock. These Late Devonian–age river sediments are now exposed as red mudstones containing fish fossils near Eden, Australia.

events. Fishes now inhabit the most diverse range of environments of all vertebrates, from abyssal ocean depths and subfrozen mountain rivers and lakes, to the extremes of subterranean cave pools and flows of water trapped inside cracks in rock deep below the ground.

This book aims to provide a general overview of the main events that define the evolution of fishes and to introduce some of the most spectacular discoveries of prehistoric fishes, with particular reference to the origins and radiations of the major groups, living and extinct. The book follows a journey from the most primitive fishes 500 million years ago to more complex forms that eventually become the first land animals. At this point the story ends by highlighting the strong links between fishes and humans. To fully comprehend the origins and evolution of fishes one must be first familiar with the basic concepts of vertebrate anatomy, geological time, the principles of evolutionary theory and the geological processes that preserve fossils. In this chapter, such background information is briefly explained as a guide for readers without specialist knowledge.

Life Histories in Stone

Fossils, from the Latin *fossilere* (to dig), are the remnants of past life of the Earth. They can be bones, leaves, petrified wood, or tracks and traces of where life once moved. They are usually organic remains that have been altered by chemical processes and compaction through a long period of burial in the ground. Let us focus on skeletons, as these are the kinds of fossils that most often represent ancient fishes. For the remains to be preserved as fossils, they must be buried rapidly, before decay by bacteria destroys them, and in optimum conditions for the preservation of their bones. This means that the chemistry of their entombment must be conducive to preserving the skeleton rather than dissolving it. A typical scenario for the preservation of a fish fossil might be as follows.

When a fish died, it might have been preyed on by scavengers and its remains scattered. Or it might have died through a mass kill event, such as caused by oxygen depletion in a lake or ocean basin, or a lake drying up completely, or by volcanic ash settling on the lake and killing the fauna. Such cases would result in a large number of complete dead fish being buried in one place, either in a marine or freshwater environment. Sediments washing in from the nearby land or settling out of oceanic waters bury the dead fish little by little. Eventually, the build-up of this sediment forms a thick pile above the dead organisms, and the weight of the sedimentary layers compresses the remains. In rare cases, when conditions are right for rapid precipitation of calcium carbonate, these chemical processes can entomb the skeleton rapidly before this compaction occurs.

In such cases, the rock enclosing the fossil skeleton can be chemically dissolved using special acids to reveal perfect three-dimensional bones (e.g., the fossil fishes from Gogo in Western Australia, or Taemas-Wee Jasper regions in New South Wales in southern Australia, which are featured throughout this book). However, most fish fossils known from complete specimens are preserved as compressed remains between flat layers of sedimentary rocks, such as shales, mudstones, or

How fossils are formed.

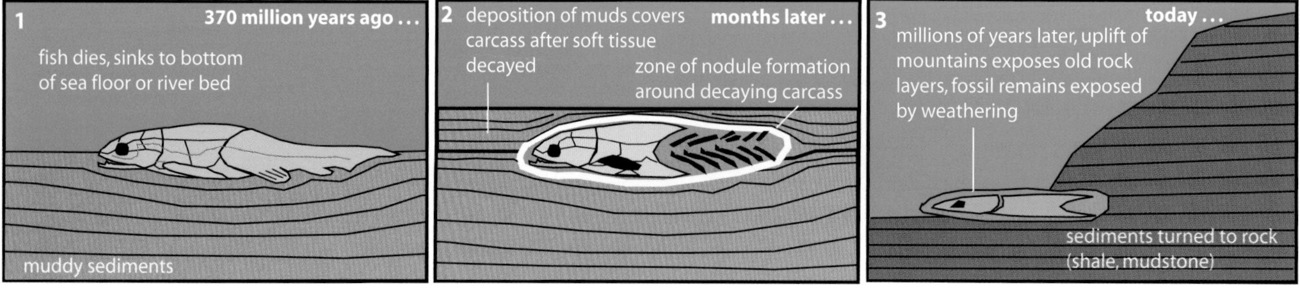

The Facts of Evolution

In November 1859, Charles Darwin released his groundbreaking book, *On the Origin of Species*, and created a worldwide sensation. In this book, which sold out on the day it was published and has been reprinted and sold ever since, Darwin unveiled his grand idea that species have evolved from one form to another over generations and that natural selection was the principal mechanism for directing this change.

Since Darwin's day, we have greatly advanced the study of evolution to incorporate new scientific disciplines such as molecular biology and genetics, plus there are countless pieces of new data from fossil discoveries that enable scientists to flesh out his original idea into a solid framework underpinning the entire diversification of life on Earth. Evolution is today viewed by scientists around the world as a grand unifying theory in biology that explains global biodiversity and intimately links it to the many transitional life-forms preserved over the past nearly 4 billion years in the fossil record.

Understanding evolution in depth requires a reasonable grasp of biology, anatomy, geology, genetics, and some mathematics, and it is for this reason that some people may have difficulty comprehending the basic tenets of evolutionary theory. Others choose to discount evolution entirely because it does not follow their belief systems. Nonetheless, evolution is a solid theory in the scientific sense: it is based on countless pieces of evidence (myriads of facts and observations) that hold the theory together and enable us to use it for predictions in the science of Earth's past, or paleontology, and in medical science. If evolution were not a totally dependable way to explain the diversity of life on the planet and how species change and mutate, then medical and biological science would have no firm foundation for understanding the rapid mutations that develop in viruses that become pandemics or how diseases evolved and thus how they can be cured.

Modern molecular biology has breathed new life into old bones. It has added the dimension of interpreting the evolutionary closeness of living species of animals and plants by comparing parts of their genetic codes. All life on the planet can be demonstrated to be interrelated through the universal commonality of molecular blueprints for forming life, complex molecules known as DNA (deoxyribonucleic acid). For example, we share about 96% of our DNA with our closet living ancestors, the chimpanzees. Examination of skeletal pat

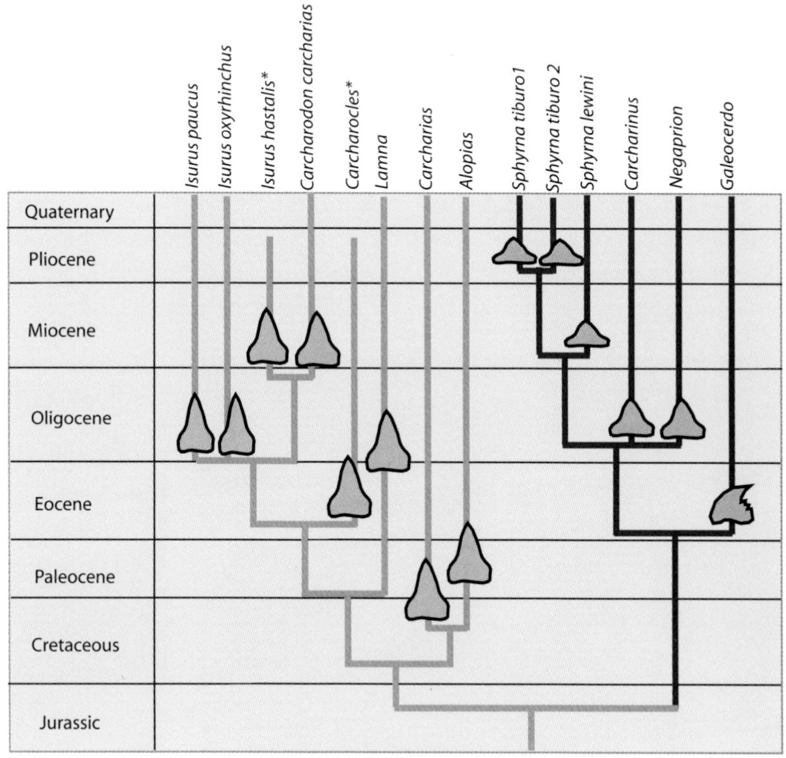

Mitochondrial DNA within the cells of animals changes with regular speciation events. This means that scientists can estimate the time of divergence from one ancestral state to another. The mitochondrial DNA divergence times in living shark lineages can be compared against the first appearance of fossil forms (here denoted by the shark's tooth icon). In this case, there is a very close correspondence, suggesting that the fossil record of these sharks gives an accurate representation of their evolutionary history. Asterisks denote extinct lineages. (After Martin, Naylor, and Palumbi, 1992)

The Facts of Evolution, cont'd.

terns of both humans and apes shows that we are more closely related (share a common pattern) than we are to any other animals. These simple tests reinforce the validity of the theory that apes and humans at some stage in the distant past descended from a common ancestor.

By analyzing the rates of variation in molecular mutations, scientists can estimate the time of such divergences and then look for evidence in the fossil record to verify or modify the assumption. An excellent study by Martin, Naylor, and Palumbi (1992), for example, calculated the divergence times based on mitochondrial DNA for living species of lamnid sharks and then correlated these with the geological dates for the fossilized teeth of these sharks. This study demonstrated that the fossil record was closely matched in this case with predictions from modern molecular biology.

It so happens that the molecular divergence time for when human ancestors diverged from chimpanzee and other great ape ancestors is around 6 million years. The fossil record supports this hypothesis, as some fossil apelike animals dating to around 4.4 million years ago, such as *Ardepithecus,* show more features in common with humans than to other apes, confirming that there are intermediate forms that would ultimately lead to the evolution of modern humans. The fossil record further shows how forms like *Ardepithecus* gave rise to *Australopithecus* around 3.5 million years ago. That evolutionary change would increase its brain size and allow it to walk upright, eventually giving rise to the first species of *Homo* (human) about 2.5 million years ago. But let us now journey back much farther and reflect on the very beginnings of all life.

The building blocks of life came together from various sources, as organic molecules such as amino acids (which form proteins) are found in meteorites that landed on Earth from space. Somehow the primeval chemical cocktails that existed in the early seas underwent transformations that lead to some simple molecules being able to replicate themselves. This defines life in its most basic form. The oldest life-forms appeared on Earth around 3.6 billion years ago, as simple cyanobacteria that lived in moundlike colonies with a host of other simple microbes to leave layered structures called stromatolites.

As stromatolites are still around today, we can confirm through study of transparent thin sections of the rocks under microscopes that microbes did indeed make these fossil structures. Life then proceeded slowly with the next big step being the advent of single-celled organisms with a nucleus (eukaryotes). Complex life consisting of many cells that worked together as one organism took a lot longer to evolve and was not common on the planet until about 600 million years ago. Around 540 million years ago at the dawn of the Phanerozoic Era (see geological timescale also in this chapter), we find the first abundant remains of life in the sedimentary rock record of animals that developed hard shells to protect them. Animals are either vertebrates (having backbones, such as fishes, reptiles, and mammals) or invertebrates, which comprise the vast majority of animals on the planet (such as worms, starfish, jellyfishes, sponges, corals, insects, crabs, and snails). The origins of the first vertebrates from such creatures involved their improved ability to move in water from development of muscles that attached to firmer sup-

Stromatolites are microbial colonies dominated by cyanobacteria that trap sediment to build up layered structures. They are examples of the oldest living record of life on Earth. These fossilized stromatolites from the Pilbara district of Western Australia are 3.45 billion years old (*A*). These living stromatolite colonies are from Shark Bay, Western Australia (*B*).

port structures inside their bodies, such as stiffened cartilage rods that would later help them form bone (see the box on larvae and evolution in Chapter 2).

Evolution can be observed in modern times. The apple maggot fly *Rhagoletis pomonella* was originally native to hawthorn trees of North America but has evolved in the last 200 years with the spread of modern agriculture to adapt to different fruit maturation times. Under laboratory conditions, it has been shown than an ancestral hawthorn fly takes 68–75 days to mature and breed. Recent races of the fly that infest apple trees may take only 45–49 days to mature and breed, while races that infest dogwood trees may take as long as 85–93 days. These times all correspond to the exact period of time it takes for fruit on the host plant to mature. Thus within 200 years different races of *Rhagoletis* have evolved to adapt to seasonal variations in the timing of fruit ripening.

Another fascinating aspect of modern evolutionary theory is how developmental changes within the growth of a single individual organism can model evolutionary changes. A change is proportion between the length of our arms and legs and size of our trunks and heads is noticeable in baby humans compared with fully grown adults. If a change in development occurs in a species, it can give rise to a descendent species that is either more like its juvenile ancestor (paedomorphosis) or by adding extra growth stages to make it look overly mature (peramorphosis). Such methods of evolutionary change are called *heterochrony*. In reality, heterochronic changes are not usually a simple case of retaining juvenile features or adding growth stages but can be a complex mixture of turning on or off growth within the development of an organism.

We humans can compare our facial features with those of a juvenile chimpanzee, both of us having a relatively flat face. This indicates that the face of humans has retained juvenile proportions relative to apelike ancestors (i.e., paedomorphosis), yet our legs have grown much longer in proportion to our bodies relative to the ancestral-ape like condition (i.e., peramorphosis). Such mixtures of growth and developmental spurts that result in new morphologies that become new species are known as evolution through *dissociated heterochrony*. The concept of heterochrony is paramount to understanding evolution, as by observing the vast degree of morphological change within the growth of a single species, we can now link such changes to large-scale evolutionary trends in many species over millions of years of change.

sandstones. We find these fossils by splitting the rocks where they are exposed in outcrops along river banks, beaches, or quarries.

While complete fish fossils are most prized by collectors and scientists alike, the harsh truth is that the vast majority of all fish fossils are fragmentary bits and pieces scattered throughout sedimentary rocks. Most abundant are isolated teeth, spines, and scales of ancient sharks, the ear-stones (otoliths) of ray-finned fishes, and pieces of broken body plates from the early armored jawless fishes and jawed placoderms. Each of these parts of the fish can have intricate characteristics that enable the specialist to make precise identification of the fossil species. For example, a single ear-bone from a fish can have a number of irregular grooves and ridges that clearly define the species it belonged to. Fish ear-bones have growth lines and can vary in chemical composition enough to show the movement of the fish from freshwater to marine phases of its life cycle can be detected by using isotopic ratios of certain chemicals in the sediment. To the trained eye, even such scraps can thus be scientifically significant as many groups of early fishes have distinctive tissue types forming their skeletal plates. Geologists searching for mineral deposits below the ground can use fish paleontology to assist in determining the age of rocks in drill cores when fish bits are the only fossils found or for clues to what kinds of the ancient environment are represented in the rocks they are searching for ore deposits. So even scraps of fossil fish can sometimes have their practical uses.

The first four-fifths of Earth's history is known as the

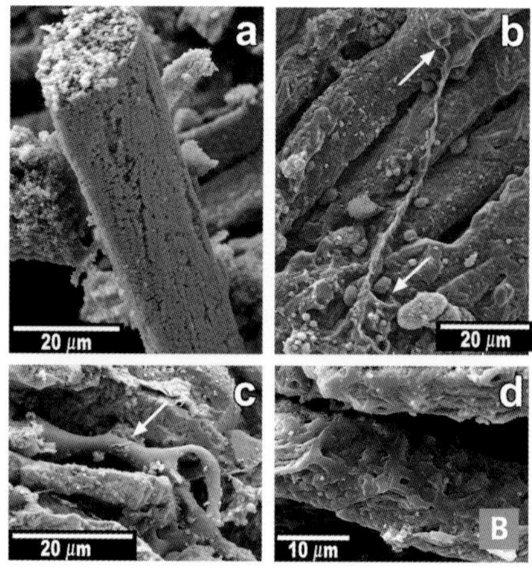

The addition of minerals has fossilized a 400-million-year-old fish bone (*A*). Soft tissues have been permineralized in a 380-million-year-old Gogo Formation placoderm (*B*). These include muscle cells (*a*) and nerve cells (*b*) still attached to the muscle cells, microcapillary (*c*), and crystals showing tissue replacement by calcium phosphate (*d*). (*A*, courtesy Gavin Young; *B*, courtesy Kate Irinajstic)

Precambrian (meaning before the Cambrian). This period spans from the Earth's molten origins 4.5 billion years ago to the explosion of life at the start of the Cambrian Period 542 million years ago.

A Guide to Basic Vertebrate Anatomy

The defining characteristic of vertebrates are that they have a stiff cartilaginous rod supporting the spine (notochord) that is replaced by a series of bony vertebrae in more advanced forms. In addition, all vertebrates possess paired muscle masses and a central nervous system. Most also have a head bearing a well-formed braincase to protect the brain and sensory organs such as eyes and nasal capsules. The most primitive fishes are the jawless forms represented by today's lampreys and hagfishes. All other fishes have jaws and teeth and can be collectively termed *gnathostomes* (all animal having jaws, including amphibians, reptiles, mammals, and birds).

The anatomy of vertebrates can be considered under the following main systems: skeletal (bones, cartilage), muscular (muscles, ligaments, tendons), nervous (brain and nerves), circulatory (arteries and veins), digestive (esophagus, stomach, intestines, etc.), endocrine and lymphatic (kidneys, glands, and secretory organs), plus the outer covering of the creature (the skin, which may develop scales, hair or fur, or feathers). As this book primarily treats fossil fishes and the evolution of their skeletons, most attention will be paid to patterns of development within the skeleton of fishes. Some exceptional fish fossils, as shown in this book, may preserve soft tissues such as muscle bundles or nerve cells. But these are extremely rare.

The skeleton of all animals, fish or landlubbers alike, is divided into the skull (inside the head) and the postcranial skeleton (body), which can be further divided into axial skeleton (spine and ribs and median fins in fishes) and the limbs, which are represented by the bones that support and form the paired fins of fishes (pectoral fins at the front, pelvic fins at the rear). The skull is the most complex set of bones in any organism. It can be divided into regions related to their prime functions, but the two main parts are the internal braincase, which protects the brain and sense organs and supports the gills, and the external dermal bones (formed in the skin or dermis), which cover the head, cheek, mouth, and underside of the head. The braincase (neurocranium) is a cartilaginous or bony box that encloses and protects the brain. In the developing embryo, it fuses together and ossifies from many separate units. The gills are supported by a series of interconnecting bones broadly called the gill arches. Both braincase and gill arch bones are formed from cartilage precursors made of endochon-

Eon	Era	million years ago	Period	Events
PHANEROZOIC EON	Cenozoic Era	23	Neogene	First hominoids (line leading to humans)
		65	Paleogene	*extinction event: end of dinosaurs*
	Mesozoic Era	145	Cretaceous	First primates; First marsupial and placental mammals
		200	Jurassic	First birds; *extinction event*
		251	Triassic	First dinosaurs, mammals and pterosaurs (flying reptiles); *extinction event: greatest loss of biodiversity*
	Palaeozoic Era	299	Permian	
		359	Carboniferous	First mammal-like reptiles; First reptiles (hard shelled eggs); First tetrapods living on land; *extinction event*
		416	Devonian	First tetrapods (amphibians) living in water; Radiation of many bony fishes (osteichthyans)
		444	Silurian	First jawed fishes, *(hard evidence)*; *extinction event*
		488	Ordovician	First fishes with jaws *(suggested by scales)*; First fishes with bony tissues
		542	Cambrian	First fishes (vertebrates), lacking bone

Before the Phanerozoic were three main periods of Precambrian time, the Proterozoic (542-1000 mya) the Archean (2,500-3,600 mya) and the Hadean (3,600-c. 4,500 mya, the birth of the Earth). During this time the first life appeared (3,600 mya) and the first cells with a nucleus evolved (c. 2,200 mya) and the first multicellular life flourished around 560 mya.

The geological time scale. mya = millions of years ago

Dating Rocks and Fossils

The Earth is approximately 4.55 billion years old, as determined by measuring the radioactive decay of certain isotopes preserved in rocks and meteorites. We know that uranium-235 decays into lead-207 with a half-life of 704 million years and that uranium-238 decays to lead-206 with half-life of 4.5 billion years. By carefully measuring the amount of decay from uranium to lead that has gone on in a rock (carefully choosing samples that have not been altered by weathering or other chemical processes to change the ratios), the uranium-to-lead ratio can determine the age of the formation of the minerals within that rock.

Measurements of meteorites from space, formed at the same time as our solar system cooled, match predictions extrapolated from Earth rocks that confirm the most likely date for Earth formation at around 4.55 billion years ago. Other isotopes like rubidium-87 decays into strontium-87 with a half-life of 48.8 billion years, so this method is accurate for determining very old rocks (usually greater than 100 million years old). In contrast, some isotopes like carbon-14 have half-lives of only 5,730 ± 30 years, so they are useful for measuring dates of organic material that have taken in a carbon-14 at time of formation and are less than around 50,000 years old. Geologists use these isotopic methods of dating to correlate the age of fossils.

As fossils occur in sedimentary rocks, they must be bracketed with radiometric dates taken from volcanic rocks above or below them. For example the town of Taggerty in Victoria, Australia, has a very thick succession of volcanic rocks, with three horizons being accurately dated using isotopic methods. However, within the thick volcanic succession there are thin beds of sedimentary rocks that formed from rivers and lakes upon the rocky volcanic terranes. Fish fossils preserved in these sediments can be accurately dated from radiometric dates on volcanic layers just above the fossil-bearing layers, so we have good evidence that they are Early-Late Devonian in age (Frasnian stage, dates are 373 ± 4 million years).

Using large numbers of reference sequences that contain well-dated layers with sediments in between them, there are global correlation charts for dating rocks from just about anywhere in the world. Thick successions of sedimentary rocks often contain lineages of microscopic fossils that evolve and change with time. These microfossils are another good way of dating rocks, as the key species that occur between certain time intervals indicate rocks of the same age from around the globe. Fossils used in this way (tied into radiometrically dated successions of rocks) are called *index fossils*. Some macrofossils, or large fossil remains, can be used in this way

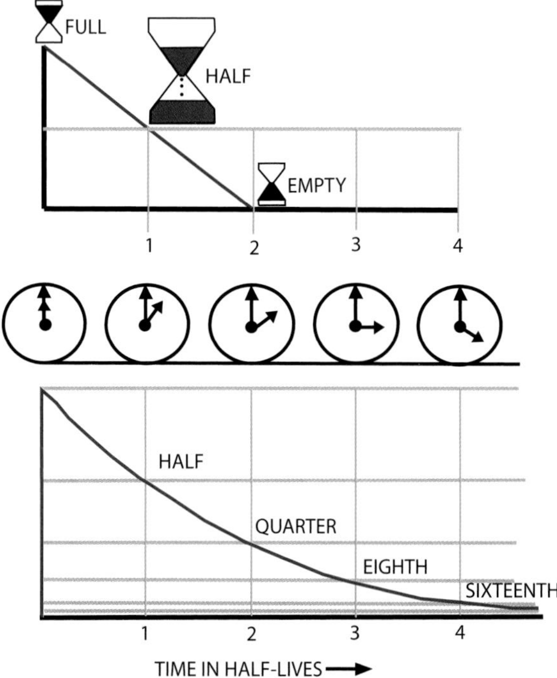

The radiometric dating of fossils involves measuring the decay rate of a radioactive isotope since it first crystallized from a molten rock. By knowing its constant rate of decay, we can then estimate its age by the amount of decay shown in the mineral structure. The clocks represent the half-lives of the radioactive isotope, or the standard length of time by which we can measure how old the mineral is, and thus the age of the rocks in which the fossils were formed.

when evolutionary lineages are diverse and widespread. For example the Devonian placoderm *Bothriolepis* occurs on every continent on Earth in rocks dated at Middle-Late Devonian age. Certain species that exist in one area can sometimes be found in nearby sedimentary basins and thus can be a useful tool to correlate age ranges. Age ranges of whole fossil assemblages can also be used when one species coexists or overlaps with a concurrent species. One manifestation of this is that in the northern hemisphere the occurrence of the placoderms *Bothriolepis* and *Phyllolepis* together indicates rocks of the latest Devonian age (Famennian stage). Also fossils of the placoderm *Groenlandaspis* occur only near the end of the Famennian stage. In Australia, once part of Gondwana, the same fish, *Groendlandaspis*, can occur in much older rocks (e.g., Middle and Late Devonian) but is generally represented by more primitive species. Dating sedimentary rocks using successive lineages of fossils is a common method known as *biostratigraphic dating*. It is how oil companies commonly identify the age of rocks when drilling into the earth. It is also a valuable proof of the reliability of evolution in a practical day-to-day activity.

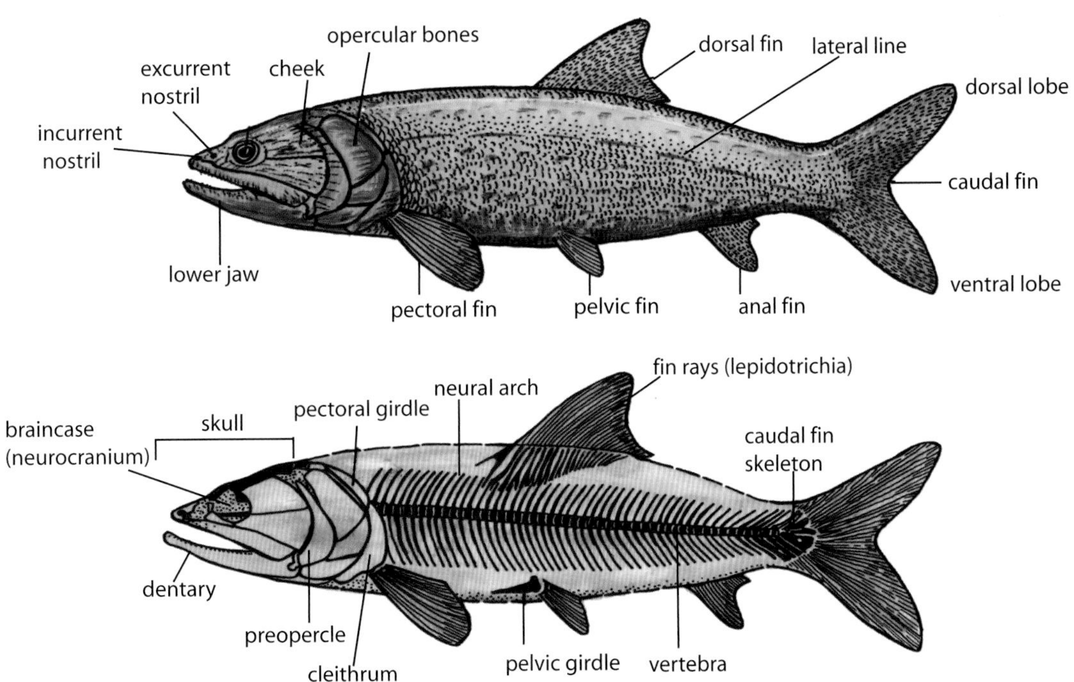

Basic features of a lower vertebrate (bony fish). *Above*, external features; *below*, major skeletal units.

dral bone or shells of thin perichondral bone that form around a cartilage core. The thicker ornamented plates of bone that adorn the outside of the head and make up the large girdle structures that support the fin are made of dermal bones. These include elements that grow and support teeth that may contain more complex tissues such as dentin and enameloids to strengthen the teeth.

Fishes typically have specialized sensory organs in the body called *lateral lines* that enable them to sense movements and chemical and temperature changes in the water around them. In some of the extinct armored fishes, we see that these sensory systems developed as linear troughs incised into the dermal outer bones or sometimes as deep pits. Most fishes have scales that are overlapping units of dermal bone that grow throughout life, but some fishes have secondarily lost scales (e.g.,

Most fishes reproduce by spawning in water, where females lay eggs that are externally fertilized by the males. But some fishes have internal fertilization that results in either live birth or large complex eggs being laid. Here, an egg case has been peeled back to show the embryo and yolk sac inside. It belongs to a lemon shark (*Negaprion*).

eels) or have microscopic scales that are not visible to the average eye (e.g., sharks and rays).

Fishes reproduce in a variety of complex ways, but most have a simple system of the females laying eggs in the water and the males fertilizing the eggs by depositing sperm over them. The eggs hatch tiny larvae that then undergo various cycles of feeding and migration until they develop into miniature versions of the adult fish. Sharks, rays, and chimaerids (class Chondrichthyes) all have internal fertilization, whereby the male inserts part of the pelvic fin called the clasper inside the female's cloaca—the single ventral body opening for emitting waste and reproductive products—to deposit sperm internally. In many sharks and rays, the young develop inside the body and then are born as a fully developed fish ready to feed and survive in the outside environment; this mode of reproduction is called *viviparity*.

Some sharks and rays develop eggs that hatch inside the mother's womb before being born to the outside world; this is called *ovivipary*. Others lay a few large eggs that hatch out well-developed juvenile fishes. While we know little of the reproductive behaviors of extinct fishes, some extraordinarily well-preserved finds from the 380-million-year-old Gogo sites in Australia have demonstrated that some of the armored placoderm fishes also gave birth to live young and reproduced by the males internally fertilizing the females. One specimen even preserved a fossilized umbilical structure showing how the embryo was nourished.

Drifting Continents and Fish Evolution

The current position of the continents is the end result of hundreds of millions of years of slow movement of large plates under the Earth's surface. These plates, which contain the major continents we know today, move just a few centimeters each year and over time have collided into each other to form supercontinents or drifted apart to form newer landmasses. This process is called *plate tectonics*. The Earth has a thin outer layer of rocks called the *crust* that sits atop a much deeper zone of molten rock called the *mantle*. Continental crust may be up to 150 km deep, but the crust forming the seafloors is just 50 km thick in places. The collision of these landmasses causes one plate to slide on top of another. The plate that is thrust downward into the Earth's molten mantle will eventually melt, and this new melt material may rise up through fissures in the crust to form volcanos. The main result of collision of plates is the buckling of the surface

Order within Nature

The system of classifying animals and plants introduced by Swedish naturalist Carl von Linné, or Carolus Linnaeus, in the eighteenth century is called *Linnean classification*. This system uses a binomial naming system comprising a genus name and species name. Such terminology based mostly on Latin and Greek word roots (but not always) means that whatever language people are speaking they can refer to a species by a universal name. Thus the great white shark alias the "white pointer" or "white death" has its specific scientific name as *Carcharodon carcharias*. Species in modern biology are usually defined by like types that can interbreed and form viable offspring, although modern molecular analysis of genetic codes (genomes) make species more easily definable especially after populations may have separated and become isolated, yet still look similar in other external features.

The Linnean system extends to containing groups of like species (as genera, singular genus) into groups called families, and groups of families into orders, and groups of orders into classes. The higher groups are called phyla (singular phylum). There may be various intermediate stages in all of these as for example, subspecies, subgenera, subfamilies, or infrafamilies, infraorders, as required when some groups are very large and have a high number of closely related forms. The human species is *Homo sapiens*, in the family Hominidae along with various fossil forms like *Australopithecus*. The family Hominidae sits within the order Primates, which includes other families such as the Pongidae (chimps and gorillas). Primates and other orders of mammals are placed in the class Mammalia, which resides in the phylum Chordata (including all chordates, creatures having a notochord).

Modern cladistic analysis of living or extinct species uses a computer program to sort out which species are more closely related, based on shared derived characters (called *synapomorphies*). Characteristics that unite species must be demonstrated to be specialized (derived), as primitive or generalized characters do not indicate closeness or relatedness. For example, nearly all fishes have fins, so fins do not unite certain orders or families of fishes unless there are special features of the fins, such has having a robust long-lobed fin with a humerus present (a synapomorphy uniting advanced lobe-fins, see Chapters 12-13). Groups of organisms that can be shown to have shared derived characteristics uniting them are called *monophyletic groups*. Those without demonstrable commonality of shared characters may be *paraphyletic groups*.

Commonly the program PAUP (Phylogenetic Analysis Using Parsimony) is used. The results indicate that there are often many new levels of classification that sit outside the normal familiar Linnean groups. Groups of closely related species or taxa (plural of taxon, meaning any group of organisms sharing certain features that unite them) are called *clades*. Cladistic classifications can use additional ranking terms such as division or cohort to denote major clusters of related organisms. Today many fossil groups are still under scrutiny as to how they are related to living groups or to other fossil groups. In cases in which we have living examples of a group (e.g., sharks or Chondrichthyes), we commonly refer to the cluster of living forms as the *crown group* and the extinct fossil forms that are at the base of the branching tree as *stem groups*. In cases where stem taxa are unresolved as to their classification, they may be left in an unclassified position pending further work.

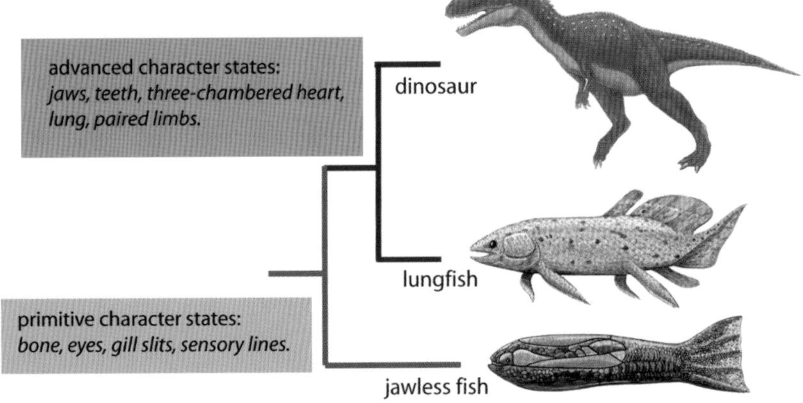

A cladogram represents the sequence of character acquisition in an evolutionary lineage. Here we see the branching sequence linking the lungfish with the dinosaur, as they share many advance features not seen in the extinct jawless fish—for example, jaws, teeth, three or more chambers in the heart, lungs, well-developed paired limbs, and many other anatomical features. Cladograms each inherently imply a new classification based on the branching nodes as higher taxonomic groups.

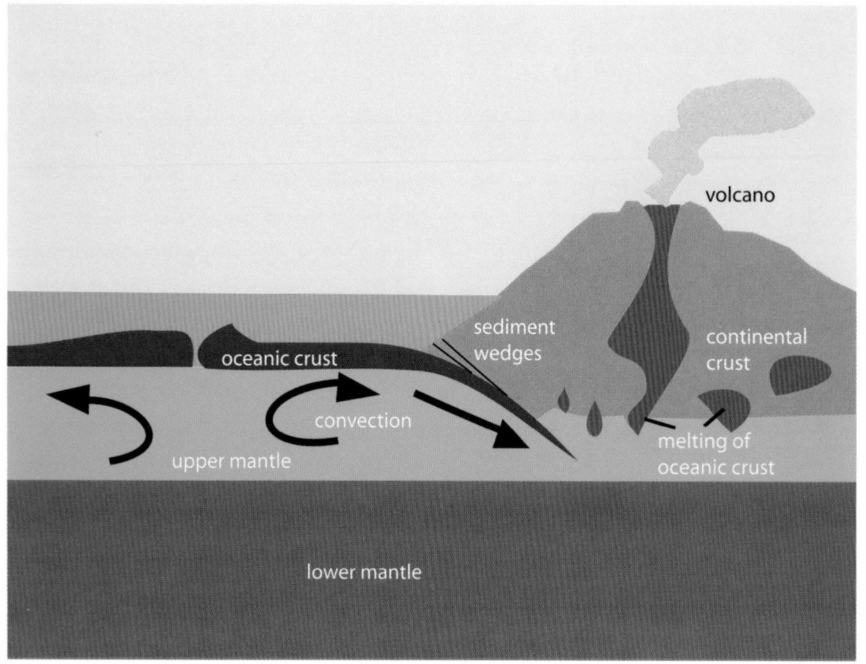

The Earth's crust is a dynamic system, continually in motion, even though the degrees of movement are impossible to detect with the human eye. Plates of the Earth's crust that carry the continents are always moving, thrusting their edges underneath or above other plate margins, as new crust is continuously formed by volcanic processes opening onto the ocean floors or erupting along the active margins of continents.

plate to form mountain chains, and this process begats a cycle of erosion and sedimentary deposition in the low-lying basin regions. Such processes form environments where fossils are commonly preserved (see box on fossil preservation above).

Around 500 million years ago when the story of fishes begins, the southern continents of Antarctica, Australia, South America, and Africa and the subcontinent of India were locked together as the newly formed supercontinent of Gondwana (accreting together around 570–510 million years ago). In the northern hemisphere, not far away, Asia and North America were fused together as another great landmass called Laurasia. Other parts of Russia and eastern Asia existed as isolated landmasses such as the South and North China terranes and the Tarim and Siberian terranes. It is important to remember the configuration of the continents in past geological times; as the continents moved throughout time, their new configurations had a profound impact on the evolution and migration of fishes. Many fishes today are spread widely throughout the oceans of the world, like the great migratory sharks and ray-finned fishes. Some groups of early fishes had widespread distributions. However, very few fossil fish species moved across great distances, as most were endemic to their regions.

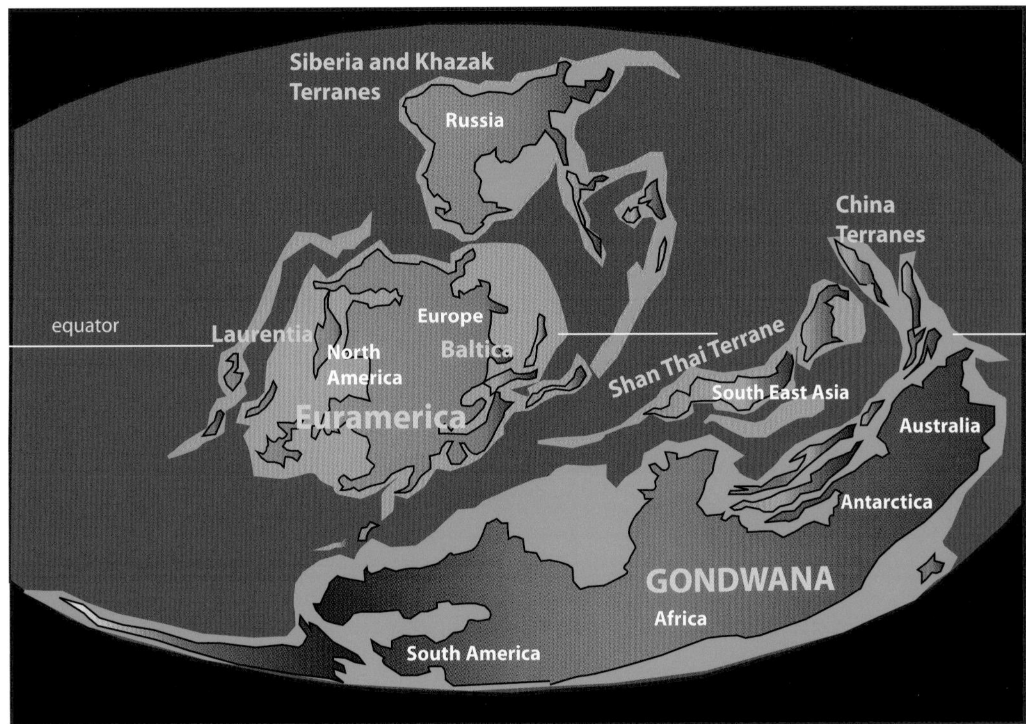

The world as it may have looked in the Devonian Period, showing continental positions and sea margins.

When fossil fishes are preserved intact inside limestone, they can be acid prepared. This technique dissolves away the limestone rock using weak solutions of acetic or formic acid. Limestone, being calcium carbonate, is dissolved by acid, whereas bone is a hydroxyapaptite mineral, so it is not affected by these acids. As bones emerge from the rock, they are hardened with glues. The process is repeated until the complete skull is freed from the rock. This specimen of the tetrapodomorph fish *Gogonasus* is shown as found (*A*), the snout emerging after a couple of weeks preparation (*B*) and mostly out of the rock (*C*). The fully prepared skull can be seen in Chapter 12.

Fossil Study and Preparation

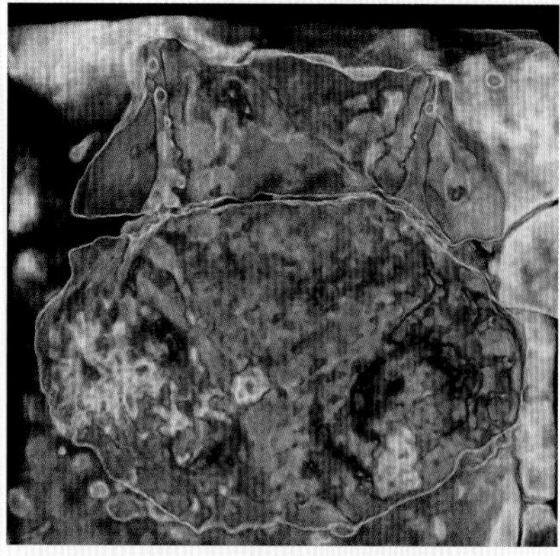

This micro-CT scan image of a Devonian fossil fish skull shows how modern technology has greatly enhanced the study of fossil fishes in recent years. Powerful x-ray beams reveal the internal sensory line canals through the bone. This is a false-colored image generated by the Australian National University's Department of Applied Mathematics. (Courtesy Tim Senden, Australian National University)

Once fossil fishes were mostly studied as the beautiful impressions of compressed skeletons preserved in rock, but today scientists employ a variety of high-tech tools to delve into their most detailed anatomy, such as micro-CT scanning, synchrotron imaging, and radiography. Advances in fossil preparation over the past few decades have seen many specimens can be etched from their limestone tombs as stunning three-dimensional skeletons. The acid-etching process is described further in Chapter 5.

Studying fossil fishes generally involves describing their anatomy using a variety of methods. Specimens that are flattened between layers of rock can be cleaned using drills and fine needles to remove rock around the delicate bones. If the specimen is in a limestone rock, it can be embedded into a slab of acrylic resin and then gently acid-prepared to remove the rock around the skeleton and reveal the 3-D form of the bones.

Traditional ways of studying the prepared fish fossils are to use photography or a camera lucida imagery to draw specimens, and in some cases applying water or ethanol can accentuate the contrast between rock and bone to clarify the boundaries of the skeleton. More advanced methods used in recent years involve x-ray radiography and microtomography to reveal the shape of bones within the rock. Such techniques work best with well-prepared specimens, such as acid-prepared skulls that can fit into a CT scanning chamber (e.g., see *Gogonasus*, Chapter 12). Use of high intensity beam lines such as produced in the synchrotron has in recent years elucidated the fine detail of fossil fish skeletal and dental tissues and, in one case, revealed a fossilized brain inside the skull of a 300-million-year-old chondrichthyan fossil.

A standard technique used to study poorly preserved fish fossils (where bone is not well preserved) in well-bedded shales and mudstones is reverse acid preparation. In this method the slab with the fossil is immersed overnight in 10% hydrochloric acid to remove the traces of bone and reveal a clear mold of where the bones were embedded in the rock layer. By gently brushing away the traces of bone, washing the rock in running water to remove traces of acid, and then casting the mold with latex rubber or silicon rubber, a detailed mold of the original bone surface can be obtained. Best results are achieved by whitening the blackened latex rubber cast with ammonium chloride sublimate, puffed from a tube heated over a Bunsen burner, as this lessens the glare off the surface and evens the contrast for highlighting fine details of the surface morphology of the cast. Good examples of this technique to reveal fine detail of skull features is seen in the Mount Howitt fish fossils such as the ray-fin *Howqualepis* (Chapter 9).

KINGDOM ANIMALIA
(animals, not plants or fungi or bacteria, etc.)

PHYLUM CHORDATA
(chordates, backboned animals)

SUPERCLASS GNATHOSTOMATA
(vertebrates with jaws)

CLASS CHONDRICHTHYES
(sharks, rays, holocephalans)

ORDER LAMNIFORMES
(lamnid sharks)

FAMILY LAMNIDAE
(mako, mackeral sharks, porbeagle, etc.)

Genus: *Carcharodon*
Species: *Carcharodon carcharias*
(great white shark)

Linnean classification is based on a hierarchical system of levels, each united by shared anatomical features reflecting a common evolutionary origin. This set of classificatory layers shows the position of the great white shark, *Carcharodon carcharias,* within the animal kingdom.

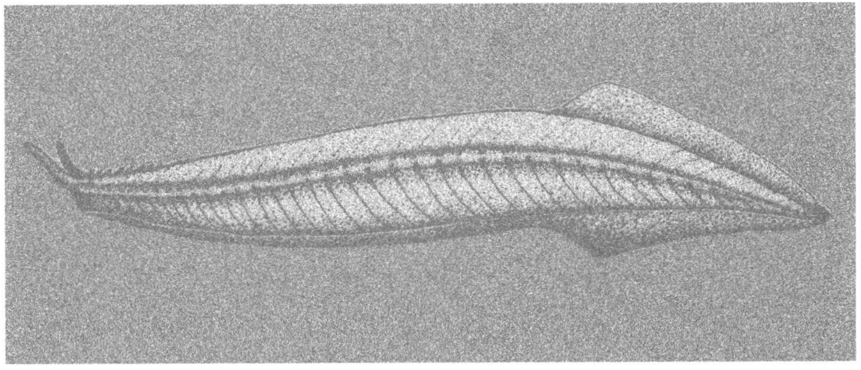

CHAPTER 2

Glorified Swimming Worms: The First Fishes

Origins of chordates and the first vertebrates

The oldest known vertebrates date back to the Early Cambrian of China. *Haikouichthys* and its kin were thought to be the first creatures to have a notochord (a stiff fibrous rod of tissue that supports the backbone and often disappears after the formation of the bone or cartilage units), muscle blocks that were V-, or chevron-, shaped, and a well-developed head with paired optic capsules for the eyes and paired olfactory organs. It was clearly a jawless, boneless creature but also had possible blocks of cartilage supporting the notochord, so it could technically be called a fish. Other enigmatic creatures having some features unique to vertebrates also occur in Cambrian and Ordovician rocks, heralding the rapid rise of true vertebrates. The conodont animal *Promissum* from the Ordovician of South Africa had well-developed chevron-shaped muscle blocks, paired eye cartilages, and possible eye muscles. Conodonts are now considered to be a more advanced fish than living lampreys, because they all possessed mineralized hard parts made up of white matter similar to cellular dermal bone, described later in the chapter. The first jawless fishes with dermal bone covering their bodies evolved seamlessly from such ancestors. The evolutionary steps leading to the first fishes are now clearly demonstrated in the fossil record.

Most people regard fishes as simple creatures with a bony skeleton that swims, has fins and gills, and is often nice to consume with a fine wine. But all of these characteristics can be found in creatures that are not fishes,

Sea squirts, or tunicates, are closest invertebrate relatives of the vertebrates. Their larvae have muscular tails that enable them to swim around. Here they are seen in the adult phase in which they have a large pharynx with gill slits. (Courtesy Clay Bryce, Western Australia Museum)

and certain fishes do not have all of them. For example, a Mexican walking "fish" (axolotl) lives underwater, has gills and arms and legs, and is actually an amphibian, not a fish. Sharks and lampreys lack a skeleton of bone (having cartilage instead), and many primitive fossil jawless fishes did not have fins, only a very simple tail. But all fishes have a well-developed head with paired eyes, sets of gills, and a notochord. The notochord also existed in many primitive fossil ancestors of the first fish; creatures having this feature are termed chordates and are placed in the phylum Chordata. Chordates thus include all vertebrates, as well as several primitive creatures that share certain advanced anatomical features with fishes.

Another characteristic of chordates is the presence of a series of V-shaped blocks of muscles along the body (called *myomeres*) that divide the tail into segments. The close corresponding numerical relationship between vertebrae and these myomeres is a feature of all chordates, including the living *Branchiostoma* (*Amphioxus*), some extinct forms such as *Cathaymyrus* from the Early Cambrian, and living tunicates (urochordates). The most significant characteristic of true vertebrates is the presence of an embryological phase containing neural crest tissue, which may develop into mineralized bone, dentin in teeth, or calcified cartilages, and epidermal placodes, which develop into the major vertebrate sensory organs, such as paired eyes, nasal capsules, and the lateral line system described in Chapter 1. Of these, perhaps the most significant characteristic of most chordates is the ability to secrete phosphatic hard tissues during the formation of the embryo, including that most advanced tissue of all, bone. Such tissues are derived from mesoderm or neural crest cells, which are not found in any living invertebrate.

Aside from the fishes and higher vertebrates with bony tissues, the primitive groups of living animals manifesting the earliest hints of becoming vertebrates include the lineage containing echinoderms (sea urchins, starfishes, etc.), hemichordates (acorn worms and pterobranchs), cephalochordates (lancelets), and urochordates (tunicates, or sea squirts). Although some paleontological studies argued for a closer relationship between echinoderms and vertebrates (e.g., Jeffries 1979), recent studies on the DNA of hemichordates show that they are more closely related to echinoderms, so they will not be further considered here in the discussion on fish origins. The most recent molecular studies confirm that tunicates are indeed the closest relative to the vertebrates.

In addition there is a mixed bag of bizarre early fossils simply termed *problematica*, some of which could

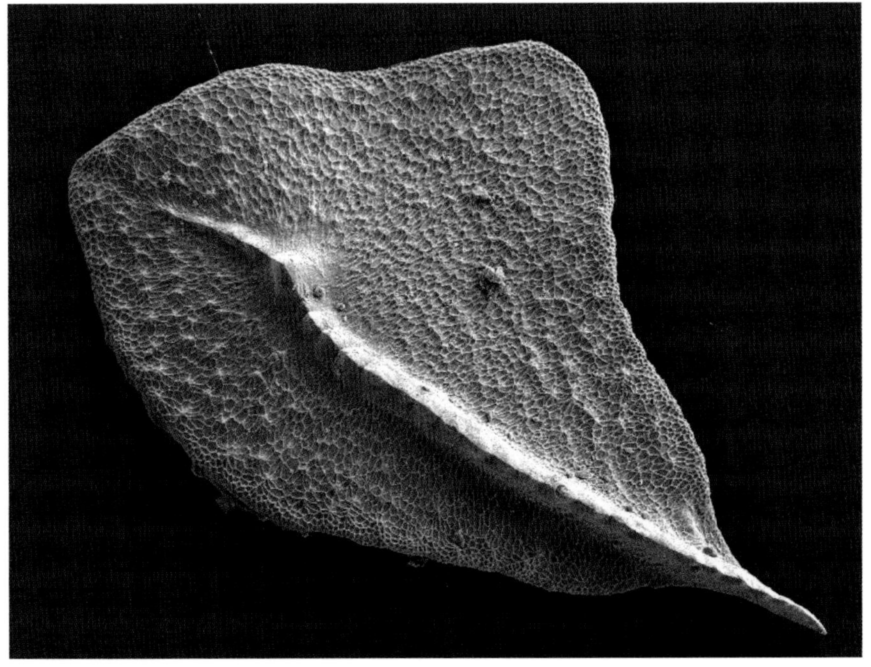

A fossil of an element of the conodont *Palmatolepis*. These enigmatic microscopic fossils are commonly used around the world to date the ages of Paleozoic marine sediments. Despite their widespread use, only recently have fossils of the whole conodont animal been discovered. These elements would have been situated in the conodont's head and may have supported filtration, food-sifting, or food-reduction structures. (Courtesy Kate Trinajstic, Curtin University)

Larvae and Evolution

The origin of the first fishes from invertebrates seems to be intimately linked with the major changes in shape exhibited by certain invertebrates during their cycle of development. The juvenile phases of fish and invertebrates are often morphologically quite different from their adult forms. Sea squirts are sessile baglike creatures that attach to the substrate and filter feed in seawater. If you pick one up, it will squirt water at you, indicating they have a muscular system inside them. Their larvae, however, are free-swimming tadpole-like forms that have a muscular tail supported by fin rods, the body is stiffened with a cartilaginous rod (notochord), and they have a hollow dorsal nerve cord and gill slits. In these features, they resemble early fishes. Changes that occur in the lifestyle of an organism may precipitate the developmental changes in their bodies, as for example when a larval sea squirt moves from open waters to the shallow and settles on the seabed.

Other invertebrate groups have similar mobile larval phases that settle into a sessile or slow-moving adult form, such as echinoderms (sea urchins, starfishes, etc.). Theories were put forward that any of these groups with free-swimming larvae could have played a role in the origins of the first fishes, simply by earlier sexual maturation to give rise to adult free-swimming forms. However, it is not so simple, as only certain kinds of invertebrates have a notochord and the ability to develop tissues that are unique to vertebrates. The derivation of such hard tissues from neural crest cells in the embryo is a feature of all vertebrates.

The larvae of the marine lamprey *Petromyzon marinus* are called ammocoetes and live in burrows in sandy river bottoms for 3 or more years before undergoing metamorphosis. The young lampreys then migrate down river to the sea, where they remain for another 3 to 4 years before becoming sexually mature. The ammocoete larva has a basic anatomical structure not too dissimilar from that of the *Branchiostoma* (cephalochordate) or tunicate larvae (urochordates), so the link to the first fishes is quite clear. Recent work analyzing the molecular similarities between these groups has determined that sea squirts are indeed closer to the origin of fishes than *Branchiostoma* or echinoderms.

have affinities to the vertebrates. All of these creatures lived in the sea, and this is undoubtedly where the first great evolutionary steps toward higher vertebrates took place. The conodonts are another major group of extinct chordates known mostly from microfossils and rare whole body fossils. They have recently been studied in great detail by UK researchers Mark Purnell and Phil Donoghue, revealing unexpected similarities with higher vertebrates and suggesting links to the advanced jawless fishes.

Lancelets (Cephalochordates)

Amphioxus and the lancelets were initially classified as slugs when first described in the 1770s, and it took almost a century before Alexander Kowalevsky showed them to be allied to vertebrates in 1866. They have since been thought of as the closest living ancestors of vertebrates, because they are small eel-like animals that have well-developed myomeres with V-shaped muscles, a well-developed pharynx with numerous gill slits, a long median dorsal fin supported by hollow tubes, and a notochord that runs from within the head to the tip of the tail. They are named lancelets because their bodies are lance-shaped, much like a primitive fish. There are only two known living genera with about 25 species, all of which are marine and reach sizes up to about 7 cm. The adults lie buried in the soft sandy shallow seafloor with their mouths protruding above the sediment to take in food from the passing seawater.

Lancelets have separate sexes and breed by shedding sperm and eggs into the water. The larvae are very fish-like in having a powerful tail but have fewer gill slits than the adult. The mouth has a circular ring of small tentacles, or cirri, which helps create a current of water around the mouth to enhance feeding.

Although lancelets lack a heart, they have a blood circulation system that is very close to typical vertebrates in possessing a large central artery, the ventral aorta. The two living genera are *Branchiostoma* (once called *Amphioxus*) and *Epigonichthys*. The fossil record of lancelets is virtually unknown. The oldest possible forms are from the early Cambrian Chengjiang fauna of southern China, such as the enigmatic *Yunnanozoon*, and *Pikaia* from the Middle Cambrian Burgess Shale of British Columbia, in Canada. Although superficially wormlike in appearance, *Pikaia* has a number of features that point to its chordate affinities. It has what appear to be a notochord and a tail fin supported by rods of cartilage. *Pikaia* has a body form and overall anatomy similar to modern lancelets; it has yet to be fully studied in detail, so its evolutionary position remains enigmatic. The discovery in the mid-1990s of an even older cephalochordate from the Early Cambrian of China, *Cathaymyrus*, shows that *Pikaia*-like creatures were around from at least 525 million years ago.

Yunnanozoon apparently had a notochord extending far forward in the head and showed the presence of sets of paired organs that were either gonads or slime glands. It was once classified as a cephalochordate but could alternatively be an early, or basal, deuterostome—that is, an animal with two openings, including all chordates and three other phlya. *Pikaia* is another "squashed slug" kind of fossil that has been reported to have a notochord, although Philippe Janvier of Paris has confided in me that he has examined all the main specimens of this animal but cannot confirm it has any chordate features. He guesses that *Pikaia* could have been a close

The lancelet, *Branchiostoma*. These primitive vertebrates are filter-feeders that lie buried in the sandy seabed. (Courtesy Douglas Elford, Western Australian Museum)

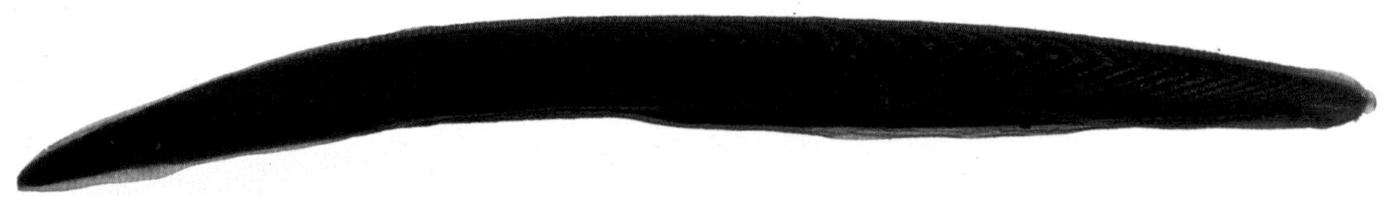

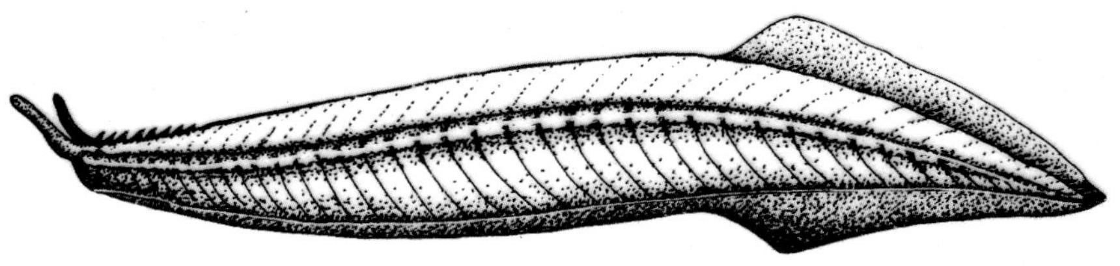

A reconstruction of *Pikaia gracilens*, an early chordate fossil, from the Middle Cambrian Burgess Shale of British Columbia, Canada.

relative of *Yunnanozoan*. *Palaeobranchiostoma* from the Permian of South Africa has some resemblance to the living lancelets but has a larger, well-developed ventral, or belly, fin, and the dorsal fin is also larger and invested with numerous small barbs. Again, poorly preserved fossils are not easily interpreted, and some paleoichthyologists have doubts about its interpretation.

Another fossil from the Early Cambrian of Chengjiang, China, *Haikouella*, has been closely related to *Yunnanozoon* in previous works. A study by Jon Mallatt and Jun-yun Chen (2003) proposed that it is more closely related to the craniates and is therefore more specialized than other cephalochordates.

The resemblance of these enigmatic animals to vertebrates, as with the tunicates, is seen mostly in the larval stages, although the asymmetry of larval cephalochordates precludes a close relationship to the main vertebrate line. The tunicates are more advanced in an evolutionary sense than are lancelets, because they have acquired the primitive phosphatic tissues that come close to being, but are not quite, true vertebrate bone.

Tunicates (Urochordates)

The tunicates are small marine animals that are commonly called "sea squirts" because, when you pick up these bloblike creatures, they often squirt water at you as their main means of defense. They feed by taking water in through the mouth and filtering it for tiny food items by passing the water through their gill slits. The name *tunicate* comes from their being embedded in a tough outer "tunic" of cellulose, the same substance that gives plants their internal support. They are also known as ascidians, from the Latin word for wineskin.

Despite the lack of similarity between an adult sea squirt and a fish, it is the sea squirt's juvenile phase, or larva, that closely resembles a primitive fish. The larva of a tunicate is rather like a tadpole in having a long muscular tail. It is supported by a notochord that starts behind the head and has a spinal nerve cord. The head end has various sensory devices that enable the creature to swim and keep its bearing with respect to gravity and to the direction of light. On finding a suitable place to settle, it anchors itself by means of three hairlike sticky structures on the head, called *papillae*, and begins its metamorphosis into the adult form. The adult remains immobile, and, as it develops, it resorbs its long tail for nourishment. Tunicates are hermaphroditic; they can develop into either sex as they mature and may even reproduce asexually by budding off new animals.

Fossils thought to be those of early tunicates are

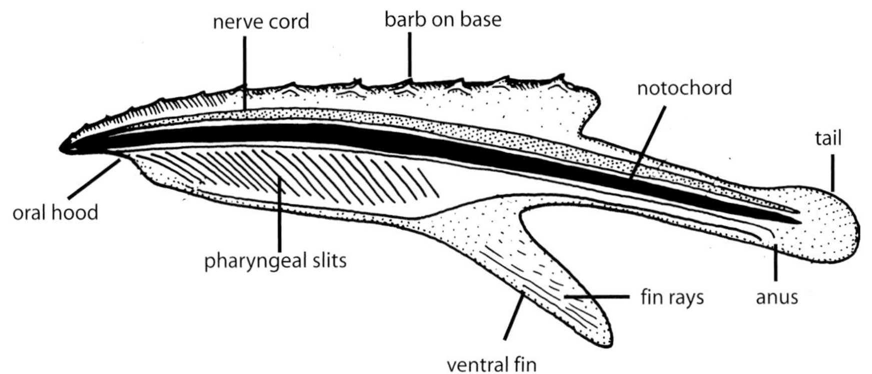

Reconstruction showing the features of *Paleobranchiostoma*, an mysterious fossil that could be a tunicate from the Permian Period of South Africa, although some scientists question this interpretation.

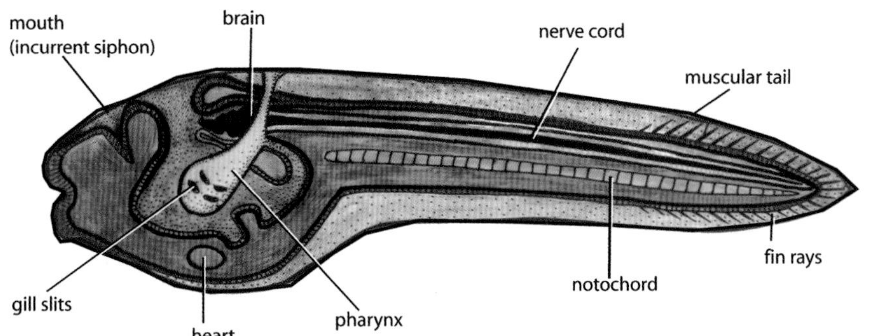

Anatomical features of the larva of a tunicate showing its vertebrate nature in the strong muscular tail, presence of a notochord, and well-developed head region.

known from the dawn of the Paleozoic Era, although there is still much debate among scientists as to whether they really are tunicates or whether they belong to completely new groups. The oldest known urochordates are possibly *Ausia fenestrata* from the Ediacaran Nama Group of Namibia and a closely related form (an *Ausia*-like genus) from the Onega Peninsula in northern Russia. Work by Pat Vickers-Rich (2007) has shown a possible affinity of these Ediacaran organisms to the ascidians. Two Early Cambrian forms are known from China, *Shankouclava shankouense* from the Lower Cambrian rocks in the vicinity of Anning, near Kunming, and *Cheunggongella* from Chengjiang. The latter form closely resembles the living tunicate *Styla*. Another fossil example is *Palaeobotryllus* from the Upper Cambrian of Nevada. Its bubble-like form makes it closely resemble the modern colonies of the tunicate *Botryllus*. Also microscopic platelets of enigmatic creatures have been compared with the spicules found in the tunics of modern tunicates, leading to speculation that the creatures are ancient sea squirts.

A problematic fossil from the Early Ordovician of China, described by Clive Burrett and myself (1989), has a phosphatic tubular exoskeleton with large blisters forming tubercles on the inside of the tube. We named this form *Fenhsiangia*, after the town of Fenhsiang in Hupei Province, China, and thought it must be somehow allied to the first vertebrates, because vertebrate bone is the only tissue to develop tubercles of this kind. Despite this comparison, the tubelike shape of *Fenhsiangia* gives no clues as to the nature of the organism or to its lifestyle. As it has tubercles on the inside of the tubes, rather than on the presumed external surfaces, it is mysterious by comparison with all other known vertebrates. Such forms give us a tantalizing glimpse into the complexities of developing primitive vertebrate tissues but tell us nothing of the internal anatomical features of their owners.

A detailed analysis from 146 genes taken from the nucleus of several invertebrate and vertebrate species by a team led by Frederic Delsuc of the University of Montreal has recently demonstrated conclusively that

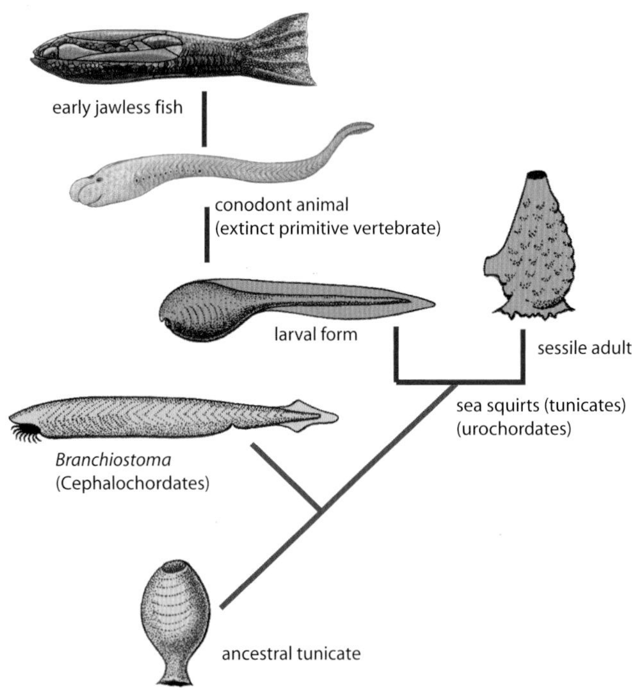

The origins of fish may be linked to early maturation of free-swimming larval phases seen in tunicates. Evolution by such processes is termed *heterochrony*, and *paedomorphosis* is a form of heterochrony in which juvenile features of one species are passed on to become adult characteristics of the descendent form.

the tunicates are the most closely related group to vertebrates. Their findings resurrect the group name "Olfactores" for a combined vertebrates plus tunicates clan, first recognized by Dick Jeffries in 1991. The embryos of urochordates have recently been shown to possess migratory cells that develop into pigment cells. In vertebrates, this is only achieved through neural crest cells, so urochordates must also possess a primitive precursor to true neural crest cells. Recent developmental studies on lancelets indicate that the midbrain-hindbrain boundary is a significant organizer region in defining vertebrates, as several important organizer genes are expressed there (Holland and Holland 2001).

To understand the links between these primitive living invertebrate forms and modern fish, we must now turn to the fossil record, where a host of strange primitive fishlike forms have been uncovered in the past decade or so.

Haikouichthys and Kin

One of the most exciting breakthroughs in fish origins has come out of China, from the early Cambrian Chengjiang deposits, dated at around 525 million years old. In 1999 Degan Shu and his colleagues, including British paleontologist Simon Conway-Morris, announced the finding of a little fishlike creature called *Haikouichthys*. Although this fossil lacked bony plates, it did possess a notochord, a long dorsal fin with supporting rays, several paired gill pouches, and eyes surrounded by possible scleral cartilages. In their subsequent paper describing new specimens of *Haikouichthys*, Shu and his team identified the possible presence of a backbone and otic, or ear, capsules, part of the developing braincase. The backbone is represented by smudges that repeat regularly, of the same shape, around the notochord, so they could be bony structures although they are not well-preserved. *Haikouichthys* and its kin (there are now four kinds of early Cambrian fish from the same layers of rock) have well-developed sense organs in the head, structures that in modern vertebrates are embryologically derived from the neural crest cells. Thus we have good evidence that *Haikouichthys* is a vertebrate at a level above or close to lampreys on the phylogenetic tree. Other forms found in the same deposit, such as *Myllokungmingia*, are more difficult to interpret and are represented by just one specimen, so their po-

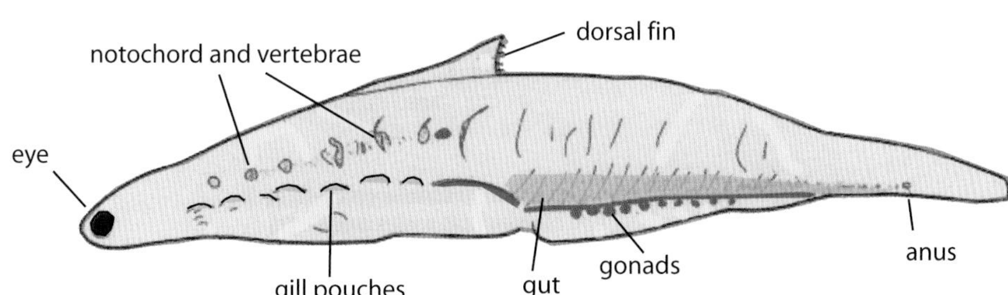

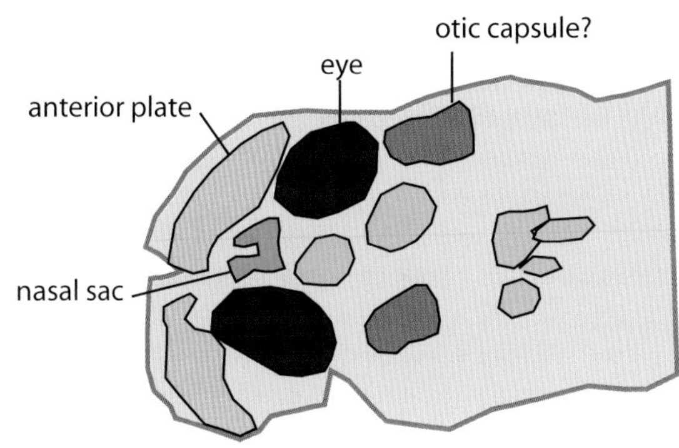

Artwork showing the body outline of *Haikouichthys* from the early Cambrian Chengjiang fauna of Yunnan, China (*above*), and features of the skull (*below*).

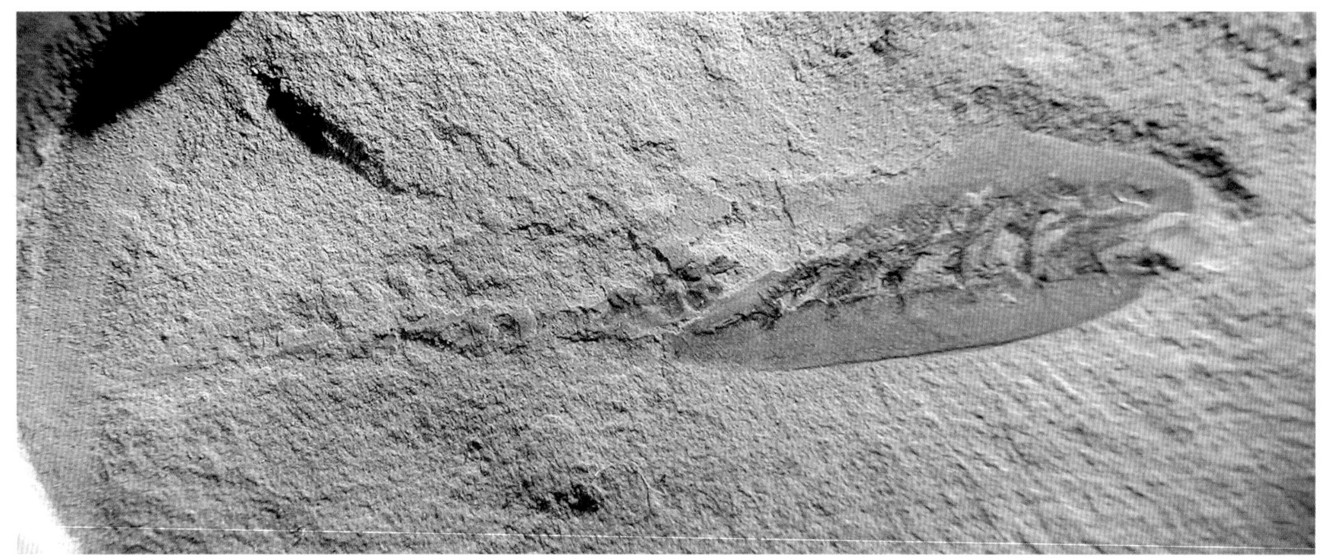

Myllokungmingia **from the early Cambrian of Yunnan, China, might well be the world's first fish. Although it lacks a bony skeleton, it has fins supported by rods, a notochord, and well-developed row of gill pouches. In essence, it is not far removed from a living hagfish.**

sition is still open to debate. Some researchers place *Myllokungmingia* at a node below lampreys but ahead of hagfishes (M. P. Smith et al. 2001).

The Conodonts: Enigmatic Fossils

The conodonts (from the Greek, meaning "cone teeth") were for many decades an enigmatic group of fossils known principally from tiny microscopic remains of phosphatic jawlike structures. Countless thousands of these fossils can be harvested from Palaeozoic limestones by dissolving the rock in weak acid and sieving off the tiny remains. Conodont elements, so-named to distinguish them from the complete animal the conodont, take the form of simple rod- or conelike forms, blades with toothlike protuberances, or complex platform shapes. Each conodont animal possessed a number of differently shaped conodont elements. M. P. Smith et al. (2001) have published sections of conodont elements showing they contained bone cells, although not all paleontologists agree with this conclusion, claiming that the mineralogy of their respective hard tissues differs considerably (Kemp 2002). The jawlike appearance of the conodont elements is deceptive, for the toothlike cusps along the ridges never appear to show much sign of wear, as teeth normally do, although some studies argue that there are wear facets on the tip of the cusps and that the cusps were used, like teeth, in the reduction of food (Purnell 1995). Purnell has suggested that some conodonts were predators.

Conodonts have been intensely studied to correlate and assess the age of rock sequences, but, until a few decades ago, we had no inkling of what kind of creatures owned these mysterious remains. In the early 1980s, the first fossilized whole conodont animals were found in the Granton Shrimp Beds near Edinburgh, of Early Carboniferous age, about 340 million years old. These show that conodonts were long wormlike creatures with tails and ossified scleral cartilages around the eyes and had supporting fin rays. In the head region, they possessed a cluster of little conodont elements, such as those described above. This, together with additional data on the hard tissue structure of conodonts, is powerful evidence that they were true vertebrates.

A 1992 paper published by Ivan Sansom and his colleagues from London in the journal *Science* suggested that true bone cells were present in some conodonts. If the possession of bone or any mineralized tissue emanating from neural crest cells is sufficient grounds to call a creature a vertebrate, then conodonts were early

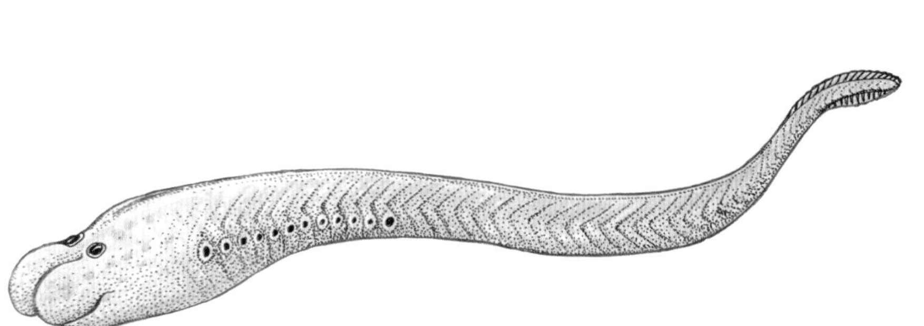

Reconstruction of a generalized conodont after the work of Richard Aldridge and colleagues. Conodonts are believed by some scientists to be deeply entrenched within the vertebrate family tree as stem gnathostomes (jawed vertebrates), whereas other scientists regard them as advanced chordates sitting below the level of the first fishes. Finding more complete fossil remains of them should one day settle the issue.

fish. Perhaps more convincing is the purported presence of cartilaginous sclerotic rings, or eye capsules, in conodonts reported from the Ordovician Soom Shale of South Africa by Sarah Gabbott and Dick Aldridge of Leicester University and Johannes Theron of the South African Geological Survey in 1995 and also identified in the few other known conodont animal fossils. The eyes of early fishes invariably lacked scleral coverage, except in the most advanced of the jawless fishes (osteostracans) and in all jawed fishes. The conodont animal called *Promissum* from South Africa was enormous by comparison to most conodonts, nearly 40 cm long. This specimen of *Promissum* shows well-preserved eye capsules and possible eye muscle tissue along with the well-preserved phosphatized conodont elements, providing proof that epidermal placodes (thickenings on the embryonic epithelial layer that develop into organs or structures) and neural crest were both present in the creature's embryonic stages. The specimen also had well-preserved fibrous muscle tissue, the oldest preserved vertebrate muscle known. Most important, they had bony protrusions called radials supporting their fins.

Robert Sansom and colleagues recently observed that the decay of lancelet and lamprey larva shows that preservation plays an important role in interpreting these early chordate-like fossils. The more the specimen decayed, the sooner it lost fundamental features that would have enabled its identification as either a stem vertebrate or stem chordate (Sansom et al. 2010).

Recent studies on the histology of conodont elements show they are made up of two main areas, the upper crown and the lower basal body. New layers were secreted on the upper crown in synch with new layers added to the base. The upper crown is composed of mainly lamellar crown tissue, which has crystallites in parallel or perpendicular orientation to the growth lines. The crown may also contain "white matter" that has a cancellous, or spongy, structure. These two tissues may even have been secreted by the same cell populations (Donoghue and Aldridge 2001). The structure of the basal body tissue is highly variable, being lamellar through to spherulitic or sometimes tubular. The lamellar crown tissue has been compared with enamel in vertebrates (Sansom et al. 1992), whereas white matter has been compared with cellular dermal bone (Smith et al. 2001). The spherulitic form of the basal body compares with globular calcified cartilage found in sharks and other chondrichthyans. Some kinds of lamellar basal tissue bear resemblance to dentin, the layer below the enamel in modern-day teeth (Sansom 1996).

As conodonts lack many of the refinements we see in the earliest fish fossils (such as having bony plates with a sculptured dentinous ornamentation), they are still regarded as suspicious newcomers to the vertebrate clan by some traditional fish paleontologists. They share some histological similarities with jawless fishes, but not enough is known of their cranial anatomy and overall features to confidently classify them above the lampreys. Some researchers place them in with the jawed vertebrates (gnathostomes), but such fishes have well-ossified braincases and jaw cartilages, neither of which has yet been identified in conodont fossils. To re-

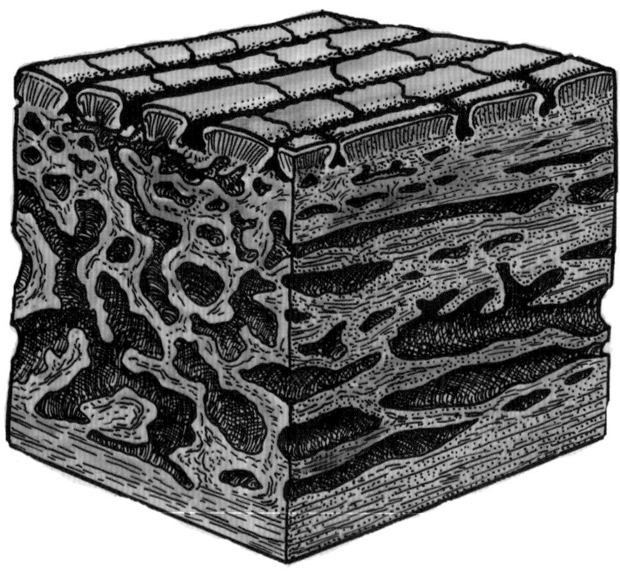

The structure of aspidin, a kind of dermal noncellular bone (bone that lacks spaces for bone cells) that makes up the shields of the earliest bone-covered jawless fishes, the extinct heterostracans. The outer sculptured surface is coated with dentinous tissues.

solve this issue, we will need some more complete and exceptionally well-preserved examples.

Bone: Its Early Beginnings

The word *fish* in common usage is usually defined as a vertebrate with a well-developed notochord, gill slits, well-formed paired sensory organs, a lateral line system, and a mineralized skeleton, be it made up of cartilage or bone. This skeleton can be either external bony plates formed in the dermis (dermal bone) or as developed later on in their evolution, internally ossified bony or cartilage units formed from cartilage precursors.

The identification of fish in the early fossil record relies heavily on the definition of what exactly bone is. Most fishes have bone in which there are cell spaces for osteocytes (bone-producing cells), and the external layers of the most primitive of all fish bones have an ornament covered by a thin enameloid layer over a dentin layer. However, some early fossil jawless forms have layers of bone that does not contain such cells, which is called aspidin.

Bone is the key to understanding the success of fishes. Bone provides a solid support for attachment of muscles, making it more efficient to use a muscular tail to propel the creature through water (well, truthfully, bone played little role in the attachment of muscles until perhaps the galeaspids and osteostracans more than 400 million years ago). Faster speeds allow the animal to escape from predators and to catch slower-moving prey. Bone not only enables greater locomotory improvements but also acts as a storehouse for phosphates and other chemicals required in daily metabolism. And, it gives protection to more vulnerable parts of the anatomy, such as the brain and heart, giving the organism a greater chance of survival after an encounter with an attacker.

Bone is very special in that it has a surface comprising a sculptured dentin layer, a spongy layer that housed bone cells (in some forms) and fibrous collagen, and a

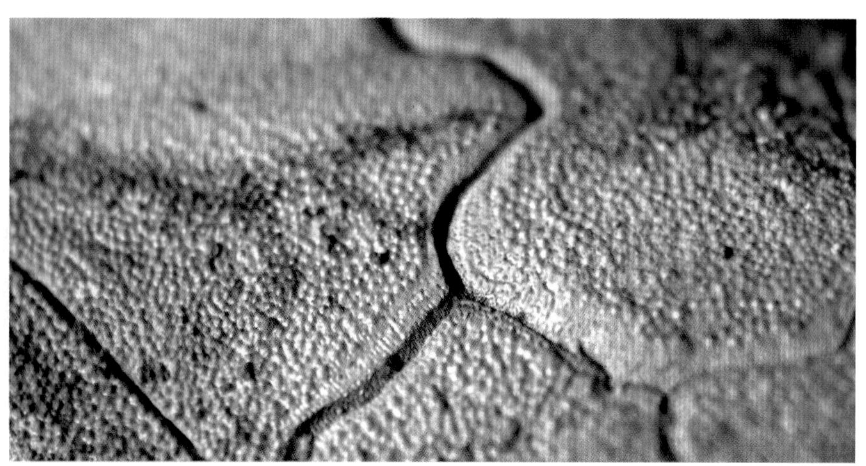

Dermal bone, shown here from a placoderm fish skull, is formed in the dermis, or skin, and is often characterized by elaborate surface ornamentation, such as ridges or tubercles. The appearance of dermal bone in the skeleton signalled the onset of the great radiation of jawless fishes beginning in the Ordovician Period.

non-cellular laminated basal layer. The first fishes had only external or dermal bone—bone formed in the dermis of the skin. Internal bony skeletons were to come much later, and this development heralded the next great explosion in fish evolution when jaws and teeth appeared. Large bony plates in fishes were thought to have formed from small centers, or plates of bone called *odontodes*, which later joined up to make large bony sheets.

After the origin of the first fish, approximately 525 million years ago, the vertebrate scene would be dominated by a remarkable array of agnathans—jawless wonders—for almost the next 100 million years.

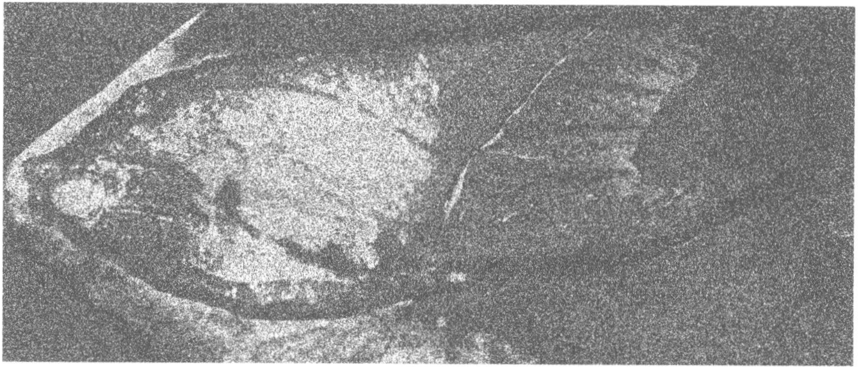

CHAPTER 3

Jawless Wonders

Hagfishes, lampreys, and their extinct armored kin

Agnathans, or jawless fishes, have a fossil record spanning about 525 million years, represented by wormlike boneless creatures, such as *Haikouichthys,* which had simple fins and paired eyes. The earliest examples of these jawless wonders are fishes that swam in seas that covered central Australia about 460 million years ago. Shortly after the Ordovician, the agnathans radiated into many diverse groups, most of them characterized by bizarre armored shields that covered the head, each with its own shape, distinctive surface sculpture, and elaborate pattern of sensory organs.

The geographic distributions of these highly distinctive groups play a major role in determining how continents were placed on the globe in Paleozoic times. The early evolution of agnathans also involved many of the great advances in vertebrate history, such as the development of cellular bone, paired limbs, intricate sensory-line systems, dentin-like tissues, complex eye muscle patterns, and the inner ear with two semicircular canals. Agnathan fossils are widely used for dating and correlating Middle Paleozoic sedimentary rocks, particularly the heterostracans from the Euramerican and Siberian regions. In spite of the wide variety of agnathans during the Silurian and Devonian, only the naked-skinned lampreys and hagfishes survived to the present day.

Agnathans take their name from the Greek *gnathos* meaning "jaw" and the prefix *a-* meaning "without," as they are a group of mostly extinct fishes

A parasitic lamprey, *Geotria australis*. Today's living jawless fishes are known only from lampreys and hagfish, rather than the many bizarre bony-shielded forms of the early Paleozoic era. (Courtesy SeaPics)

that lacked true bony jaws and teeth. The name agnathan is no longer restricted to specific types of fishes but now is instead seen to represent the basal radiation—or evolutionary starting point—of all fishes. As such, agnathans are not characterized by any unique features, so they are considered a *paraphyletic group* in modern taxonomic parlance. Although modern agnathans hold little interest to most humans, apart from being a minor food source in some countries, the study of fossil agnathans is vital to our understanding of many important anatomical transformations that took place early in vertebrate evolution. The fossil agnathans are our only window into understanding the origins of bone and jaws and of the complete organization of the standard vertebrate head pattern, so their significance to modern zoology quickly becomes clear. This aside, it is the simple beauty and mystery of the numerous bizarre-looking agnathan fossils that interests us as much as their quintessential scientific value.

Agnathan fishes even played a small role in human history. King Henry I of England died in Lyons-la-Forêt, Normandy, France, on 1 December 1135, after an excessive banquet in which he gorged himself on "lamprey stew." Lampreys are still considered a great delicacy in some parts of Europe, especially the highly prized "Lamproie a la bordelaise," cooked in a white Bordeaux with fresh onions. They are parasites that feed on other live fishes by attaching themselves with a sucker disc around the mouth. They cut into the flesh to feed on the blood of the host. Their other living relatives are the myxiniforms, or hagfishes, which are far more primitive than lampreys in many aspects of anatomy and are by and large deep-sea carrion feeders. Hagfishes are the largest extant agnathans and may reach lengths of up to 1.4 m.

The hagfish, *Myxine glutinosa*, is more primitive in many respects than the lampreys in its cranial structure and represents the most basal of all living fishes. They are mostly deep-sea forms that scavenge rotting carcasses. (Courtesy SeaPics)

The modern lamprey has a fossil record spanning back to the Late Devonian, known from the small broad-headed form *Rinipiscis* from South Africa, remaining almost unchanged throughout the past 360 million years. Lampreys are thought by some researchers to be descended from eel-like anapsid fishes. Hagfishes are considered more primitive than lampreys in many details of their anatomy, such as their lack of a head skeleton, which is present in lampreys.

The extinct fossil agnathans include eight major groups of armored and nonarmored forms, most of which had evolved by the start of the Silurian Period, about 430 million years ago. These are named the Arandaspida, Astraspida, Osteostraci, Heterostraci, Anaspida, Thelodonti, Galeaspida, and Pituriaspida. Only three of these groups—the Arandaspida, Thelodonti, and the Pituriaspida—are recorded in Australia and other Eastern Gondwanan regions. The Osteostraci and Heterostraci are unique to the ancient Old Red Continent Euramerica (Europe, Greenland, western Russia, and North America), and fossils of the Galeaspida are unique in the ancient Chinese terranes (South China, Vietnam).

Close-up view showing the horny teeth (not true vertebrate teeth) of the oral disc of the lamprey *Geotria*. Many lampreys are parasitic forms that attach to fishes and feed on their blood. (Courtesy iStock)

The Oldest Fishes with Bone: Cambrian and Ordovician Agnathans

In October 1996, Gavin Young of the Australian Geological Survey Organisation in Canberra and his colleagues published a report in *Nature* of a possible Late Cambrian vertebrate from central Queensland, Australia, dated at about 520 million years old. These bony fragments possessed an extensive pore-canal sensory system, a feature found in all true fishes. The histology of these bony pieces shows a three-layered skeleton that has a thick enameloid outer layer but lacks dentinous tissue, once thought to be a definable feature of all early vertebrate tissues. This finding supported the idea that bony sheets could well have been a primitive pattern for fishes and that enameloid tissues preceded the widespread evolution of dentin support tissues in the dermal skeleton. Young and his colleagues suggested that a system of canals penetrating bone and opening through pores to the surface (with enamel-like hard tissues) may well have preceded the formation of cellular bone with dentinous tissues in vertebrates.

This 1996 discovery was preceded by the unearthing of a curious fossil from the Late Cambrian at first thought to be the oldest known bit of fish bone. This fossil, named *Anataolepis,* was originally described from fragments found in North America by Boeckelie and Fortey in 1977. Later, it was attributed as an arthropod and then again restudied and found to be a vertebrate with dentin tissues present (Smith and Sansom 2005). It is slightly younger in age than the Australian material.

The model of bone formation as determined from the Australian fossils described in the *Nature* article could have led to the conclusion that conodonts and *Anatolepis* represent divergent lineages within the first radiations of vertebrate hard tissues and so may not be as closely allied to fishes as some researchers have so passionately argued. Alternatively, some scientists think that these remains from central Australia possibly belong to arthropods (M. M. Smith and Coates 2001).

The oldest identifiable fossil vertebrate skeleton that has undoubted real bone comes from the Early Ordovician Horn Valley Siltstone and Pacoota Sandstone of central Australia, dated at about 490 million years old. These fragments of true dermal bone point to Gondwana as the most likely place for the origin of all vertebrates. At this time the South China terrane was closely situated to Gondwana, so close biogeographic affinity of the remarkable Early Cambrian Chengjiang vertebrates (*Haikouichthys, Myllokungmingia,* etc.) also fits in with this region as a center of origin for the first fishes.

One fossil scale type from this region, named *Areyongia,* has been likened to ancient shark scales by Young (1997). Young suggests that these scales were possibly a primitive precursor to those of the chondrichthyan *Polymerolepis*; yet, as their internal structures show significant differences, he was not confident of placing the species within the sharks (class Chondrichthyes). Sansom (2001) suggested that the absence of differentiation into a crown and base precluded them from being vertebrate scales, so classifying this scale remains a topic for future study.

The oldest partially articulated fish fossils are from the Ordovician Period, with specimens known from

Basic Structure of Primitive Agnathans

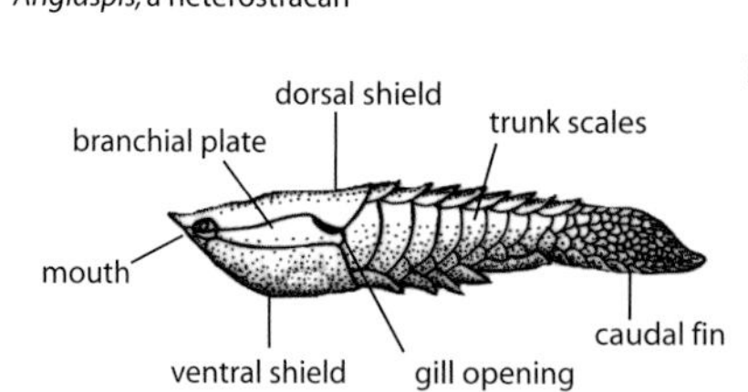

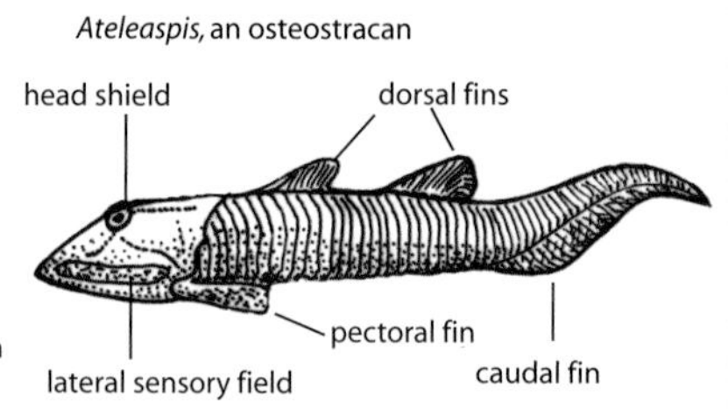

Basic features of two of the major groups of extinct jawless fishes, a heterostracan and an osteostracan.

Early agnathans are known largely from fossils of their cranial shields, isolated scales, and bone fragments. In rare cases, whole fossil agnathans have been found, showing that the early, heavily armored forms had tails covered by thick bony scales. Often they have a dorsal fin or an anal fin, or both, but the presence of paired pectoral fins is known in only three groups, the Thelodonti, Osteostraci, and the Pituriaspida.

The bone forming their shields is made up of an outer ornamental layer, often covered by a shiny enameloid layer over the dentin and perforated by numerous pores. Below this is a middle vascular layer with spaces for collagen fibers and, in some cases (e.g., osteostracans), bones cells (osteocytes). The base of the bone is made up of laminated noncellular bone (called aspidin). Typically, bone is made up of cells, such as osteoblasts and osteoclasts, that perform a variety of functions for bone health. In humans with certain bone diseases, these cells are missing. They are also not found in teleost fishes, the bones of which are said to be noncellular or acellular.

The typical form of the bony shield differs in each of the armored agnathan groups. The shield in the Osteostraci and the Galeaspida is formed from a single unit of bone perforated by holes for the eyes, nostrils, and other sensory organs and open on the underside for the mouth. In the Heterostraci, the armor is formed from several plates of differing sizes and the body is covered by large overlapping scales. In the Anaspida and the Thelodonti, there is no enlarged bony covering, only the scales covering the head and body, and also in the mouth, gill arches, and pharynx. There are large, flat scales in some anaspids; but others such as *Jamoytius* from Scotland appear to have naked bodies, approaching the condition seen in lampreys.

The braincase is well known in some fossil agnathans in which it is well ossified, such as with osteostracans and galeaspids. These show that the inner ear possessed two semicircular canals and that the brain was well formed and segmented into discrete divisions (as in higher vertebrates). A large vein drained the blood from the head, and a complex system of sensory fields and lateral lines was developed in most fossil forms. Only in the Galeaspida and the Osteostraci are the soft tissues of the braincase preserved as delicate shells of perichondral bone, and these are also the only groups to show the head vein placed dorsally near the top of the armor.

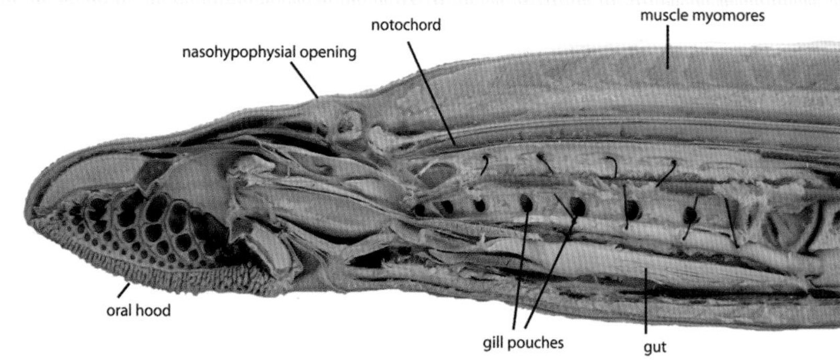

A cross-section of a lamprey head showing anatomical features.

Australia (*Arandaspis* and *Porophoraspis*), South America (*Sacabambaspis*), and North America (*Astraspis* and *Eryptychius*). These fossils all share the primitive feature of having numerous paired openings for the gills, a feature reduced in number in all subsequent jawless fishes except for some anaspids. The bone making up the shields of the North American *Astraspis* is composed of four layers of phosphatic minerals including fluorapatite and hydroxlyapatite. This suggests a close relationship to the heterostracans, a diverse group that

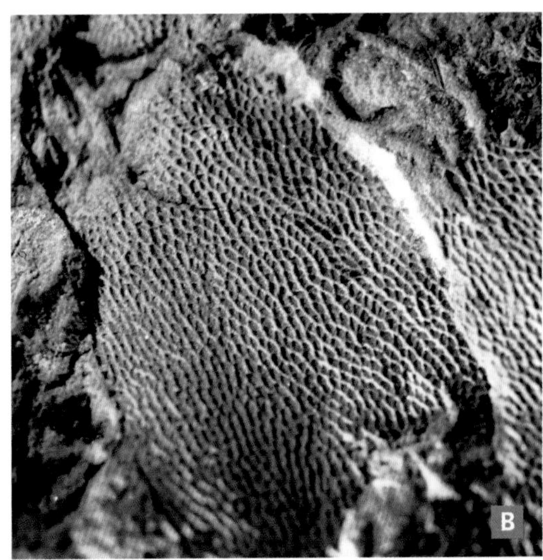

The remains of *Arandaspis prionotolepis*, one of the earliest jawless fishes having a bony external covering, from the Middle Ordovician Period of central Australia. Impression of the ventral shield (*A*) and a close-up of the impression of the surface ornamentation (*B*). The length of specimen is approximately 20 cm.

had similar shields to these Ordovician forms but that possessed only one branchial opening over the gills.

The earliest relatively complete fish fossils come from near Alice Springs in central Australia, where they occur in fine-grained sandstones dated at about 470 million years old. The rocks in which these fish fossils were found represent a shallow marine incursion that covered all of central Australia known as the Larapintine Sea. When the fossils were first found in the rocks in the mid-1960s, the strata were immediately thought to be Devonian age, because at that time Ordovician fish fossils were virtually unheard of. Further collecting at the sites by Alex Ritchie of the Australian Museum in the 1970s and 1980s yielded a number of good specimens of these early primitive fishes.

The shields of *Arandaspis* (named after the Aranda tribe of Aboriginal people) are not preserved as bone but as impressions in the ancient sandstones. These tell us exactly the shape of these armored agnathans and what their body scales were like. *Arandaspis* had a simple dorsal and ventral shield with up to 14 or so paired branchial plates covering the gills. The eyes were tiny and situated right at the front of the head, like the headlights on a car, and there were two tiny pineal openings on the top of the dorsal shield, probably light-sensory organs. The format of the tail is largely unknown, except for the fact that it bore many rows of long trunk scales, each ornamented by many fine parallel ridges of bone, making them comb-shaped. *Arandaspis* occurs

A reconstructed model of *Arandaspis* as a living fish. (Model by Kirsten Tullis, Western Australian Museum)

The Stairway Sandstone where the fossils of *Arandaspis* and other early jawless fishes were found in the Northern Territory, Australia. (Courtesy Alex Ritchie, Australian National University)

with *Porophoraspis*, distinguished by different ornamentation of its plates.

The discovery of the world's first nearly complete Ordovician fish fossils in central Bolivia by French-Canadian Pierre-Yves Gagnier in the mid–1980s caused worldwide scientific interest when preliminary results were first published in *National Geographic* magazine. These fish, called *Sacabambaspis* after the town of Sacabamba in Bolivia, are slightly younger than the Australian fossils (around 450 million years old) but are much better preserved. There a two villages, about 30 km apart. One is Sacabamba, where the holotype was found by Gabriela Rodrigo in 1985, and the other is Sacabambilla, where Gagnier found the articulated specimens 2 years later. These fossils show the entire articulated armor and give the overall body form of the fish. Like *Arandaspis, Sacabambaspis* had a large dorsal and ventral shield with numerous rectangular branchial plates, small eyes at the front of the skull, and paired pineal openings. It also had a rounded plate on each side near the front of the head. The body was covered with many fine elongated scales, and, although it lacked paired or median fins, the tail was quite well developed with a long ventral lobe.

The North American fishes *Astraspis* and *Eriptychius* were for many years known as the earliest fish fossils, having been first described by Charles Doolittle Walcott in the late nineteenth century. Their abundant remains come from the Harding Sandstone of Colorado and are preserved as isolated small fragments of bone. An

A cast of *Sacambambaspis*, the oldest relatively complete fossil fish known. It comes from Bolivia and is of Late Ordovician age. Total length of the specimen is just under 30 cm.

almost-complete specimen of *Eriptychius* was described in the late 1980s by David Elliott. The body looked similar to *Sacabambaspis,* although it had much coarser, diamond-shaped scales covering the tail and its shield was made up of many polygonal units, called *tesserae.* The significance of these fossils lies in their excellent bone preservation, allowing them to play an important role in discovering how primitive bone evolved. The bone of *Astraspis* has four layers: an outer thin layer of enameloid capping a second layer of dentin, which forms the ridges and tubercles of the outside of the plates, a third layer of cancellous or spongy bone, and a fourth basal layer of aspidin, a layered hard tissue that lacks bone cells.

Astraspis is one of the oldest vertebrates known from North America and was initially described from fragmentary plates of bone from the Late Ordovician Harding Sandstone of Colorado by Charles Doolittle Walcott in the 1890s. In recent years, more complete material has been found, enabling this new reconstruction of the fish. (After the work of David Elliott)

Although these primitive Ordovician agnathans closely resembled the Heterostraci (the major Silurian and Devonian group of agnathans, described below), they lacked one distinct feature of that group—a single external branchial opening for the gills. In many other respects, such as having a shield formed of numerous plates, notably large median dorsal and ventral plates, and similar bone structure, they were very similar to the Heterostraci. The heterostracans most likely evolved from such ancestral stock (Donohue and Aldridge 2001).

The bony shield of _Liliaspis_, a heterostracan from the Early Devonian of Russia. Note the distinctive raised ridge ornamentation. (Courtesy Oleg Lebedev, Borissiak Paleontological Institute, Moscow)

The Heterostraci: A Great Radiation

The Heterostraci (meaning "different shield") were a diverse and wonderful assemblage of ancient jawless fishes that thrived during the Silurian and Devonian Periods. They are easily recognized in their fossilized state by the several plates of bone forming the shield, their single branchial opening for the gills, and their characteristically elaborate bone surface patterns. Typically there are two large bony plates covering the top and underneath of the head (the dorsal and ventral shields) and a single plate covering each of the gill openings along the side of the armor (the branchial plates or plate, where they are fused into one bone). Smaller separate plates may form around the eyes (such as the orbital, suborbital, and lateral plates), and the mouth may also have unusual oral plates lining its opening. Sensory lines criss-cross the plates as linear or curved grooves in the bone or as visually clear delineations between the surface ornamental ridges and pustules.

The heterostracans underwent their major radiation early in the Silurian Period and were common in Euramerica (including Siberia) throughout the Devonian, reaching a peak of diversity in the Early Devonian. The largest heterostracans were the giant flattened psammosteids, reaching estimated lengths of around 1 m, but most heterostracans were small fishes about 10 to 20 cm long.

The small traquairaspids and cyathaspids from Arctic Canada and Britain are easily distinguished by their relatively simple-shaped shields with highly elaborate surface ornament. They were primitive heterostracans that lacked the elaborate spines developed in later lineages such as the pteraspidiforms, and the tail has only a few large scales. Some of these, such as the Silurian *Aethenaegis* (the name meaning "Athena's shield") from the Delorme Group of the Northwest Territories of Canada, are beautifully preserved as whole fishes. *Aethenaegis* was a small fish about 5 cm long, which had a V-shaped leading edge on the lower lip of the mouth that may have been used for plankton or detrital feeding. Other cyathaspids, such as *Traquairaspis, Corvaspis, Tolypelepis,* and *Lepidaspis,* had very distinctive surface ornaments, consisting of many polygonal units of elaborately sculptured bony ridges. The cyathaspids flourished during the later half of the Silurian Period but were extinct by the early part of the Devonian.

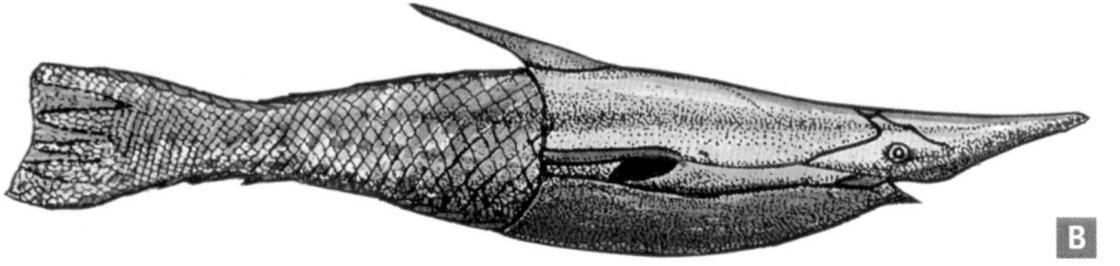

Lack of jaws did not inhibit the diversification of agnathans. Many had elaborate armored head plates and thick bony tail scales, like *Errivaspis waynensis*, an Early Devonian heterostracan, from the Wayne Herbert Quarry, Herefordshire, United Kingdom (*A*). A reconstruction of *Errivaspis* (*B*).

(*Above*) Pteraspid fossils are most commonly represented by parts of the shield, as in the dorsal surface of the armor of this specimen of *Protaspis transversa* from the Early Devonian of Wyoming (*A*) and the ventral surface and tail scales of another specimen (*B*). The concentric growth lines can be clearly seen in each of the plates forming the shield of the second specimen.

Shield of *Anglaspis*, a cyathaspid from the Early Devonian of England.

One of the most successful groups of Devonian heterostracans were the pteraspidiforms (Greek *pteros*, meaning "wing," and *aspis* meaning "shield"), so-named because of the winglike pointed spines at the sides of their armor, called *cornua*. Pteraspidiforms have a more complex shield than do cyathaspids, with separate rostral, pineal, and dorsal discs forming the upper part of the armor. Some forms, such as *Doryaspis* from Spitsbergen, Norway, evolved bizarre pointed processes, or protuberances, at the front of the armor (the rostrum of *Doryaspis* is ventral to the mouth and widely flared lateral wings). Others, such as *Unarkaspis* from Canada, also had wide lateral spines on the armor as well as high dorsal spines. What was once seen as one genus, *Pteraspis*, was recently subdivided into a number of distinct genera by French paleontologist Alain Blieck. His studies of pteraspidiforms have shown them to be very useful in age determination of Devonian rocks in Spitsbergen in Norway and other parts of Europe, in western Russia, and in North America. Earlier work by British paleontologist Errol White first established a detailed stratigraphic zonation using heterostracan fossils. Some of the well-known pteraspidiforms include *Errivaspis* (named in honour of White) from Britain and France, *Rhinopteraspis* from Europe and North America, which has a long elongated rostrum, and the large flat-

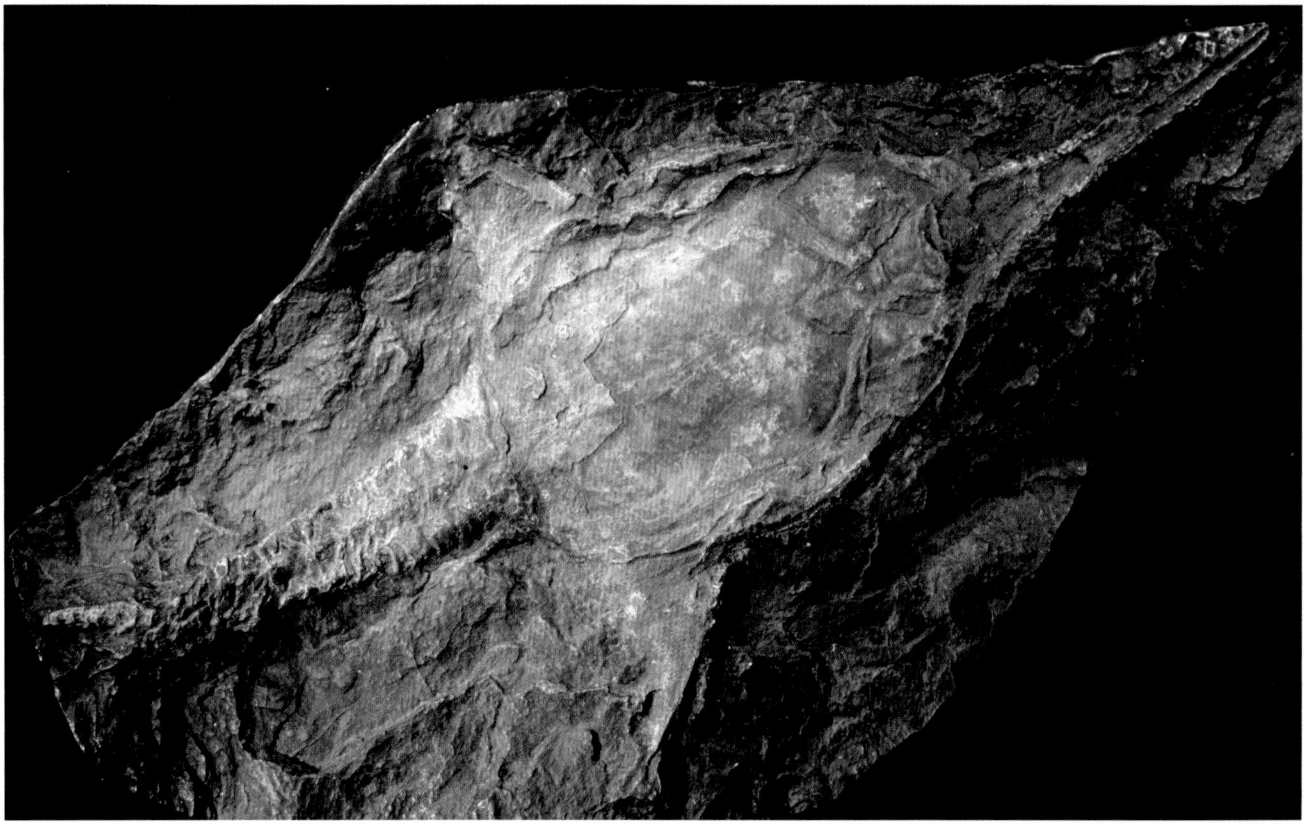

***Unarkaspis* (previously *Lyktaspis*) *nathorsti*, an unusual pteraspid with a long rostrum from the Early Devonian of Spitsbergen, Norway (cast).**

tened form, *Drepanaspis*, from the Hunsrück Slates of the Rhineland, Germany.

Other heterostracans having strange armor include a group unique to the Russian terranes, the amphiaspids. These had wide, rounded armor made of a single piece of bone, and some of the shields resemble flying saucers. Most had shields about 10 to 18 cm long, the largest forms having shields about 40 cm long. *Lecaniaspis* and *Elgonaspis* had bony feeding tubes or scoops at the front of their heads that may have functioned as a pump to suck in small organisms from the mud. The eyes were very small or entirely absent in the amphiaspids, as they lived in muddy habitats and instead relied on hiding themselves in the seafloor away from the gaze of predators. Their exquisite lateral-line systems would also have served them well for sensing when danger was near. Some of the amphiaspid fossils have traces of healed bites on them, suggesting that they regularly survived attacks from the larger jawed fishes that inhabited the same shallow seas.

The Anaspida: Forerunners of Lampreys?

The Anaspida (meaning "no shield") were simple, laterally compressed eel-like jawless fishes that may or may not have had a covering of thin elongated scales on

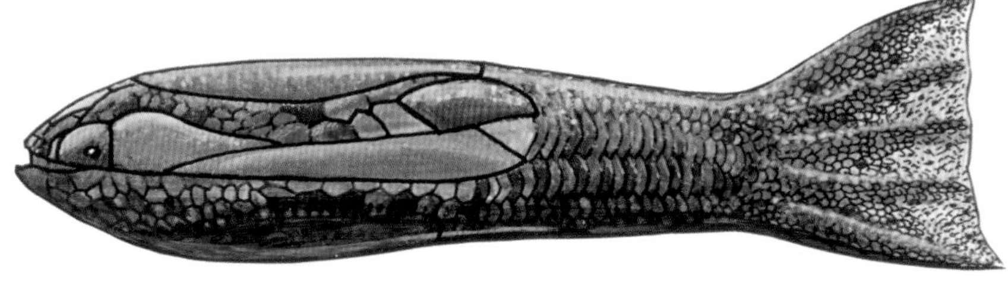

Reconstruction of *Drepanaspis*, a large flattened heterostracan from the Early Devonian of Germany, seen in side view.

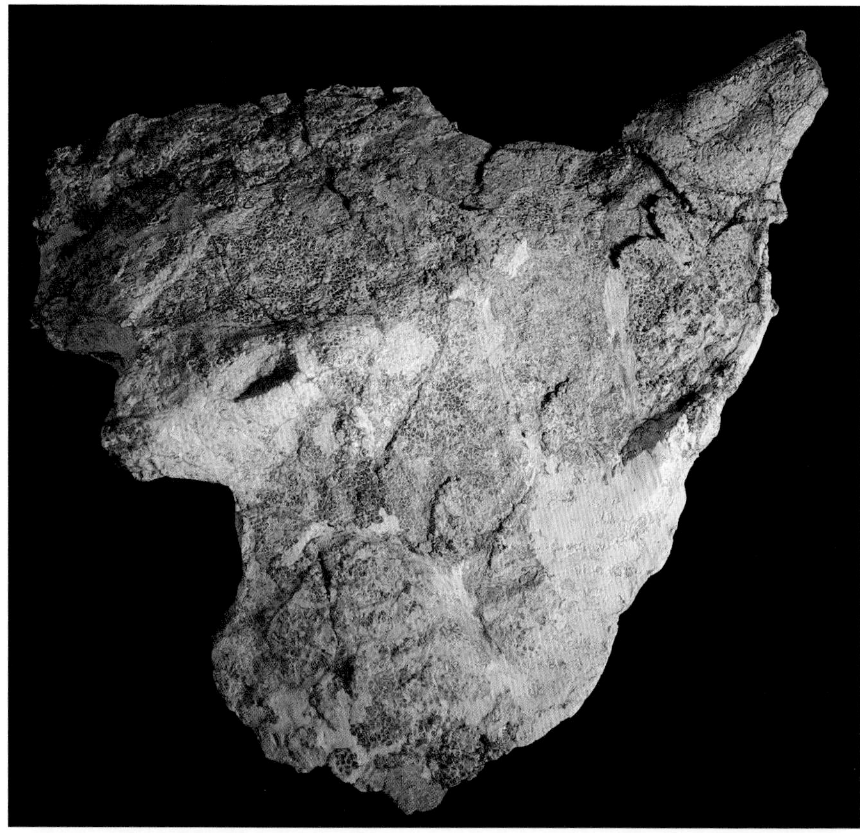

***Elgonaspis*, an unusual tube-mouthed heterostracan from the early Devonian of Russia. It was probably a filter-feeder that used its tube to suck in plankton and organic debris.** (Courtesy Oleg Lebedev, Borissiak Paleontological Institute, Moscow)

***Olbiaspis*, a saucer-shaped agnathan from the Early Devonian of Russia.** (Courtesy Oleg Lebedev, Borissiak Paleontological Institute, Moscow)

their bodies. They were mostly small, rarely exceeding 15 cm in length, although there are some huge anaspid scales known from in the Early Silurian of Canada. The group flourished during the Silurian and early part of the Devonian, and descriptions by Alex Ritchie provided much information on its anatomy (Ritchie 1964, 1980). They had simple fins that developed along the dorsal and ventral ridges of the body, and some forms,

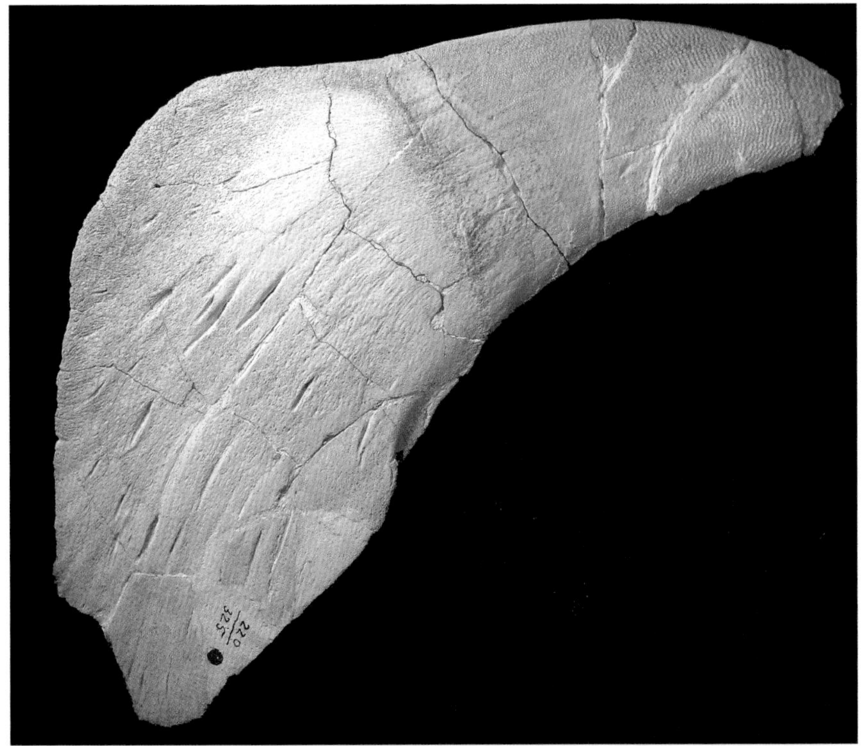

(*Left*) Cornual plate from a large agnathan, *Psammosteus*, from the Late Devonian of the Russian platform. These giant jawless fishes may have reached lengths of nearly 1 m and were the last surviving family of heterostracans. (Courtesy Oleg Lebedev, Borissiak Paleontological Institute, Moscow)

(*Below*) *Birkenia*, an anaspid covered by thin scales, from the Silurian of Scotland. Fossil impression (*A*); whitened cast showing details of the scale cover (*B*). (Both images courtesy Henning Blom, Uppsala University, Sweden)

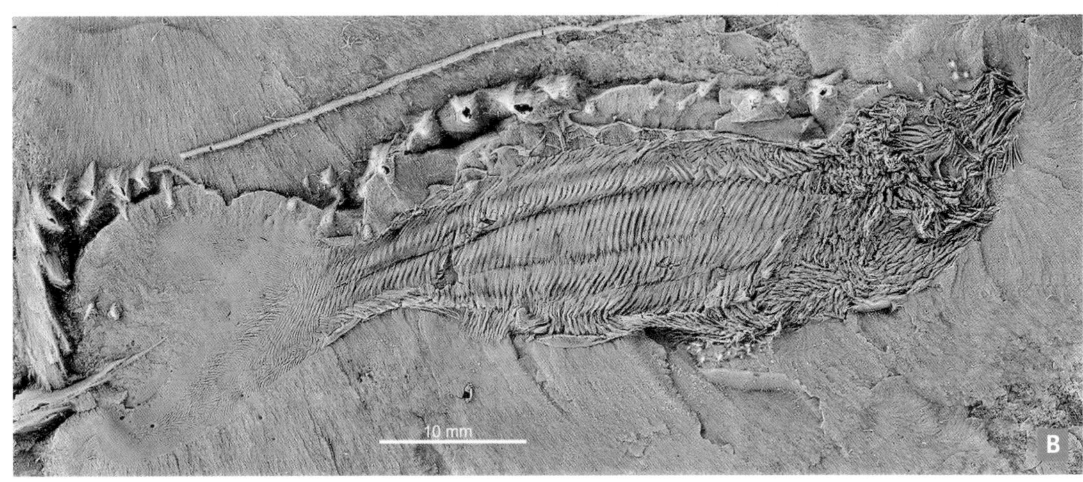

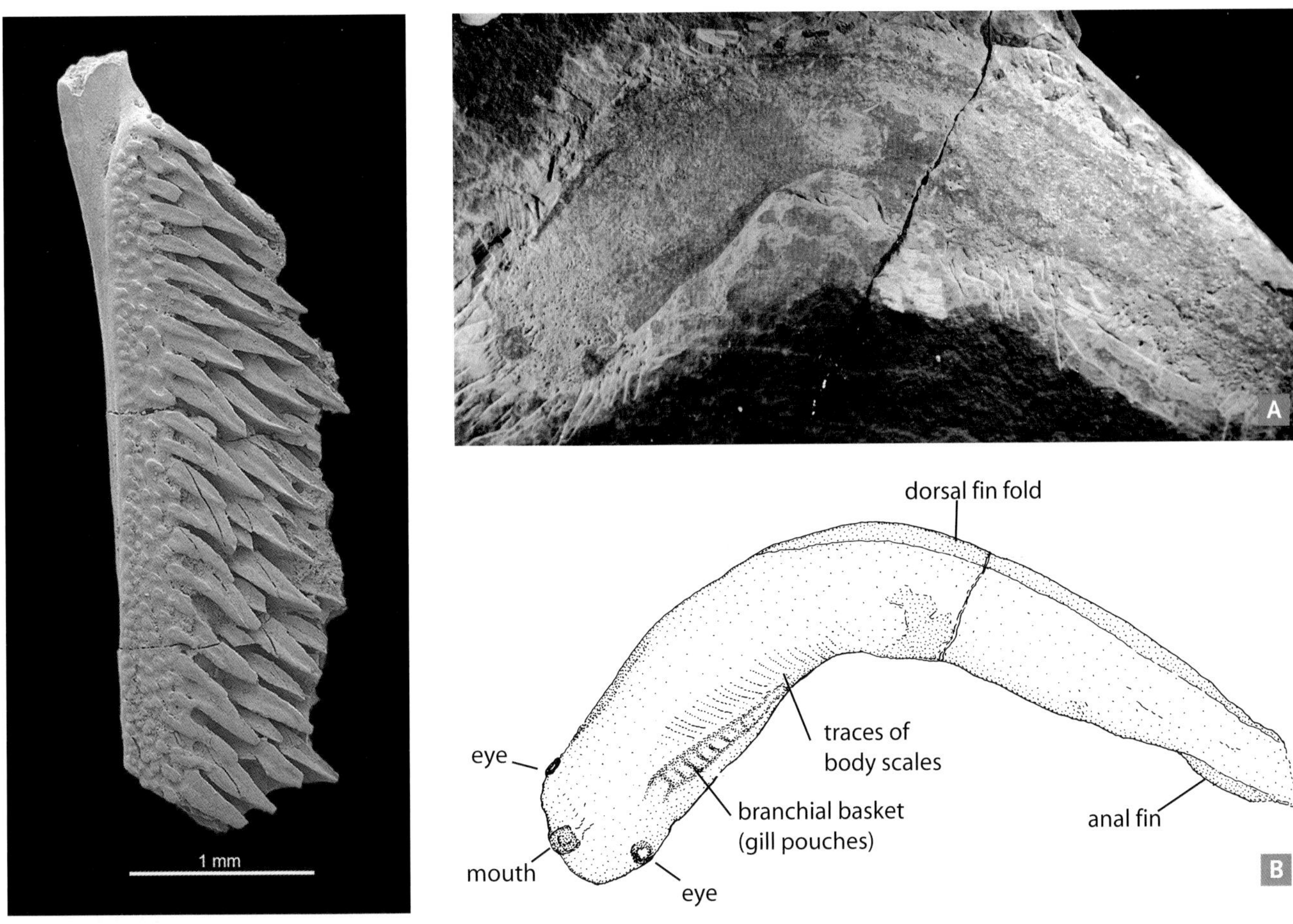

Scanning electron microscope image of a trunk scale of an anaspid, showing detailed surface sculpture. (Courtesy Henning Blom, Uppsala University, Sweden)

The fossil remains of the anaspid *Jamotius* from the Late Silurian of Scotland (*A*) with diagram (*B*) showing main features of the specimen.

such as *Jamoytius* and *Pharyngolepis*, had well-formed lateral fins that were supported by radials. *Cowielepis*, from the Middle Silurian of Scotland, showed that all anaspids probably had a long ventrolateral fin along each side of the body (Blom 2008). The tail was supported by the body axis directed downward and had a thin dorsal (epichordal) lobe. *Birkenia* and *Lasanius* had elaborate arrangements of dorsal scales along the ridge of the body. Like the osteostracans and lampreys, they had a single nasohypophysial opening on the top of the head. The gills opened as a row of holes along the side of the animal, varying in number from 6 to 15 pairs of openings. All known fossil anaspids lived in the ancient Euramerican continent, and their remains are best known from sites in Scotland, Norway, Estonia, and Canada.

Recent discoveries of anaspid-like creatures called *Endeiolepis* and *Euphanerops* from the Late Devonian of Scaumenac Bay, Canada, lead Philippe Janvier and Marius Arsenault to suggest that anaspids are the closest fossil ancestors of modern lampreys. Janvier suspects that *Endeiolepis* could possibly also be the same as *Euphanerops*, as the "ventrolateral scales" of *Endeiolepis* are now known to be the internal cast of the branchial basket of *Euphanerops*. The fossil forms from Canada show that they both possessed a long row of gill arches that may have stretched almost to the tail, and in *Euphanerops* this may have been as many as 30 pairs of gills.

A major revision of the Northern hemisphere birkeniid anaspids was completed by Henning Blom (2007). This monographic study described some 15 new species, 10 new genera, and 2 new families of anaspids

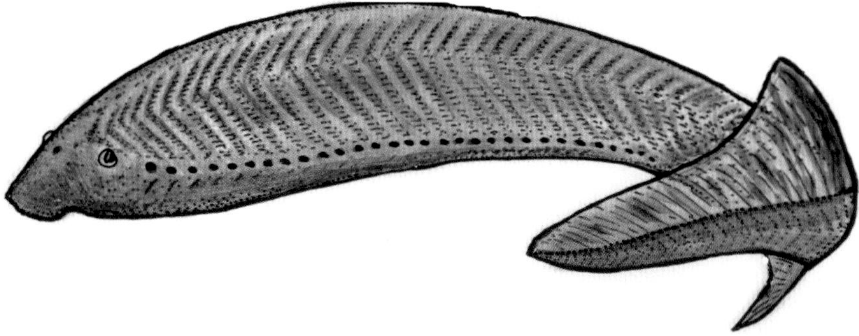

Reconstruction of *Endeiolepis*, a link between the early fossil anaspids and the living lampreys. (After the work of Philippe Janvier and Marius Arsenault)

***Euphanerops*, an anaspid from the Late Devonian Escuminac Formation of Quebec, Canada. This specimen shows part of the body and tail outline.**

and provided a detailed phylogenetic framework for the group.

The anaspids were probably much like lampreys in their lifestyle, being either parasitic feeders on live fish or detrital feeders. However, their remains have been found mostly in marine sediments, except for the Scottish species, which may have come from freshwater deposits. This, of course, does not preclude the possibility that, like some lampreys, they may well have spent a major phase of their life in marine conditions.

The Thelodonti: The Scales Tell the Tales

The Thelodonti (meaning "nipple tooth") are known mostly in the fossil record from their characteristic scales, which have a distinctive crown made of shiny dentin on a bony base perforated underneath by a large pulp cavity. Thelodont scales vary in morphology according to which part of the fish they come from. Short squat head scales give way to elongate trunk scales, and there are many shapes of fin scale and internal pharyngeal scales. The tissues making up the scales are also quite variable. Rare whole thelodont fossils show that most were flattened fishes with broad winglike pectoral fin folds lacking radials as seen in anaspids. They also feature large heads with ventral rows of gill openings (for example, *Turinia*). Thelodonts ranged in size up to nearly 1 m in length, but most were small fishes, generally less than 15 cm long.

Finds of whole, well-preserved thelodonts from the Northwest Territories of Canada, described by Mike Caldwell and Mark Wilson (1993), show that the group actually radiated into many different forms, some of which, the Furcacaudiformes, had deep bodies with

Thelodus, a complete thelodont fish from the late Silurian Lesmahagow site of Scotland.

large forked tails and small triangular dorsal fins, exemplified by *Sphenonectris*. The most remarkable feature of these deep-bodied forms, such as *Furcacauda*, is that they show the presence of a large stomach—an organ thought to be absent in jawless fishes, as it is lacking in the living forms such as lampreys.

The oldest fossil thelodonts, known from scales found in Siberia, are of Late Ordovician age, and the group became extinct by the end of the Frasnian stage of the Late Devonian. Most genera of thelodonts had died out by the end of the early part of the Devonian in Euramerica, but those in Gondwana, belonging to the genera *Turinia* and *Australolepis*, survived later. Their scales range in size from about 0.5 to 2 mm, and each thelodont would have had several thousands scales covering its body and lining the insides of the mouth and gill slits. This resulted in many different scale shapes for each individual fish; thus the job of sorting out species from isolated scales is really the domain of thelodont specialists.

The thelodonts probably lived a variety of lifestyles. Sue Turner, one of the world's foremost experts on thelodonts, believes that the larger, flat forms such as *Turinia* may have been slow-moving bottom feeders, much like the modern angel sharks (*Squatina*), perhaps either grubbing the mud for invertebrates or waiting in ambush for small items of passing prey. The deep-bodied thelodonts from Canada were likely to have been more

Trunk scales of the thelodont *Turinia antarctica*, from the Middle Devonian Aztec Siltstone of Antarctica, showing crown view (*A*) and basal view with pulp cavity (*B*). The scale is about 2 mm in size. (Courtesy Ken Walker, Museum Victoria)

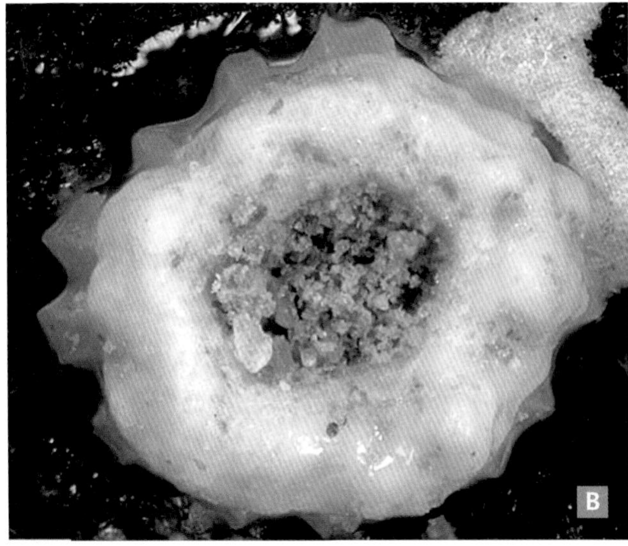

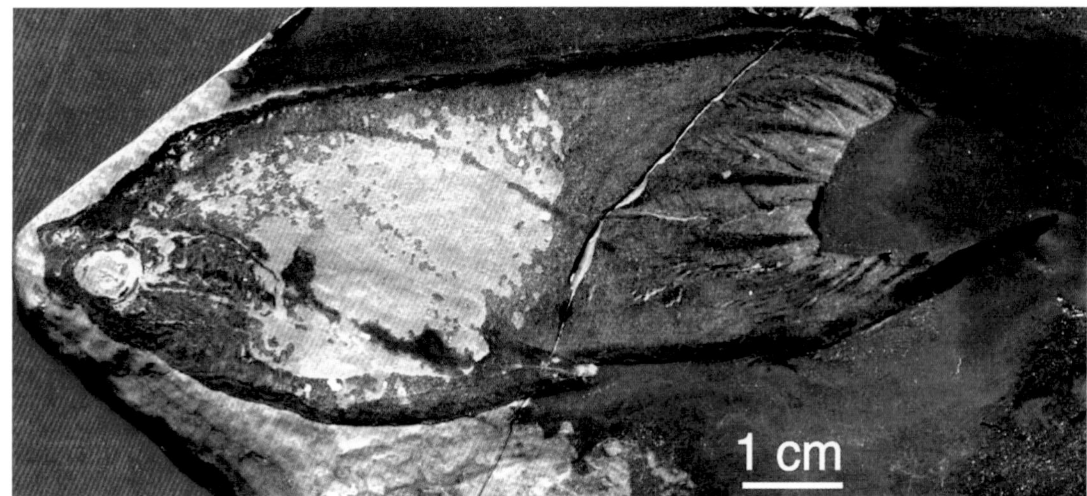

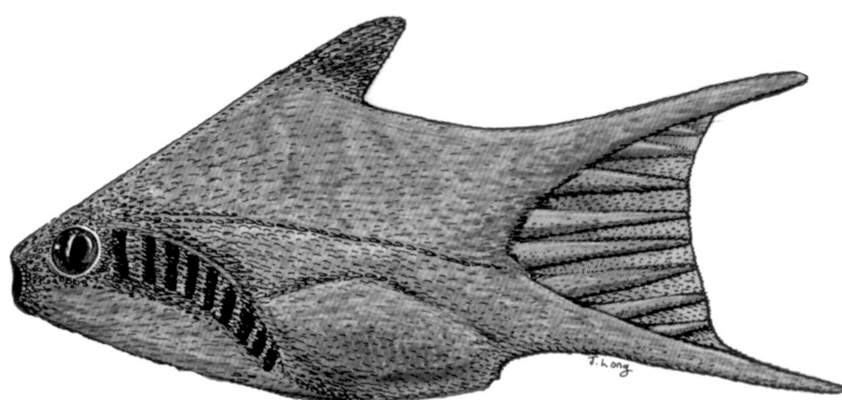

(*Above*) The fork-tailed thelodont *Sphenonectris turnerae*, from the Early Devonian MOTH—for Man on the Hill—locality of Canada. (Courtesy Mark Wilson, University of Alberta, Canada)

(*Left*) Reconstruction of a fork-tailed thelodont, *Furcicauda*, from the Early Devonian MOTH Locality of Canada.

active swimmers that lived by filter-feeding or catching free-floating prey. Some thelodonts like *Lanarkia* had strongly pointed body spines that may have been able to move from laying flat along the body to pointing outward defensively if the fish puffed itself up, as modern pufferfish do today (suggested by Turner 1992).

Fossils of whole fishes found in Canada support a suggestion that thelodonts were far more advanced than previously thought and may be either a closer link to jawed fishes or close relatives of the heterostracans. The link to jawed fishes is based on the primitive scales of early thelodonts being very similar to those of early sharks and even primitive thelodonts having well-formed stomachs (although a stomach is present in the anaspid *Euphanerops* as well). Also, some thelodonts (*Loganellia*) have denticles in the pharynx, like jawed vertebrates. The link to heterostracans is based on the similarity of the unique fork-tailed caudal fin structure, also seen in the strange family of irregulariaspidid heterostracans from northwest Canada (Pellerin and Wilson 1995). In fact, thelodonts are perhaps not naturally monophyletic—a group sharing the same ancestor and characterized as having predominant features in common—with some being close to heterostracans, for example, and others closer to jawed vertebrates.

The Galeaspida: Mysterious Oriental Agnathans

The Galeaspida (meaning "helmet shield") were a group of extinct jawless fishes unique to the ancient terranes that today make up southern China and northern Vietnam. Perhaps because of the long isolation of these ancient continental blocks, the galeaspids evolved a completely different style of armor from other agnathans, some being the most bizarre-looking remains of all known fishes. Some, like *Dongfangaspis*, had up to 45

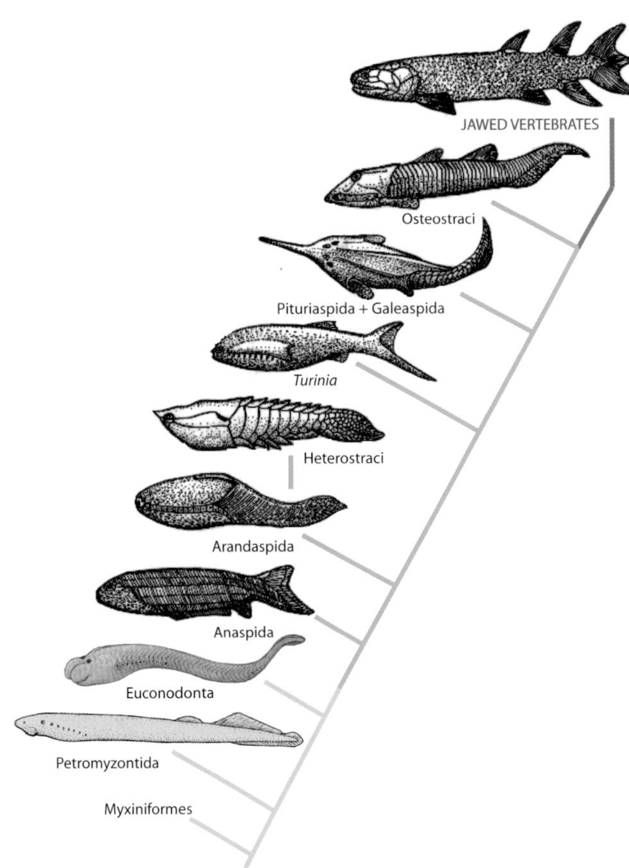

Relationships of the major groups of living and jawless fishes, after recent work by Philippe Janvier and others.

pairs of gill openings. The armor of galeaspids is made up of a single bony shield without separate plates, except on the ventral side of the head, as in osteostracans and pituriaspids. The unique feature of galeaspids is that the armor has a large median hole in front of the paired eye holes, called the median dorsal fenestra. This is very large in most galeaspids, opening directly below to the paired nasal cavities. In a recent phylogenetic analysis of the group it was discovered that this fenestra evolved independently in two lineages, once within the polybranchiapsids and once within the huananaspidiforms (Zhu and Zhikun 2006).

Galeaspids were a diverse group, with more than 80 known species, and the fine preservation of tubes of laminated perichondral bone around soft tissues of the head tells us much about their soft anatomy. They had a complex brain and well-developed inner ear with two vertical semicircular canals. The group had evolved by

The fossil impression of the shield of *Pituriaspis doylei* (*A*), a representative of a "class" of vertebrates newly described by Gavin Young from central Australia in 1992, in which all the original bone has weathered away. A restoration (based on the latex peels) of *Pituriaspis* (*B*).

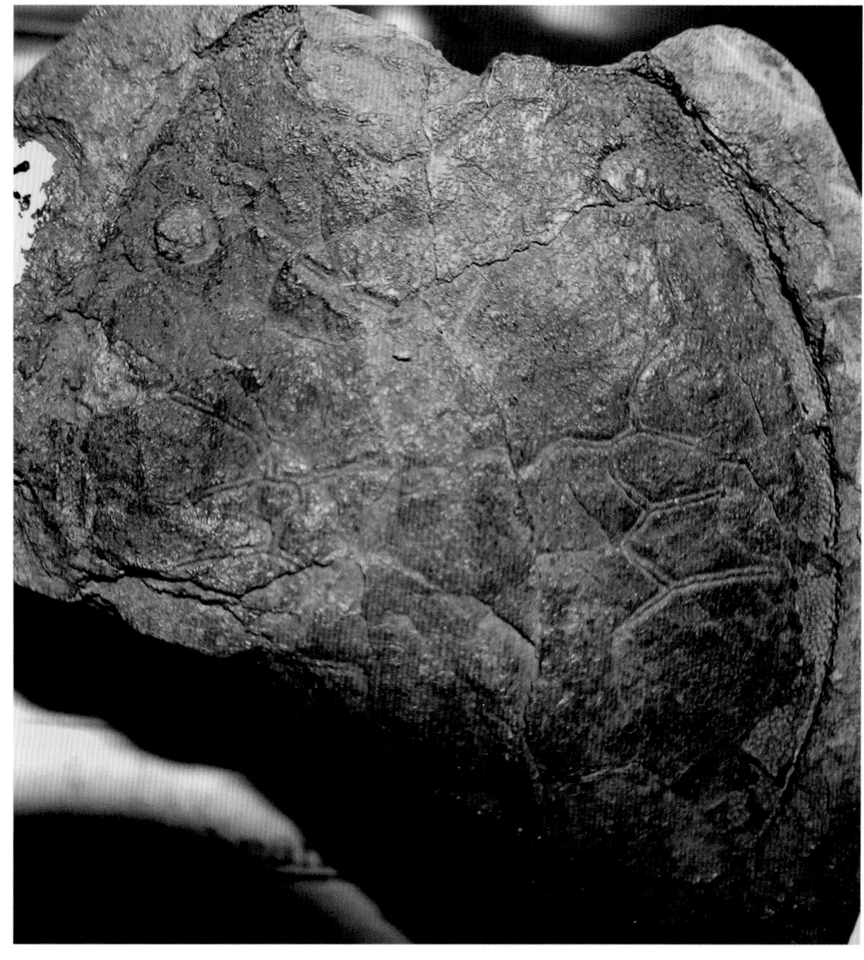

An impression of the bony shield of *Polybranchiaspis*, a common galeaspid agnathan from the Early Devonian of China and northern Vietnam.

This impression of a galeaspid shield from the Early Devonian of China shows clearly the circular holes for the eyes and the elongate median dorsal opening as well as numerous raised lines representing the pathways of sensory canals.

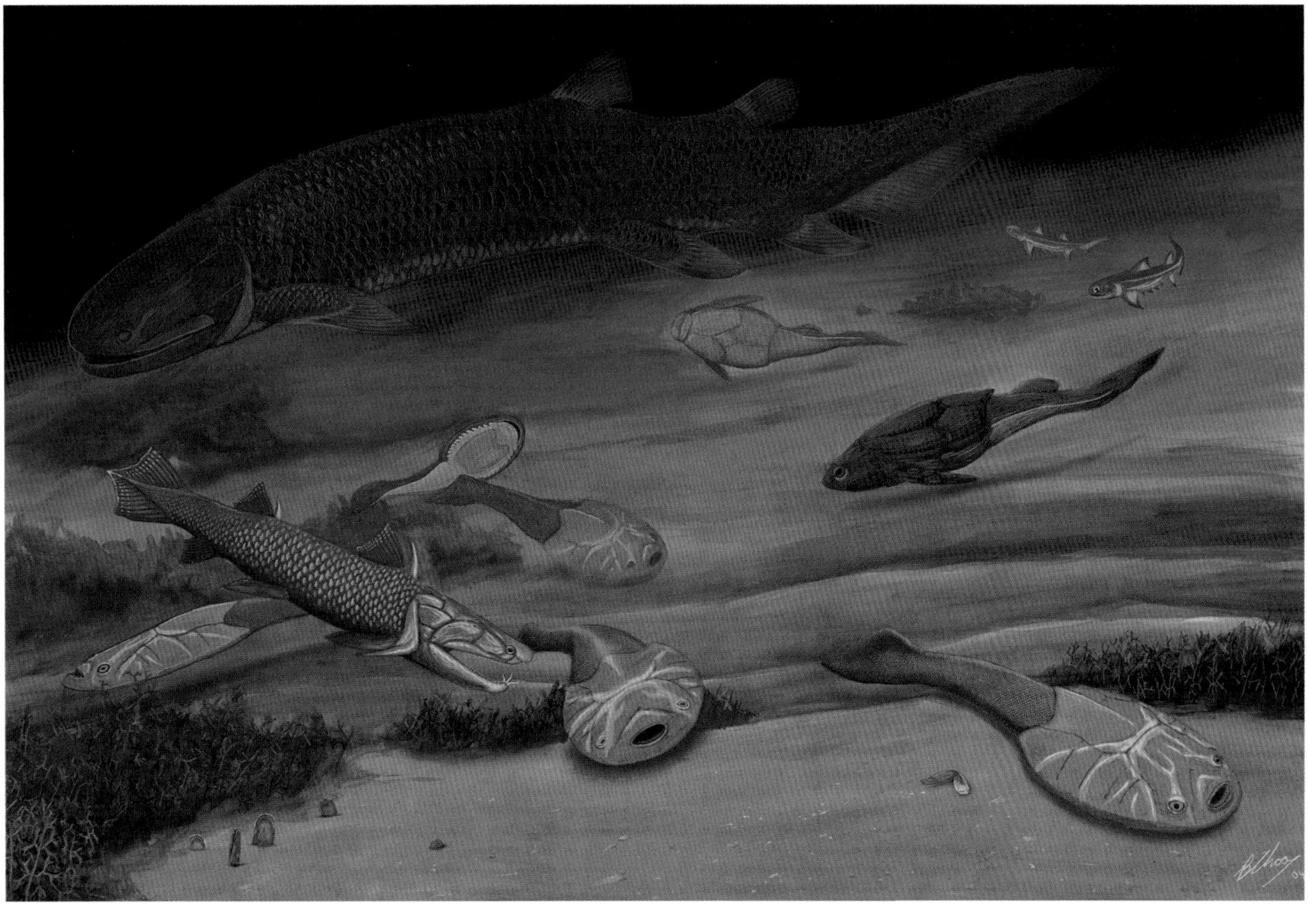

the beginning of the Silurian, reached a peak of diversity in the Early Devonian, and most had become extinct by the Middle Devonian (only one species known from the Emsian and one from the Eifelian stages; Zhu 2000). Finally one indeterminate form of galeaspid is known from the latest Devonian in Ningxia, northern China, before the group became extinct at the end of the Devonian Period.

Average galeaspids, like *Polybranchiaspis,* had simple ovoid shield with well-developed radiating patterns of sensory-line grooves adorning the surface of the top. Triangular shields are represented by fishes such as *Tridensaspis,* and a broad, almost semicircular shield shape is seen in *Hanyangaspis* (this genus is now regarded as one of the basal forms; Zhu and Zhikun 2006). Extreme shape development is seen in *Huananaspis* and *Lungmenshanaspis,* for example, which possessed drawn-out narrow process of bone at the front and sides of the armor. It is also seen in others such as

A scene showing fishes of the Early Devonian Xitun fauna of Yunnan, China. Several galeaspid agnathans can be seen on the seafloor in the foreground chased by the large osteichthyan *Youngolepis* (*rear*) and *Psarolepis* (*left*). Small yunnanolepid antiarchs can be seen on the seabed at the back right, and in the far distant view are two early acanthodians. (Courtesy Brian Choo)

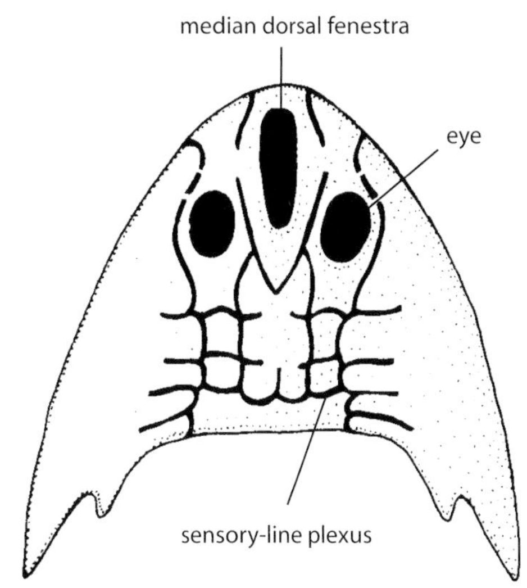

Sinogaleapis **shield showing head structure of a galeaspid.**

Sanchaspis, which had a forward-projecting tube with a bulbous enlargement on the end. *Polybranchiaspis* was one of the first named galeaspids and has been found throughout China and in the north of Vietnam.

The Pituriaspida: A Class of Their Own

Pituriaspids take their unusual name from an Australian Aboriginal word *pituri*, the name of a plant containing a narcotic drug (can I get some for my garden?) sometimes used by central Australian Aboriginals. The fossil was so weird that the discoverer, Gavin Young, thought he may have been hallucinating when he saw it. Pituriaspids come from the Toko Range in southwestern Queensland, Australia. They lived close to the start of the Middle Devonian and represent the only body fossils of agnathans of Devonian age from Australia. The unique feature of pituriaspids is that they have a long bony armor, creating a tube around the head and trunk region, and this armor has a large opening below the eye holes. The front of the armor features a long, forward-projecting extension of bone called a *rostrum*. Two forms are known, *Pituriaspis* and *Neeyambaspis*, the latter having broader, shorter armor. The pituriaspids are believed to have had well-developed pectoral fins, as seen by paired openings on each side of the armor, and a strong shoulder of bone that would have protected the front edge of the fin.

At present we know less about the anatomy of the pituriaspids than we do other fossil agnathans, but from their general appearance we can place them close to the osteostracans. Young (1991) classified the pituriaspids at the base of the radiation containing osteostracans and galeaspids. The main feature of pituriaspids is that they show that a cephalaspid-like shield that can exist without any dorsal nasohypophyseal opening.

The Osteostraci: The Pinnacle of Jawless Achievement

The Osteostraci (meaning "bone shield") were once called ostracoderms. They were a diverse group of solid-

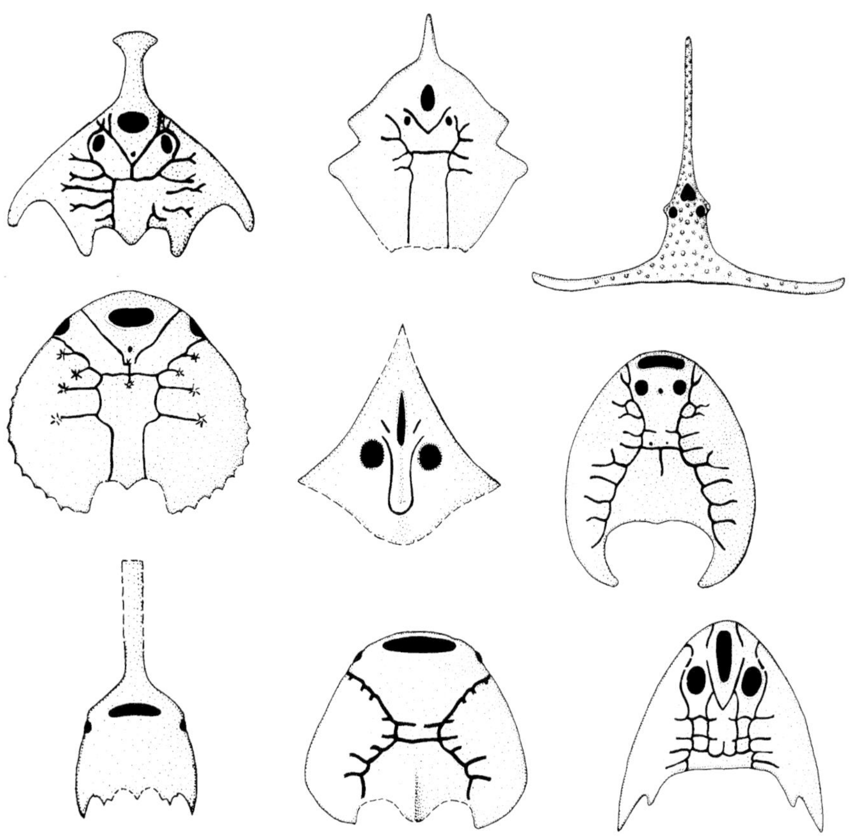

Diversity of galeaspid head shield shapes. Top row, left to right: *Sanchiaspis, Nanpanaspis, Lungmanshanaspis;* middle row: *Cyclodisaspis, Tridenaspis, Changxingaspis;* bottom row: *Sanquiaspis, Hangyangaspis, Sinogaleaspis.*

A complete specimen of the osteostracan *Cephalaspis pagei*, from the Early Devonian of Britain. (Courtesy Rob Sansom, Leicester University, and The Natural History Museum, London)

shielded agnathans restricted to the ancient Euramerican continent, with their fossil remains well known from the Old Red Sandstone outcroppings in Britain, in Spitsbergen in Norway and in other parts of Europe, in western Russia, and in North America. They had a large bony shield with two round eye holes, a key-shaped smaller opening for nasal organs, and a tiny pineal opening between the eyes. The sides of the shield had large areas of probable sensory function, and a similar sensory field sat on the top of the shield. The shield of many species had well-developed cornua, generally projecting rearward. The shield grew from tesserated smaller plates that fused to form a solid shield at maturity. The pectoral fins were well developed in osteostracans and attached on simple internally ossified shoulder girdle bones (scapulocoracoids), on which articulated a simple, paddle-shaped, cartilage support.

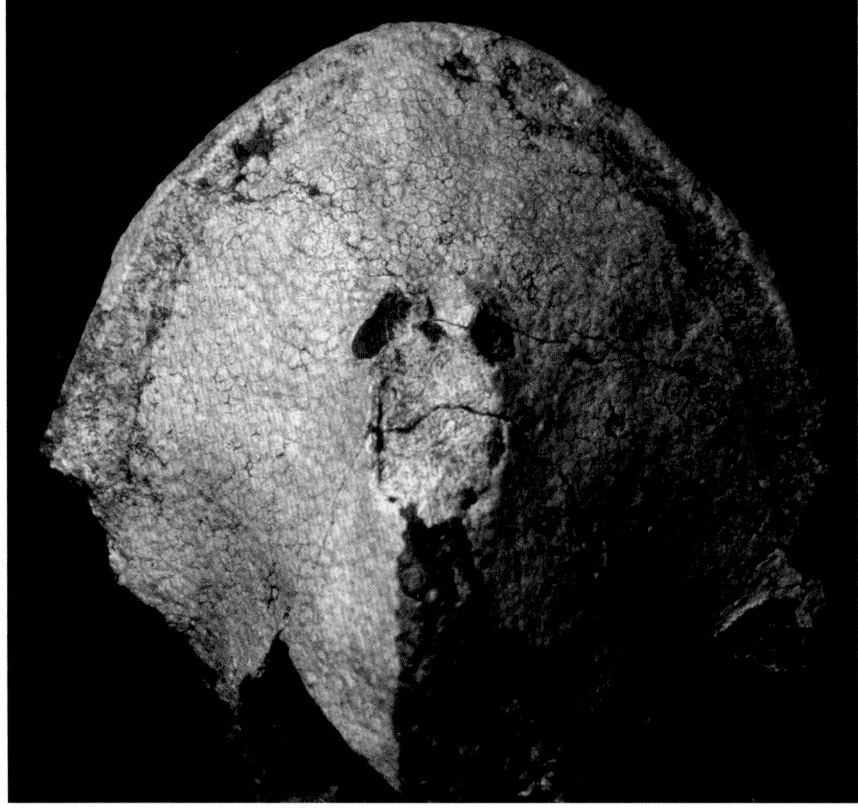

The head shield of an osteostracan, *Zenaspis selwayi*, showing how the shield is made up of many polygonal units of bone. Note the central nasohypophysial opening between the eyes.

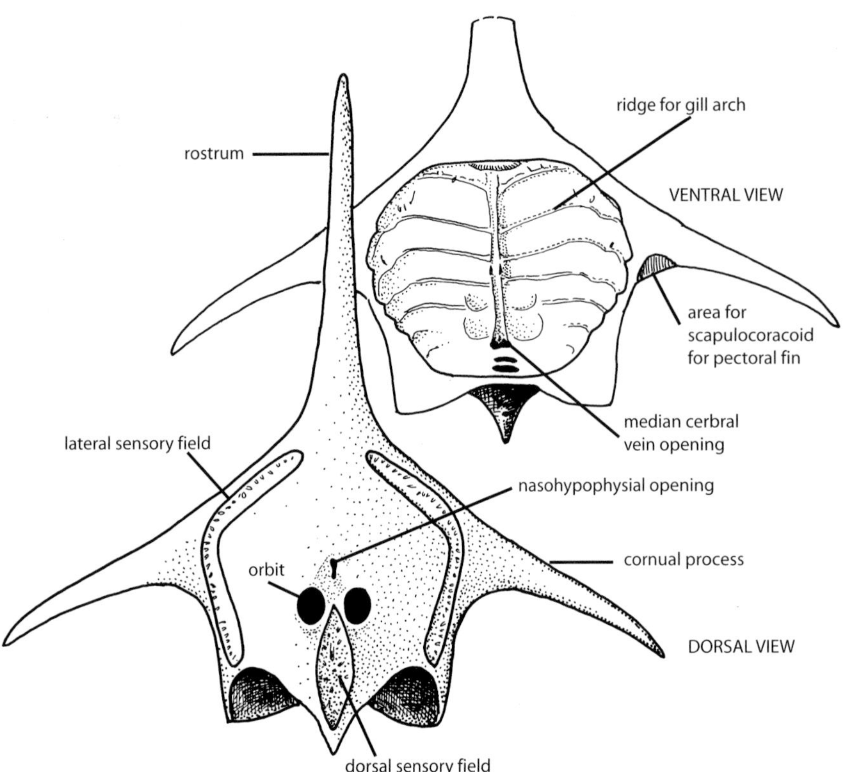

The basic structure of an osteostracan head shield (*Boreaspis*).

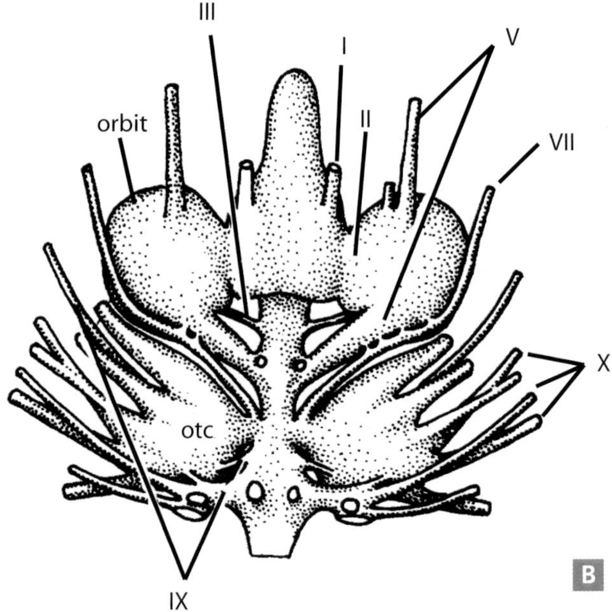

The inside of an osteostracan head shield (*A*), showing delicate perichondral bone preserving pathways of the soft tissues around the brain (*Norselaspis*). Restoration (*B*) of an osteostracan brain (*Boreaspis*). Both are from the early Devonian of Spitsbergen, Norway. (Courtesy of D. Servette—MNHN-Paleontologie)

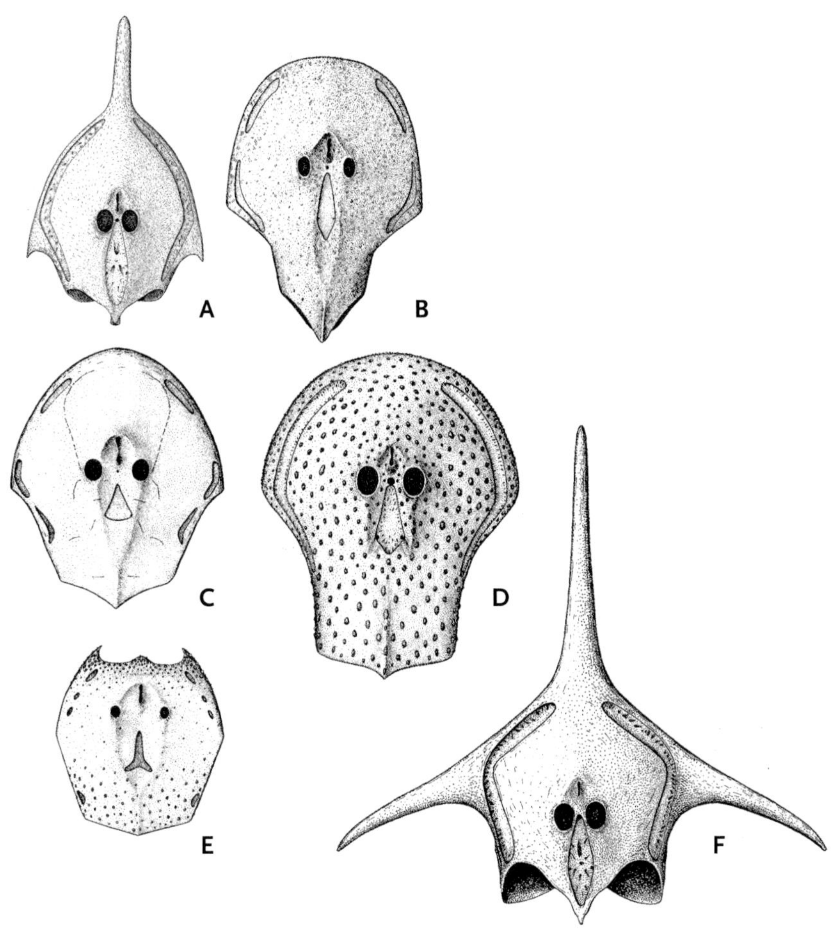

Different shapes of osteostracan head shields show the range of variation seen in this group: *Belonaspis* (*A*), *Axinaspis* (*B*), *Nectaspis* (*C*), *Norselaspis* (*D*), *Gustavaspis* (*E*), *Boreaspis* (*F*).

In some osteostracans (e.g., *Tremataspis*), the paired fins secondarily disappeared, giving the shield the shape of an olive. They had one or sometimes two dorsal fins on the body, as shown in *Ateleaspis* from Scotland. The underside of the fossil shield is largely open beneath the cavity for the mouth and gills, but in life was covered by a mosaic of many small platelets, as shown in rare well-preserved fossils. Up to 10 pairs of gill slits were present, indicated by small paired openings beneath the shield. In some osteostracan fossils, impressions of the brain cavity can be seen, with pathways of nerves, arteries, and veins preserved by thin perichondral bone. From such fossils, we know that the fishes had only two semicircular canals forming the inner ear, as opposed to the three found in higher vertebrates, and that the general plan of the cranial nerves and vascular supply to the head was similar to that of the larvae of lampreys (Janvier 1985). Osteostracans also had well-developed sclerotic bones around the eyes, and some had small tuberculated oral areas at the front of the mouth, clearly an aid to seizing food.

The lifestyle of osteostracans has been reconstructed by Afanassieva (1992) as being of two kinds. The first was largely bottom-dwelling and was heavily shielded without paired fins and, hence, less mobile. Forms such as *Tremataspis* could move only by flits of the short tail. Their heavily ossified shields gave them weight to lie easily on the bottom of the seafloor feeding in the mud. The other group of osteostracans were stronger swimmers with well-developed pectoral fins and long tails (as in *Cephalaspis* and kin). These fishes sometimes had bizarre wide shields, as in *Parameteoraspis*, for protection against predators. They were clearly capable of moving away from danger quickly if the need arose.

***Tremataspis*, a long-shielded osteostracan from the Late Silurian of the Island of Oesel, Sweden. The length of specimen is just under 4 cm.**

The pore-canal system in the bones of all osteostracans was probably a mucus-secreting organ that facilitated easier swimming by reducing drag as the fishes swam or moved along the water bottom. All the osteostracan fossils with well-preserved tails show the presence of thick scales, often arranged as a series of vertically oriented rectangular units, which are capped by a series of smaller ridge scales along the back and meeting another series underneath on the belly.

The osteostracans underwent a major radiation during the Early Devonian resulting in a great diversity of forms, ranging from those with simple semicircular head shields (such as *Cephalaspis*) to others with prominent dorsal spines (*Machairaspis*) or elongated shields that cover much of the trunk of the fish (*Thyestes, Dartmuthia, Nectaspis*). The phylogeny of osteostracans has been greatly elucidated in recent years by the work of Robert Sansom (2009). The osteostracans died out early in the Late Devonian.

Links to the First Jawed Fishes

Two groups of extinct jawless fishes—the thelodonts and an osteostracans—are contenders for the title of possible ancestor of the jawed fishes, or gnathostomes. Thelodonts were once favored as a sister group to gnathostomes because well-preserved fork-tailed specimens from Canada (Furcacaudiformes) showed the presence of a well-developed stomach, a feature currently only seen in living jawed fishes and higher vertebrates. Furthermore, the thelodonts show broad-based pectoral fins and a well-developed caudal fin, which could be argued to be a later development than the osteostracan tail. The resemblance of thelodont scales to the teeth of primitive jawed fishes and the presence of denticles in the pharynx are other features that could unite them with early gnathostomes, in particular to sharks.

The Osteostraci are the group currently favored by most paleontologists as being closer on the evolutionary line that leads to jawed vertebrates. Philippe Janvier (2001, 2007) elegantly summarized the shared characteristics of osteostracans with jawed fishes as follows: the development of paired fins (including pectoral fins with an ossified scapulocoracoid and a cartilaginous fin skeleton), the open endolymphatic duct on the head, ossified bones around the eye (sclerotic bones), cellular bone in both the external and internal ossifications, two dorsal fins, an epicercal tail, and slit-shaped gill openings. Wilson et al. (2007) also demonstrated the suprabranchial paired fins in osteostracans were closer to gnathostome pectoral fins than to other fin structures seen in jawless fishes. The Galeaspida also share with osteostracans and gnathostomes a perichondrially ossified (or calcified) endoskeleton, externally open endolymphatic ducts, a large dorsal jugular vein, and an occipital region developed on the braincase that encloses the exit for the vagus nerve. Although pituriaspids are poorly known at present, they are believed to be related

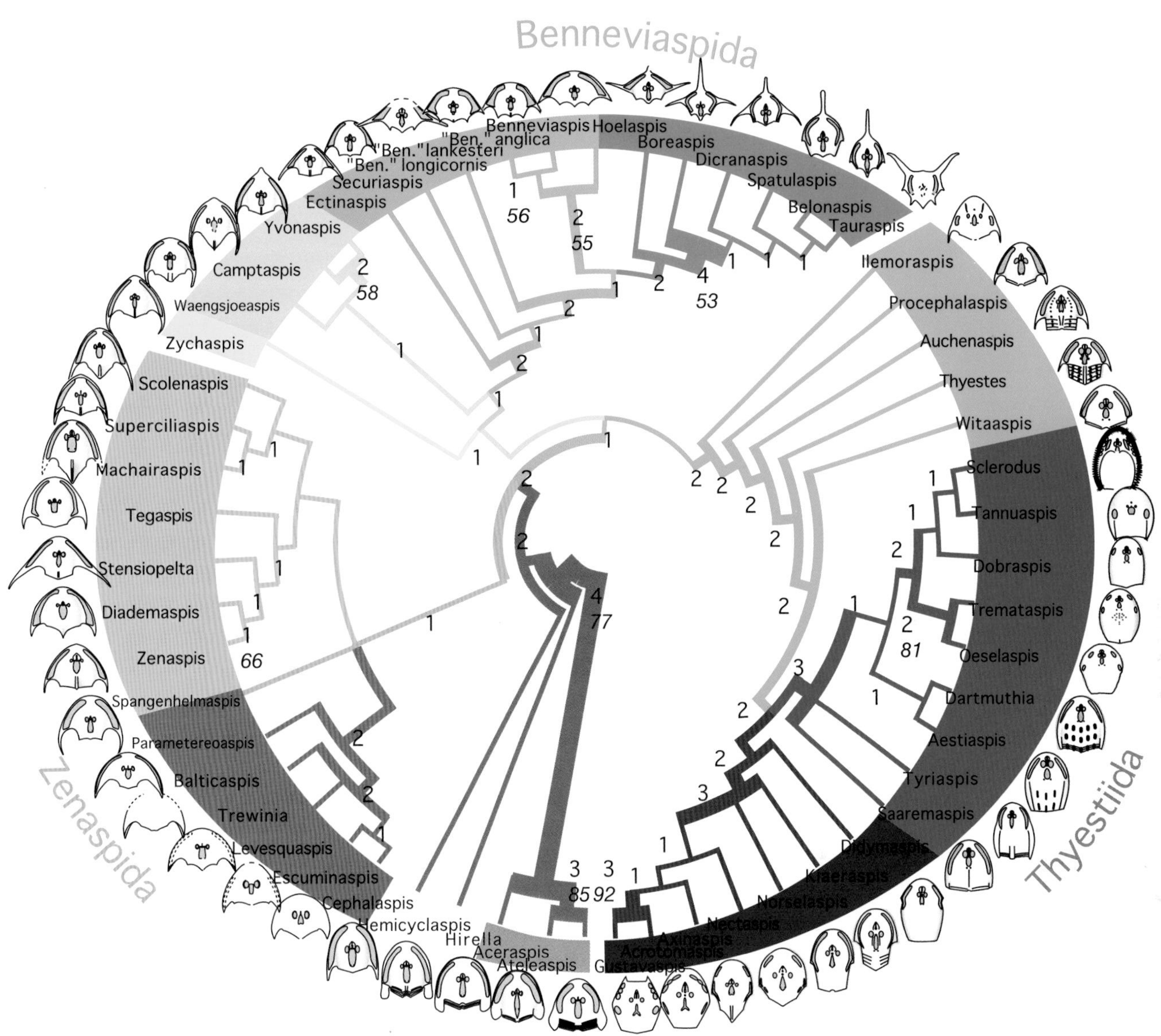

A recent scheme of interrelationships of the osteostracan fishes by Rob Sansom. The different shading and numbers refer to distinct clades (see Sansom 2009 for details).

to osteostracans (and possibly also to galeaspids) because of overall similarities in the head shield and the presence of pectoral fins.

A vital connection between the appearance of eye bones and the origins of jaws is seen through discoveries in developmental biology. Through studies of the developing chick embryo, Brian Hall of Halifax University, Canada, determined that the tissue forming the buds that develop into the sclerotic bones covering the outer margin of the eye are also integral to initiating the formation of the lower jaw. Perhaps it was the developmental pathway for the by-product of jaw ossification. Once sclerotic bones had formed, a series of connector genes might have accidentally begun making a lower jaw cartilage, perhaps as a means of strengthening the existing mouth parts. Such a developmental model would indicate that toothlike structures could have evolved within the mouth cavity well before supporting jawbones did. Diverse groups of fossil agnathans show that some did have functioning bony plates in the mouth area for feeding on carrion or possibly live prey.

The initiation of an ossified support for these mouth parts, the Meckel's cartilage, may therefore have arisen as a result of protecting vital sensory organs, the eyes. It is interesting to note that among all jawless forms only the osteostracans have well-ossified scleral bones, and, therefore, some latent potential for jaw development.

The earliest jawed fishes were once presumed to be chondrichthyans, such as the sharks, as their scales date back farther than for any other group, to the Late Ordovician / Early Silurian of North America. The earliest evidence for prismatic calcified cartilage, one of the two chondrichthyan characteristics, is late Lower Devonian but certain isolated teeth from the Lochkovian come from chondrichthyans. All presumed Early Silurian or Late Ordovician "shark" scraps can be anything, although certain scales, such as the Late Silurian Elegestolepis are likely to be placoid, the type found on sharks. The earliest clear evidence for gnathostomes is from Late Ordovician acanthodian type scales, but confirmation in the form of complete body fossils is still lacking. Further evidence of Early Silurian placoderms from China also places that group back in the running for being the first gnathostomes. The view supported by many researchers who study early vertebrates is that placoderms are most likely the most basal jawed vertebrates (see Chapter 4 for further discussion).

The problem of whether the early sharklike scales referred to as "mongolepids" belonged to sharks with teeth and jaws has not yet been resolved, as the occurrence of shark teeth and scales in the same deposits is not known until the Early Devonian. Thus the high degree of anatomical complexity seen in the osteostracans was to set the stage for the next great revolution in fish evolution, the rise of the jawed fishes. With jaws and teeth the predator-prey arms race was set to begin, and the war for survival or extinction in the fish world was to keep raging throughout the Devonian Period and beyond.

The End of the Jawless Empire

By the start of the Late Devonian, most of the many families of primitive jawless fishes had become extinct. The handful of survivors included four species of osteostracans found in the Escuminac Formation of Canada, one possible (but indeterminate) galeaspid from Ningxia in northern China, one genus of thelodont from Australia (*Australolepis*), the lamprey-like anaspids also from the Escuminac Formation of Canada (*Euphanerops*), and a few species of the large, flattened psammosteid heterostracans from Europe. A possible reason for the steep decline in agnathan diversity at this time is the rapid increase in the diversity of jawed fishes, which may have either displaced some of the agnathans or fed on them or their eggs or larvae directly.

In the Silurian, the agnathans ruled as the dominant fish type both in diversity and biomass, as the early jawed fishes were comparatively rare. By the start of the Devonian, all the major groups of jawed fishes had appeared, with some like the stem gnathostome acanthodians reaching levels of high diversity by the last phase of

A restoration of the high-shielded osteostracan *Meteoraspis*.

the Early Devonian. By the Middle and Late Devonian, many of the jawed fish groups reached a peak of diversity, such as placoderms, porolepiforms, tetraopodomorph fishes, and dipnoans.

It is clear from their similar body shapes that some of these placoderms could have taken over the niches of agnathans that had succumbed to increasing predation pressure. For example, the flattened phyllolepid placoderms appeared straight after the extinction of the flattened psammosteid agnathans, as we know from their fossils in the same succession in the east Baltic region. Long-shielded heterostracans were probably outcompeted by the long-shielded early placoderms, which could either defend themselves better against predators or perhaps swim faster. The detrital bottom-feeding agnathans may have been put out of business by the many new forms of bottom-feeding placoderms, such as antiarchs. By the close of the Devonian, 355 million years ago, only the lampreys and the hagfishes, unburdened by bony armor, were left to carry the flag for the once-mighty jawless vertebrate empire.

CHAPTER 4

Armored Fishes and Fishes with Arms

The armored placoderms, kings of the Devonian seas, rivers, and lakes

The placoderms were an unusual group of armor-plated jawed fishes that first appeared early in the Silurian Period about 430 million years ago and dominated the waterways of the Devonian Period, becoming extinct at the end of that age. They were some of the most bizarre-looking vertebrates that ever lived and included the antiarchs, characterized by external bone-covered "arms," and the gargantuan dinichthyids, the world's first megapredators, reaching sizes of 8 m or more. Most placoderms sit on the evolutionary tree somewhere after the jawless osteostracans and before the first sharks. Their rapid evolution and diversification meant that species generally occupied short time spans. Their fossilized bones can often be used to accurately determine the precise ages of Devonian rocks. Some of the world's best-preserved placoderms came from Australia. Beautiful specimens of primitive placoderms from the Early Devonian Limestones of southeastern Australia show extraordinary detail of the anatomy of these primitive jawed fishes. A great diversity of advanced placoderms, perfectly preserved in three dimensions from the Gogo Formation of northern Western Australia, show the complex adaptations evolved by these fishes for life in a reef ecosystem. Some of these Gogo placoderms have revealed the secrets of their sex lives, showing that many reproduced by copulation and some gave birth to live young.

The word *placoderm* comes from the Greek, meaning "plated skin," al-

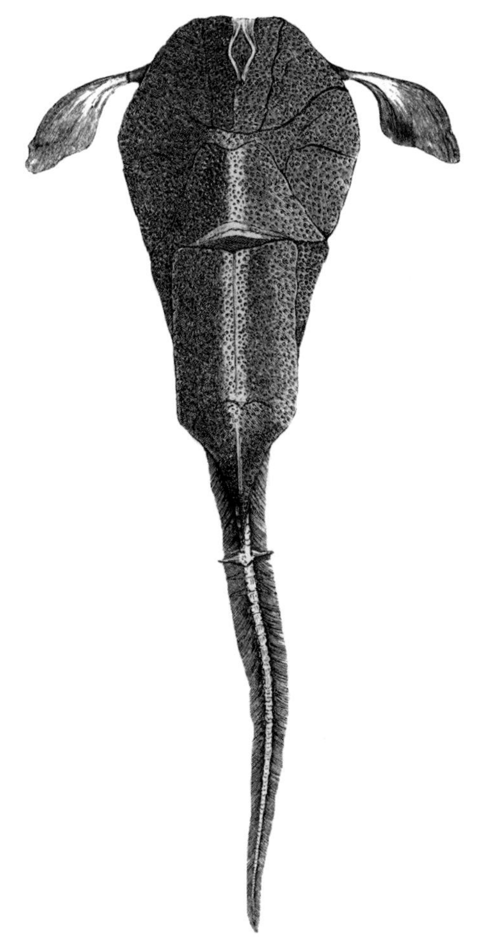

Hugh Miller's 1841 "crude drawing" of the Scottish Old Red Sandstone placoderm *Coccosteus*. The depiction of the pectoral fins coming out of the notches for the eyes indicates the poor state of knowledge of the group at the time. Miller's discoveries were seen by Swiss researcher Louis Agassiz, who became the first scientist to make a detailed study of the group.

luding to their characteristic feature of having the head and trunk covered in a mosaic of overlapping bony plates. Placoderm fossils have been known for hundreds of years, most notably those from the Old Red Sandstone outcrops of Scotland. Some of the earliest reconstructions of placoderms depict them as jawless fishes with peculiar winglike arms and unusually large heads. Hugh Miller, in particular, is famous for his "crude drawings," one being an 1838 composite restoration confusing two different Scottish placoderms, *Coccosteus* and *Pterichthyodes.* After the specimens were studied further, the facts emerged that placoderms had true jaws and fins much like other fishes. The earliest detailed study of placoderms was by noted French paleontologist Louis Agassiz, who published a five-volume set entitled *Recherches sur les poissons fossiles* (Research on Fossil Fishes) between 1833 and 1843, illustrating many forms of placoderms. Some early naturalists even regarded the antiarch placoderms as invertebrates akin to large beetles or as kinds of fossil turtles because of the boxlike shell on their backs.

Placoderm Origins and Relationships

The revolution in placoderm studies came in the early 1930s, when Swedish professor Erik Stensiö of the Natural History Museum in Stockholm began looking at placoderms in great anatomical detail. Stensiö took specimens of placoderm skulls with braincases intact in the rock and slowly ground them away, recording each cross-section of a tenth of a millimeter in wax templates, each magnified by 10 times the original section

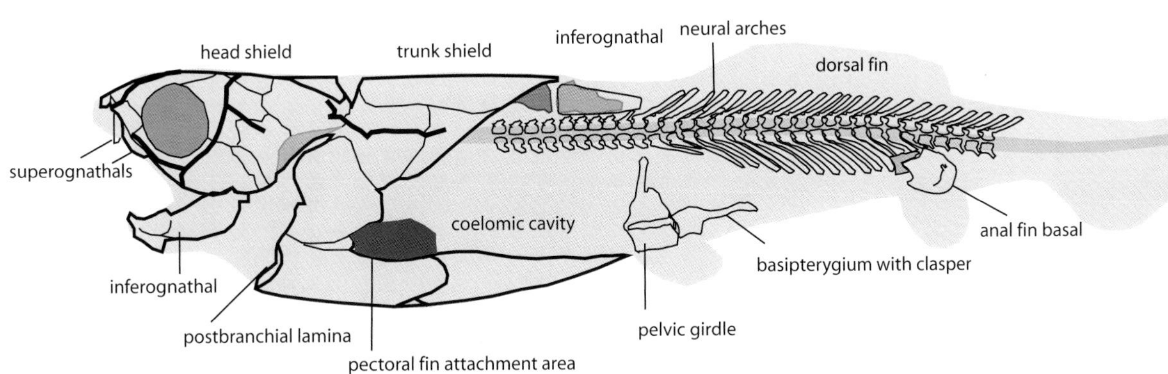

Basic anatomy of the placoderm *Incisoscutum*, from the Late Devonian Gogo Formation of Australia.

Fossil fish researchers at the Natural History Museum, Stockholm, Sweden in 1968. ***Left to right:*** **Hans-Peter Schultze (Germany), Tor Ørvig (Norway), Hans Bjerring (Sweden), Erik Stensiö (Sweden), Gareth Nelson (USA), Meeman Chang (China), Raymond Thorsteinsson (Canada), Erik Jarvik (Sweden), Emilia Vorobjeva (USSR), and Hans Jessen (Germany).** (With the permission of the Swedish Museum of Natural History)

size; when these were assembled, they made a large model of the three-dimensional skull clearly showing canals for nerves, arteries, and veins. This method, using acid etching to remove the layers, was called Sollas' grinding technique after Englishman William Sollas. By taking the negative of where bone was missing, Stensiö could then make a model of the brain cavity with soft tissues shown emerging from parts of the brain. His detailed study initiated a bold new era in placoderm research and fossil fish studies in general, by describing their soft anatomy and establishing placoderms as jawed fishes that were possibly related to modern-day sharks.

In recent years, this acid-etching technique has enabled us to prepare specimens out of the rock in three-dimensional form, thus confirming much of Stensiö's work and discovering much new information on placoderm anatomy and relationships. Today, placoderms are seen as an important group for solving many questions about geology. As stated above, they are useful as index fossils for giving age determinations for Devonian sediments, and some groups have distinct biogeographic ranges, and therefore tell us about the positions of certain continents in past geological times.

The affinities of the placoderms have long been a topic of debate among paleontologists. Swedish scientist Erik Stensiö favored a close relationship between placoderms and sharks, based on his many reconstructions of the anatomy of placoderms from the detailed serially ground sections. However, many people involved in research on placoderms today have claimed that Stensiö based his anatomical reconstructions of placoderms on a shark model, thereby artificially making them always

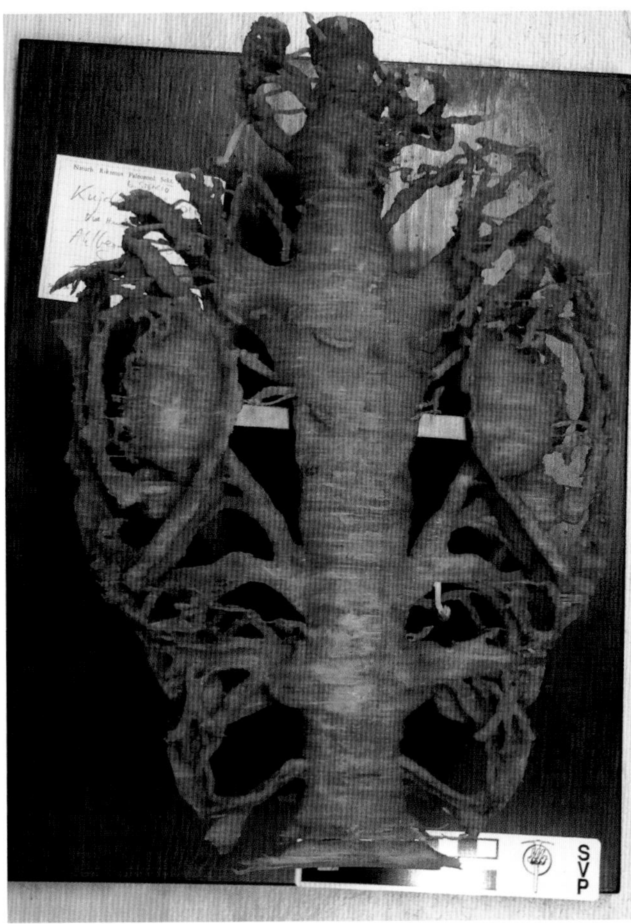

A wax model made up of scaled up slices taken from a series of ground down slices through the skull enlarged in wax sheets, showing the brain cavity of an Early Devonian placoderm, *Kujdanowiaspis*, made by Erik Stensiö of Sweden in the 1930s.

appear more sharklike. This led some scientists such as Brian Gardiner to argue that the placoderms could be closely related to the ancestors of the true bony fishes (osteichthyans). Others like Bob Schaeffer and Gavin Young have argued that placoderms evolved well before sharks, acanthodians, or bony fishes and are the primitive ancestral group to all jawed fishes.

This debate over placoderm relationships has heated up in recent years, with a number of new viewpoints being published. Some agree with Stensiö that placoderms should be classified with sharks and rays (in a superclass called the Elasmobranchiomorphii). Others believe placoderms are closely related to the bony fishes, or Osteichthyes. The latest opinion shared by many is that the placoderms occupy a position more primitive than either of these groups, at the base of the radiation of all jawed fishes. For over a century, it has been held that placoderms were a natural monophyletic group sharing unique characters, a view that has championed recently in papers by Gavin Young and Daniel Goujet.

In early 2009, Martin Brazeau argued that placoderms are a paraphyletic group of species, that is, some but not all descendents of a common ancestor. Some placoderms, in his view, should be regarded as stem gnathostomes, while others, like the arthrodires, form monophyletic groupings. Indeed some placoderms, like antiarchs, completely lack pelvic girdles and fins, and this is strong evidence that the acquisition of this important feature could unite some placoderms with chondrichthyans and osteichthyans, making the antiarchs stem gnathostomes.

Recent work by my colleagues and me has shown that some groups of placoderms had complex reproductive behavior, copulating like modern sharks and rays and some giving birth to well-developed live young (Long et al. 2008, 2009). Although this may not directly link them to the sharks, it suggests that the evolution of complex reproductive strategies might well have appeared before the first sharks.

Some evidence is now emerging that shows the complex reproductive behavior in some placoderms was more widespread than previously thought. It includes sexual dimorphism of arthrodires and ptyctodontids, with males internally fertilizing females using claspers similar to those found on modern sharks but covered with dermal bone. This suggests that external claspers were a specialized feature that evolved in sharks and possibly all placoderms and implies that claspers primitively present in all placoderm groups could have been independently lost or have not yet been discovered in some groups of placoderms.

Anatomical features that once supported an affinity between sharks and placoderms—such as the presence of an eyestalk connecting the eyeball to the braincase; the structure of the pectoral and pelvic fins, which are fleshy broad-based structures (except in antiarchs, in

Gogo Fishes and the Acid-Preparation Technique

The world's best placoderm fossils come from the Late Devonian reef deposits of on the Gogo and Christmas Creek stations in the Kimberley district of north Western Australia. The Gogo Formation represents the quiet deeper waters well away from high energy reef fronts. The great diversity of fishes found in these deposits is shown throughout this book, preserved in exquisite detail because of the lack of later geological activity, which kept the region free of large crustal movements that usually deform and compress fossils of this age.

In the deeper inter-reef basins of the ancient reef system, muddy lime-rich sediments slowly accumulated over the dead bodies of organisms, many of which may well have lived on or around the reef when alive. After death (by whatever means), fish carcasses tend to float for a while, then sink to the muddy seafloor and may be scavenged at any time during this period. Thus the remains of Gogo fishes found in the fine mudstones comprise complete skeletons, isolated bones, or pieces of a carcass. The Gogo fishes were rapidly encased in fine limy mud, which sets as hard calcite crystals formed not long after burial. This process protected the delicate skeletons from being crushed by the weight of accumulating overlying sediments.

Searching for Gogo fishes in the field is a matter of hard work, hitting thousands of limestone nodules with a hammer until a lucky find is made. Sometimes a keen eye can spot a bit of bone weathering out of a concretion, and in such cases the limestone nodule doesn't have to be broken. If a specimen of a complete fish has been split through the middle or broken by hammer into a handful of pieces, the specimen may be prepared using several techniques. One method involves embedding each side of the split fish in an epoxy resin slab and then acid-etching each half (see p. 15). This reveals the two sides of the fish with all bones in articulated position, as the fish was buried. Alternatively the pieces can be glued together with acid-resistant epoxy resin and then the whole nodule dissolved in acid. This gives a whole, undamaged skeleton, except for visible lines in some bones where the original breaks occurred.

The acid-etching process uses weak acetic or formic acid (about 10% solution) and, because each treatment requires extensive washing in running water, the whole preparation procedure may take several months for each specimen. During the preparation, after each treatment in acid and washing in water, the exposed bones are left to dry in the air. They are then impregnated with a plastic glue that is absorbed into the bones and gives them internal strength. When the bones have soaked up as much plastic glue as necessary, the specimens can then be placed back in acid for further dissolution of the enclosing rock.

Once the fossil fish plates are freed from the rock, they can then be assembled (like building a model airplane) to restore the placoderm's external skeleton in perfect three-dimensional form. This chapter uses many examples of the superb Gogo placoderms to show their overall appearance and anatomical features.

which they are bony props); and the similar shapes of the braincases—are now widely regarded as either generalized gnathostome characteristics or as convergent features. The presence of little cartilage rings around the nasal area, called annular cartilages, was proposed as an important similarity between sharks and placoderms by Stensiö. Although he never saw such a structure in any fossil, he based his theory on similarities in the general appearance of the snout region in the two groups. A superbly preserved placoderm from Gogo, *Mcnamaraspis,* has shown that annular cartilages really did exist in some placoderms. Furthermore, the general shape of the bodies with broad pectoral fins in predatory placoderms, like in *Coccosteus,* is reminiscent of sharks and suggests a similar swimming style and internal anatomy (the absence of a swim-bladder, a balloon-like organ that fills with gas to increase buoyancy) in the two groups. The overall feeling by most paleoichthyologists today is that placoderms are indeed very primitive jawed vertebrates that sit at the base of the tree below chondrichthyans and osteichthyans (Goujet and Young 2004).

There are some 240 known genera of placoderms, divided into seven major orders, the largest order of placoderms being the Arthrodira ("jointed necks"), comprising more than 60% of all known placoderm species.

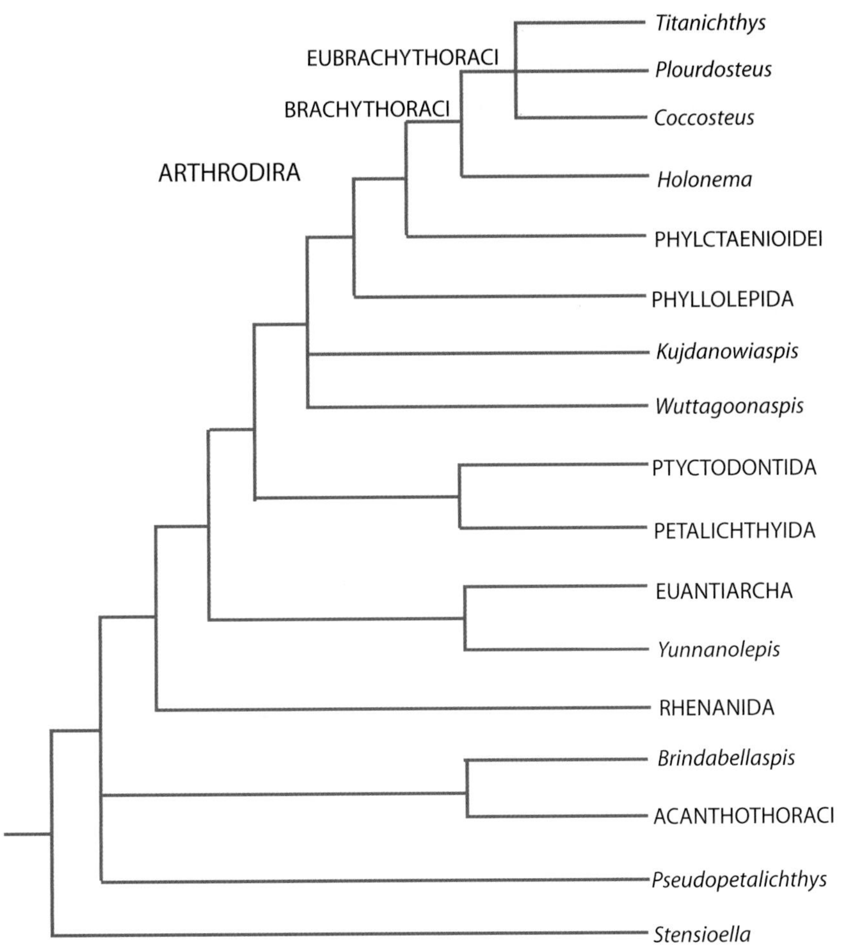

A scheme of relationships of the major placoderm groups, based on recent work by Gavin Young, Daniel Goujet, and Vincent Dupret.

All these groups had appeared by the start of the Devonian, with antiarchs and arthrodires now known from the Middle Silurian, most surviving through to the end of the Devonian. The different orders are recognized by a combination of plate patterns making up the external armor and by the unusual and often complex development of bony superficial ornament on the plates of some groups.

A long-held view championed by Roger Miles and Gavin Young (1977) is that the ptyctodontids, rhenanids, and acanthothoracids are the most primitive placoderms, whereas the antiarchs and arthrodires (including phyllolepids) are considered the most advanced orders. The other orders would slot somewhere in between. Several schemes of interrelationships have been proposed in recent years, but there is still ongoing debate about the precise relationships of certain orders, especially the antiarchs and petalichthyids. Recent work suggested that antiarchs could alternatively be the most primitive of all placoderms, as they lack pelvic fins and girdles (Brazeau 2009). However, Young (2008) showed convincing evidence that antiarchs are highly specialized in their pectoral fin anatomy. Robert Carr and colleagues (2009) presented a recent phylogeny of placoderms that accepts they are a natural (monophyletic) group and yet still places antiarchs near the base of the placoderm family tree. This scheme of relationships has been reproduced below as a representative placoderm phylogeny.

Stensioellida and Pseudopetalichthyida: Early Enigmas

The Stensioellida and Pseudopetalichthyida were primitive placoderms with little bony armor developed.

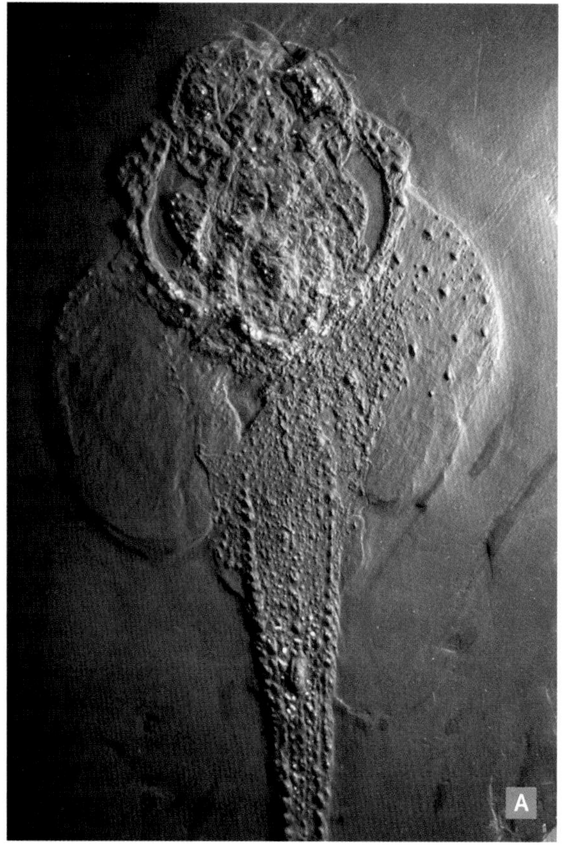

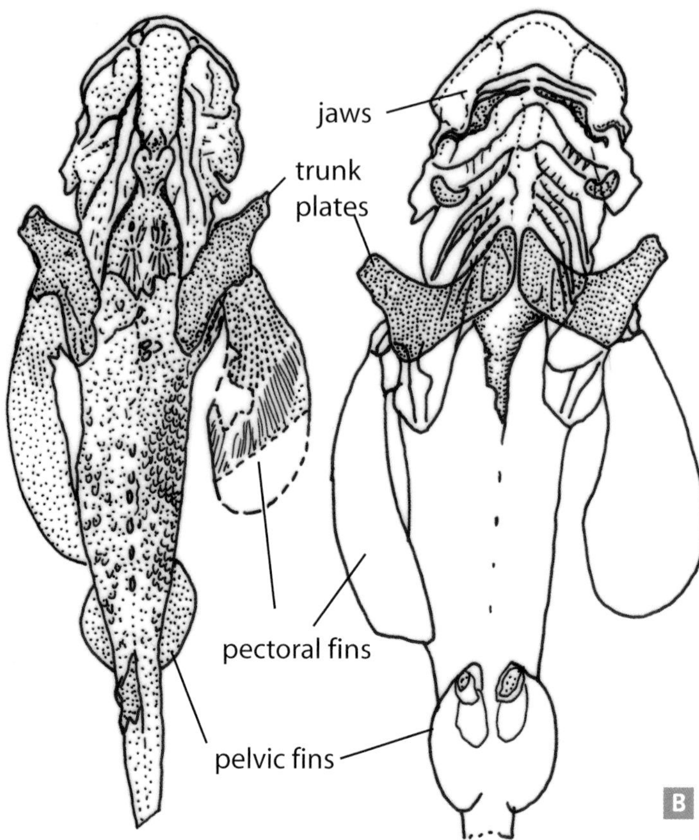

(*Above*) The Early Devonian placoderm *Gemuendina* from the Hunsruck Shales of Germany is regarded as one of the most basal of all placoderms (*A*). A sketch showing major features of *Stensioella* (*B*).

(*Left*) Gavin Young with Early Devonian Taemas limestone in New South Wales, Australia, the site for many significant Early Devonian fish discoveries, in the background.

Their broad winglike pectoral fins are reminiscent of those of rays, and their bodies have many small ornamented denticles, or placoid scales, set in the skin. The relationship of these groups is still not clear. Philippe Janvier (1996) regards them as possibly related to the holocephalan chondrichthyans. Others regard them as stem gnathostomes outside of the placoderm group. Most researchers consider them to be the most primi-

The Basic Structure of Placoderms

Most placoderms are characterized by their peculiar armor made up of an overlapping series of bony plates with extensive flat overlap surfaces that form a protective cover around the head (the head shield) and enclosing an immobile ring of bone around the front of the fish's body (the trunk shield). The head and trunk shields in most placoderms are articulated by bony knobs and grooves, although in some rare exceptions the two parts may be fused as one composite shield.

There are seven major orders of placoderms, each categorized by its own pattern of bony plate shapes that make up its armor. Each armored plate can be identified by its combination of shapes, overlapping areas for neighboring plates, presence or absence of sensory-line canals, and an external surface texture, called the *dermal ornament*. Generalized placoderms all show an external ornament of simple wartlike structures called tubercles, whereas more specialized groups, such as the phyllolepids, may develop complex linear or reticulated network patterns.

The braincase of all early placoderms was well ossified with layers of laminar perichondral bone, but in later species, such as in the Late Devonian Gogo forms, it was made entirely of cartilage, possibly to reduce weight. The jaws are nearly always simple rods of bone that may have pointed teeth for gripping prey. In those species whose diets involve crushing hard-shelled prey, the jaws may have areas of thickened tubercles or be smooth in

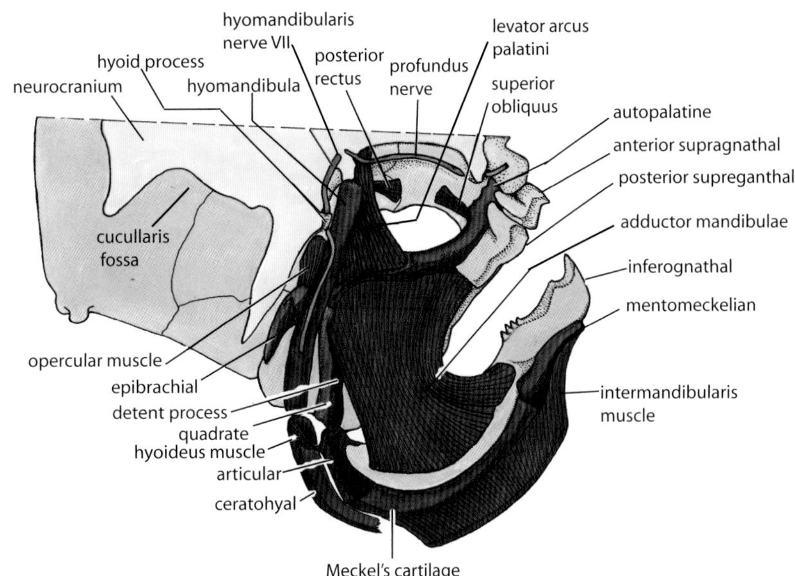

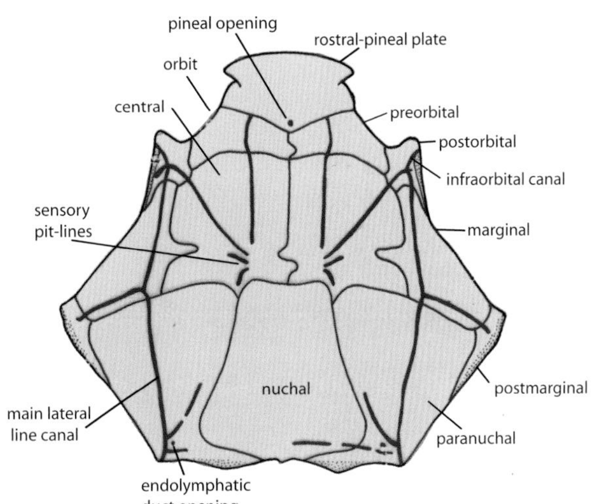

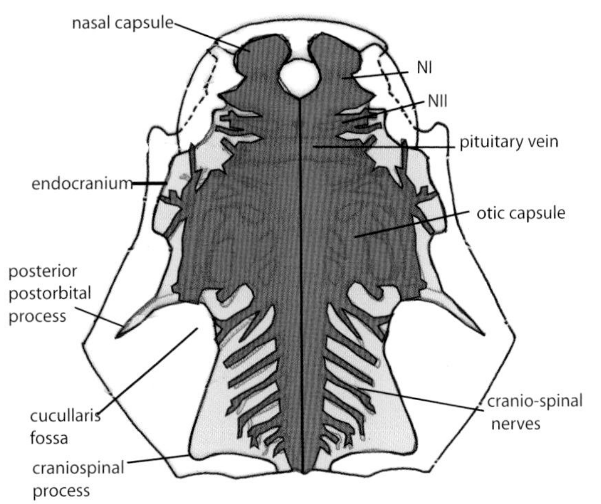

(*Above*) Anatomical features of the arthrodire skull with some of the soft tissues restored, from *Mcnamaraspis* (After Long 1995).

(*Left*) The skull of the Early Devonian arthrodire *Buchanosteus*, showing external dermal bones (*above*) and endocranial features (*below*).

The Basic Structure of Placoderms, cont'd.

presumed filter-feeding forms (e.g., *Titanichthys*). The jaw joint is simple: in arthrodires, the lower jaws articulate against a knoblike quadrate bone that is commonly fused to the cheekbones but is a separate ossification in ptyctodontids. The eyes are surrounded by a simple ring of bone, made up of three to five sclerotic plates, and are connected to the braincase by an eyestalk. The head shield may exhibit a variety of bone patterns, some with eye notches on the side of the head, others having eyes and nostrils together in the center of the skull via a median opening (as in antiarchs and rhenanids).

The body of placoderms is generally torpedo-like, as in sharks, although notable exceptions are seen in the flattened groups, the phyllolepids and the rhenanids. There is only one dorsal fin developed—despite many erroneous reconstructions made by Erik Stensiö, in which two dorsal fins were always assumed to be present. There are paired pectoral and pelvic fins and a single anal fin, with heterocercal caudal fin shape in arthrodires and acanthothoracids, whereas antiarchs seemed not to have ever developed pelvic fins. Ptyctodontids and some arthrodires (phyllolepids, *Incisoscutum*) show sexual dimorphism with males having elaborate bony claspers that they inserted into the females. The body is primitively covered with thick bony platelets that resemble miniature versions of the dermal bones, often having similar ornament on each scale. In advanced lineages of placoderms, the body scales may be reduced or absent, corresponding with the overall trend seen throughout the skeleton to reduce weight.

tive of all placoderms, as they lack a number of bones found in all other groups. This group is only known from a few species, and all known specimens are from the Early Devonian Black Shales (Hunsrück-Schiefer) of the German Rhineland.

Rhenanida and Acanthothoraci: The Elaborately Armored Ones

The Rhenanida were a group of flattened placoderms similar to stensioellids in having very large winglike pectoral fins and variable amounts of dermal bone covering the skull, mostly composed of polygonal platelets. The skull roof pattern thus doesn't resemble the larger patterns of bones making up the skulls of other placoderms. The trunk shield is very short, and the tail has many bony platelets of varying sizes. The group is known only from Europe and North American marine sediments of Early Devonian age. Rhenanids like *Gemuendina* are now often grouped close to the Acanthothoraci (meaning "spiny trunk shield"), a group characterized by heavily ossified armor with elaborate ornamentation. Indeed, the fossil plates of some of these fishes have the most beautiful surface patterns ever seen in the dermal skeleton of any vertebrate. Acanthothoracids are characterized by their own patterns of skull bones and short trunk shields.

Australia has an excellent representation of fossil acanthothoracids, from the Early Devonian Limestones around Taemas and Wee Jasper, in New South Wales, and near Buchan in Victoria. These are three-dimensional preserved skulls and trunk plates and often include surprising preservation of structures like the bony covering around the placoderm eyeball. One such fossil, a sclerotic capsule, found near Taemas, dem-

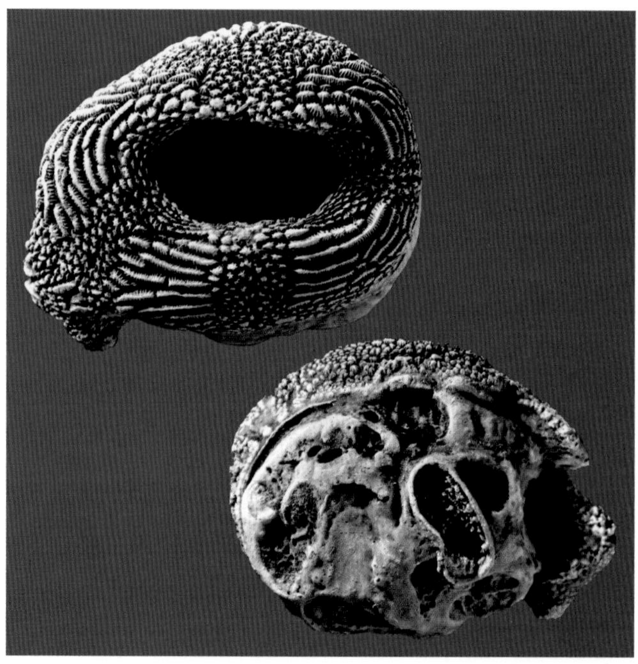

Early Devonian acanthothoracid placoderm eye capsule (*Murrindalaspis*) from Taemas, Australia. (Courtesy Gavin Young)

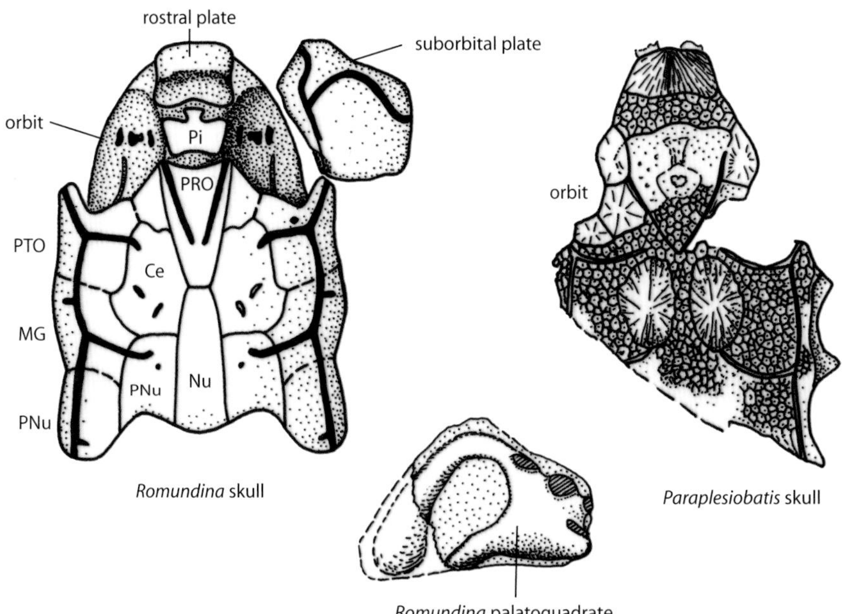

Acanthothoracid placoderm skull anatomy. Ce = central; MG = marginal; Nu = nuchal; Pi = pineal; PNu = paranuchal; PRO = preorbital; PTO = postorbital.

The skull of the long-snouted acanthothoracid *Brindabellaspis stensioi*, from the Early Devonian of Taemas, New South Wales, Australia.

onstrated that the eyes of placoderms had an unusual pattern with seven external eye muscles unlike the six standard muscles seen in lampreys and all living jawed fishes (Young 2008). Also unlike most modern fishes, however, the eyes were heavily invested with bone, and each eyeball was connected to the braincase by a pedicle of bone or cartilage called the eyestalk. The eyestalk is another feature seen in both sharks and placoderms.

The best-known acanthothoracid fossils from southeastern Australia are the remains of the long-snouted *Brindabellaspis* and the two high-crested forms, *Weejasperaspis* and *Murrindalaspis*. These fossils have exquisite preservation of the cavities surrounding the brain and cranial nerves and reveal much about the soft anatomy of primitive placoderms. Other well-preserved acanthothoracid fossils occur in the Early Devonian sediments of Eastern Europe, Russia, and Arctic Canada.

The Antiarchi: Always at Arms Length

The antiarchs were little placoderms usually about 20–30 cm in length and reaching a maximum size of about 1 m. They are characterized by having their pec-

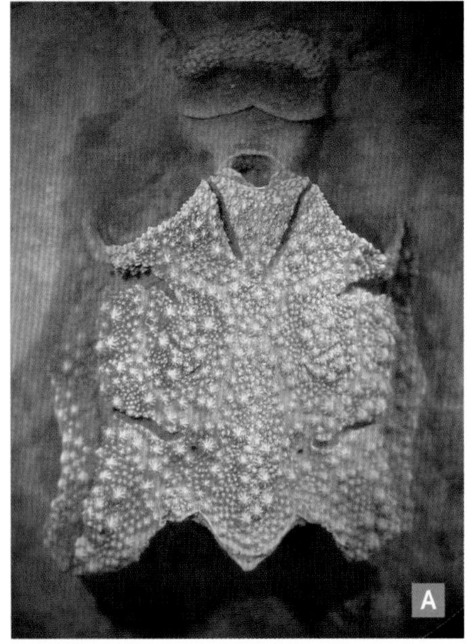

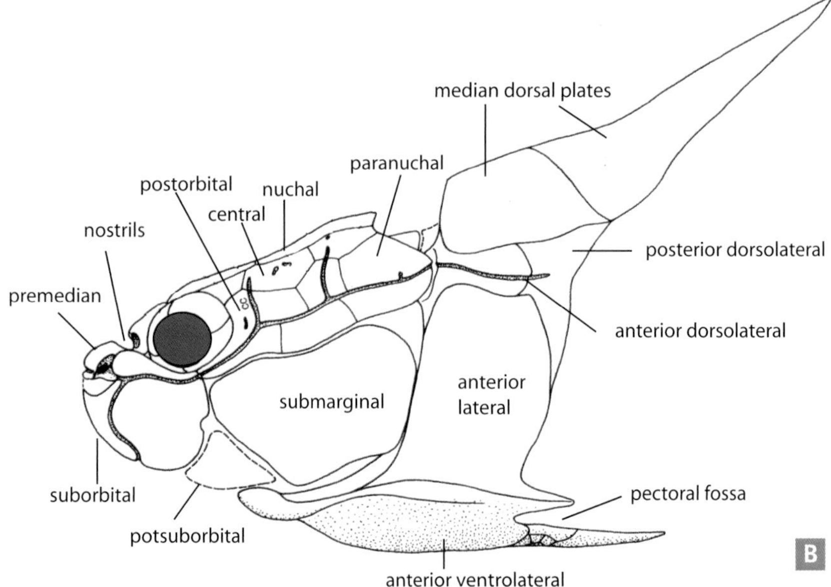

The skull roof of the acanthothoracid *Romundina stellata* from the Early Devonian of Arctic Canada (*A*). The restoration of the dermal armor of the same species in side view (*B*). (Courtesy of Daniel Goujet)

toral fins enclosed in bony tubes, referred to as pectoral appendages (sometimes called "arms"). Most of the advanced antiarchs had segmented arms, although some like *Remigolepis* had short oarlike props. The head shield of antiarchs featured a single opening in the middle for the eyes and nostrils and pineal eye, and the trunk shield of all antiarchs was very long relative to the overall length of the fish. Antiarchs had a trunk shield with two plates along the back called median dorsal plates. The group first appeared in the Silurian of China (e.g., *Silurlepis*) and flourished in the Early Devonian of China, becoming widespread by the Middle Devonian and reaching a peak of species diversity during the Late Devonian.

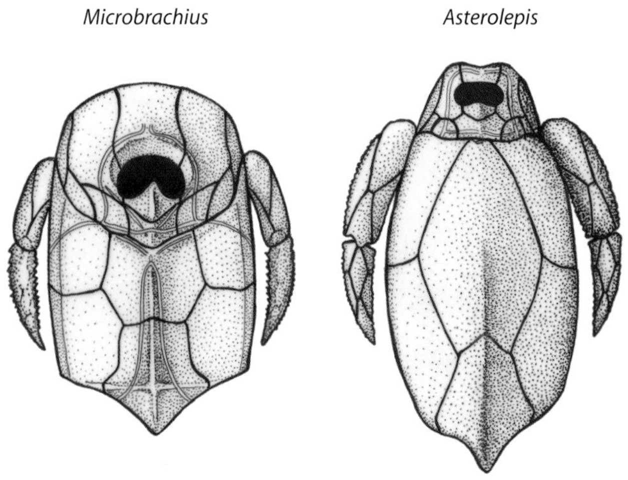

The dermal armors in dorsal view of two asterolepidoid antiarch placoderms.

The earliest and most primitive antiarchs are well represented in Silurian -Lower Devonian rocks of Yunnan, China, where they were first discovered and where their name, the yunnanolepids, comes from. They had a single proplike pectoral fin, which had not developed the ball-and-socket articulation seen in most other antiarchs. The armor of these little fishes, usually less than 5 cm long, had a covering of wartlike tubercles. Some more advanced forms had an incipient pectoral joint (in *Procondylepis* for example), and the true antiarch shoulder joint evolved from this basic structure.

The sinolepids (meaning "Chinese scale" forms) were a peculiar group of big-headed antiarchs with long, segmented pectoral fins. Until recently they were only known by one genus, *Sinolepis,* from the Late Devonian of China. Several other sinolepids like *Dayoushania* have since been recognized from numerous sites of Early-Late Devonian age in China and northern Vietnam. Sinolepids resemble the more primitive yunnanolepids in their head structure but have reduced trunk shields and

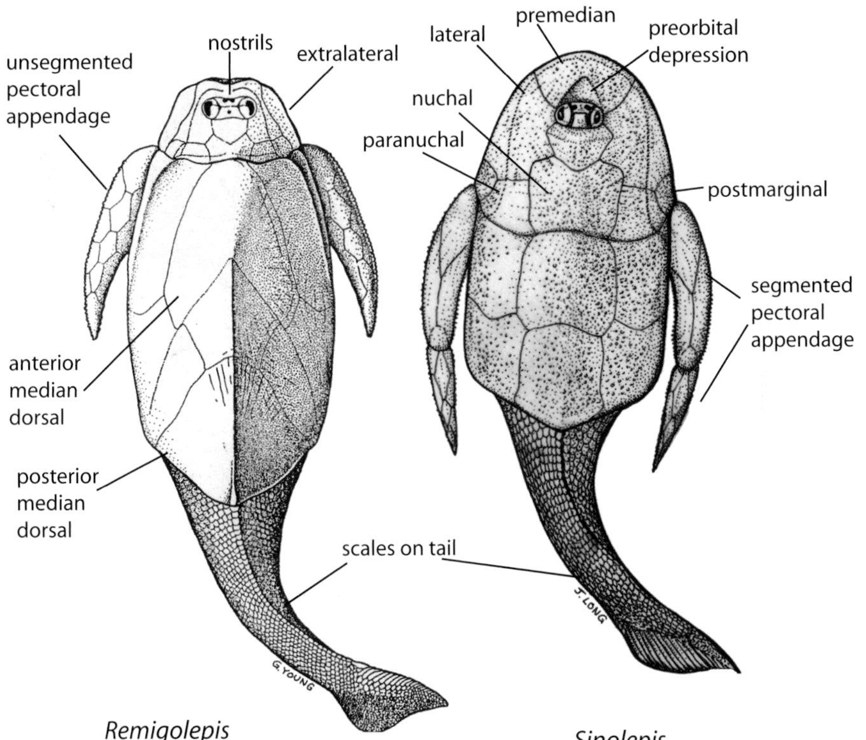

(*Left*) **Dermal armors of two antiarchs, a sinolepid and a bothriolepid, in dorsal view.**

(*Below*) **The skull of *Yunnanolepis*, a primitive antiarch from the Early Devonian of China and Vietnam.** (Courtesy Daniel Goujet)

long, segmented pectoral appendages. The sinolepid *Grenfellaspis* has been discovered near Grenfell, New South Wales, Australia. This discovery provides strong evidence that the Chinese and Australian terranes came into close proximity, enabling migration of fish faunas, during the Late Devonian.

The most successful groups of placoderms were the bothrolepids and the asterolepids, groups that flourished in the Middle and Late Devonian around the world. Asterolepids had long trunk shields with small heads and robust, short, segmented pectoral appendages. The best known examples are *Asterolepis* (meaning "star scale"),

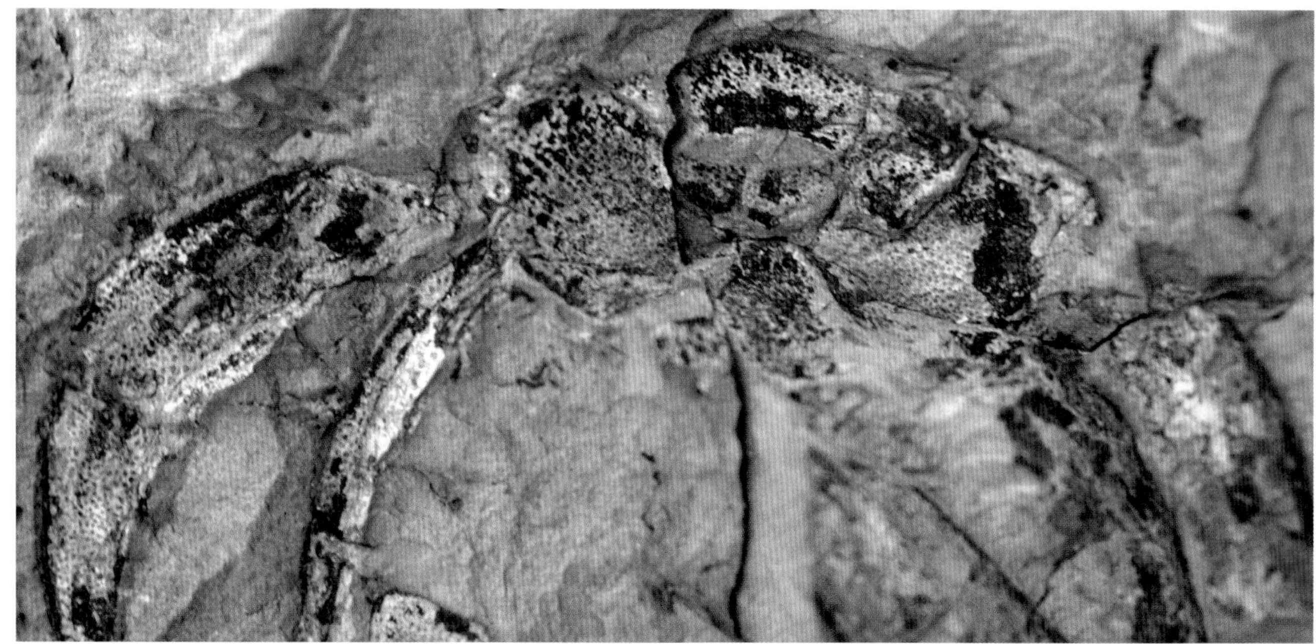

The skull and upper trunk armor of the sinolepid *Dayoushania*, from the Early Devonian of China.

known from Europe, Greenland, and North America, and *Remigolepis* (from the Greek meaning "oarsman"), known from Greenland, China, and Australia. *Remigolepis* is particularly unusual in having a short, stout pectoral appendage that lacks a joint. Many other asterolepids are known from around the world, such as the Middle Devonian genera *Pterichthyodes* from the Old Red Sandstone of Scotland and *Sherbonaspis* from Australia. The asterolepids first appeared in marine environments and soon after invaded freshwater river and lake systems. By the end of the Devonian, they were pushed out of the highly competitive marine realm and became freshwater river and lake dwellers.

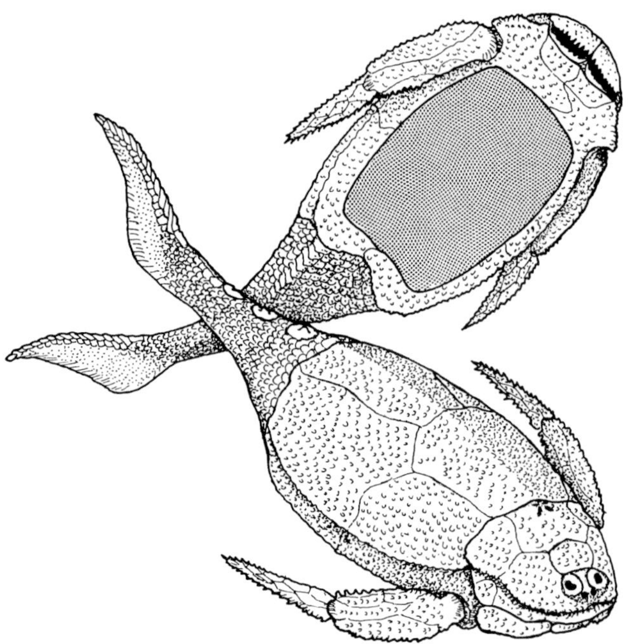

Reconstruction of the sinolepid antiarch *Grenfellaspis* from the Late Devonian of Australia. (Courtesy Gavin Young)

The most successful placoderm of all time was undoubtedly *Bothriolepis* (meaning "pitted scale"), a genus of little antiarchs having long, segmented arms and known from more than 100 species found in Middle and Late Devonian rocks of every continent, including Antarctica. *Bothriolepis* had a special feature of its skull that may have been the key to its success: a separate partition of bone below the opening for the eyes and nostrils, enclosing the nasal capsules called a preorbital recess. Serially sectioned specimens of *Bothriolepis* show that inside the armor the fish had paired "lunglike" organs and a spiral intestine, preserved full of organic sediment, differing from the sediment type surrounding the fossil. Other interpretations suggest these organs could alternatively be lymph glands.

Bothriolepis was probably a mud-grubber that in-

gested organic-rich mud for its food. Its long pectoral appendages could also have been used to push itself deeper into the mud for feeding. Another suggestion is that the fish used its long arms to walk out of the water and its "lungs" to breathe as it crawled out of the water to invade new pools free from predators and full of rotting vegetation. *Bothriolepis* is known mostly from freshwater deposits, as well as rarer marine sites such

(*Top*) *Asterolepis* was a wide-ranging genus of antiarch found in the Middle Devonian of Europe, Asia, and North America. This specimen is from the famous Lode Quarry of Latvia. (Courtesy Ervins Luksevics, Latvia)

(*Bottom*) The pectoral fin of *Remigolepis*, from the Late Devonian near Eden, Australia.

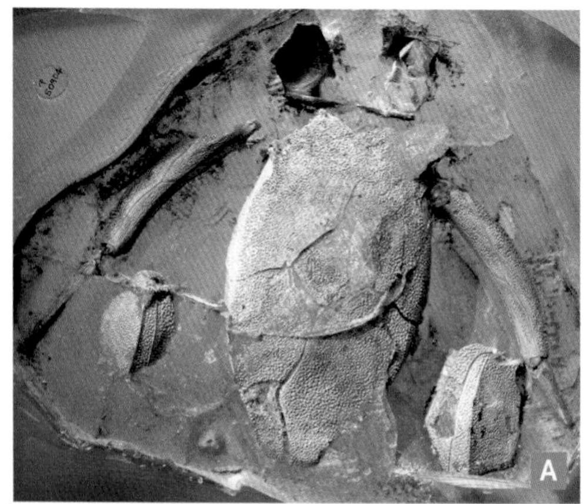

as the Devonian reefs at Gogo, Western Australia. Its ability to disperse around the Devonian world was via shallow seaways, from which it could invade river systems. Perhaps, like many modern fishes, it spent a large part of its life in the sea, moving upstream to breed and die. The fossil record suggests that most *Bothriolepis* species, irrespective of where they lived, died in freshwater habitats.

(*Top*) *Bothriolepis* was one of the most successful of all Devonian fishes, represented by over 100 species and occurring on every continent. These superb 3-D specimens come from the Gogo Formation of Western Australia. Specimen showing ventral view (*A*); Specimen showing armor in side view with pectoral fins missing (*B*).

(*Bottom*) *Bothriolepis canadensis* from the Escuminac Formation of Quebec, Canada.

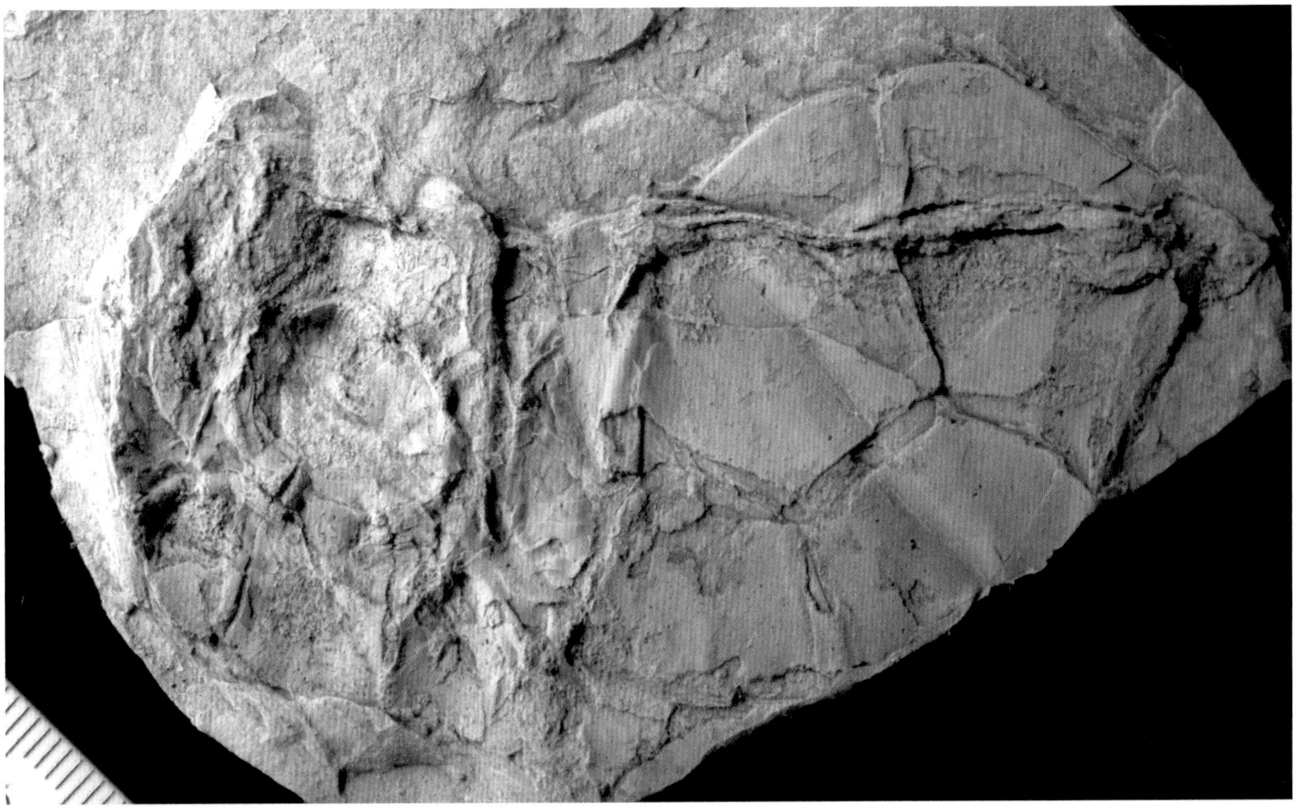

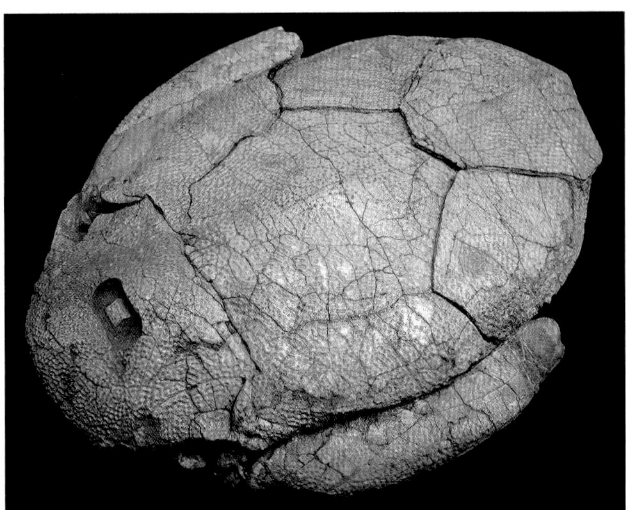

(*Top*) This latex peel of a fossil shows the high-crested *Bothriolepis gippslandiensis* from Victoria, Australia, in side view.

(*Bottom Left*) *Bothriolepis maxima* from the Baltic States was one of the largest known antiarchs, reaching around 1 m in length. (Courtesy Oleg Lebedev, Borissiak Paleontological Institute, Moscow)

(*Bottom Right*) A reconstruction of the elongate bothriolepid *Vietnamaspis*, from the Middle-Late Devonian of Vietnam.

Ptyctodontida: Crushing Teeth and Spiky Claspers

The Ptyctodontida (meaning "beaked tooth") were an unusual group of placoderms with strong crushing plates in the jaws. They had long bodies with whiplike tails and large heads with big eyes and in many respects resembled the modern-day chimaerids and whipfishes.

They had very short trunk shields and reduced head bone cover and were specially adapted for feeding along the bottom of the seas on hard-shelled organisms. They are the only placoderms to show sexual dimorphism—the males having dermal clasping organs used to internally fertilize the females. Embryos found inside mother fishes of the genera *Materpiscis* and *Austroptyctodus* confirmed this (Long et al. 2008, 2009).

Ptyctodontid fossils in Australia are known from isolated tooth plates of Early Devonian age from the limestones near Taemas-Wee Jasper in New South Wales and from extremely well-preserved articulated skeletons of Late Devonian age from the reef deposits at Gogo, Western Australia. There are three forms from Gogo: *Austroptyctodus gardineri* was about 20 cm long, with shearing tooth plates and a relatively low trunk shield. *Campbellodus decipiens* was about 30 to 40 cm maximum length and had an unusual trunk shield with three plates forming a high spine on its back, preceding the dorsal fin. Specimens of *Ctenurella* from similar-age rocks in Germany have whole bodies preserved that reveal the outline of the fish and the position of the fins, and these have been used to give the restoration of *Campbellodus* shown in this book. It is interesting that *Campbellodus* shows the body covered by fine overlapping scales, whereas other specimens from around the world lack this body scale cover. Other Gogo specimens, such as *Materpiscis* and *Austroptyctodus,* also include well-preserved parts of the braincase, indicating that it was highly modified for placoderms—not a single bony box with holes for nerves and arteries but a complex unit made up of several ossifications.

Ptyctodontid fossils are also well known from Middle and Late Devonian deposits in Europe, North America, and Russia. Whole skeletons of the little ptyctodont *Rhamphodopsis* have been described from Scotland and were the first studied examples showing the sexual dimorphism in placoderms by English palaeontologists D. M. S. Watson and Roger Miles (Miles 1967a). In many cases, ptyctodontids are often represented solely by their characteristic tooth plates. Such isolated fossils from North America indicate that the largest species of ptyctodontids (*Eczematolepis,* Late Devonian of New York) had tooth plates about 15–20 cm long, suggesting a total fish length of about 2.5 m.

Petalichthyida and Their Odd Relatives

The Petalichthyida were unusual little placoderms with widely splayed pectoral fins and all their dermal bones ornamented with characteristic linear rows of little tubercles. Their bones have thick tubes carrying the sensory-line nerves, and these tubes clearly stand out on the inside surface of the skull bones. Petalichthyids are known from Europe, North and South America, Asia, and Australia. They reached their peak of diversity in the Early Devonian, and only a few species sur-

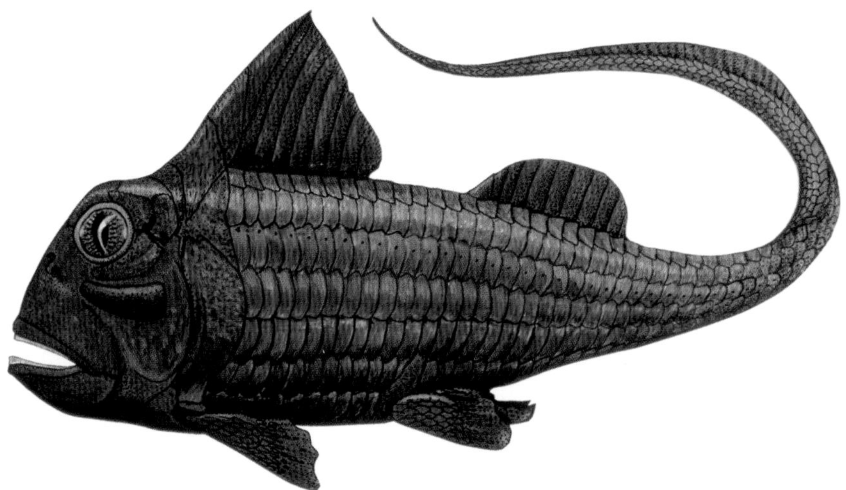

The ptyctodontid *Campbellodus decipiens*, which is known from relatively complete 3-D specimens from the Gogo Formation of Western Australia. This restoration shows a male displaying pelvic claspers.

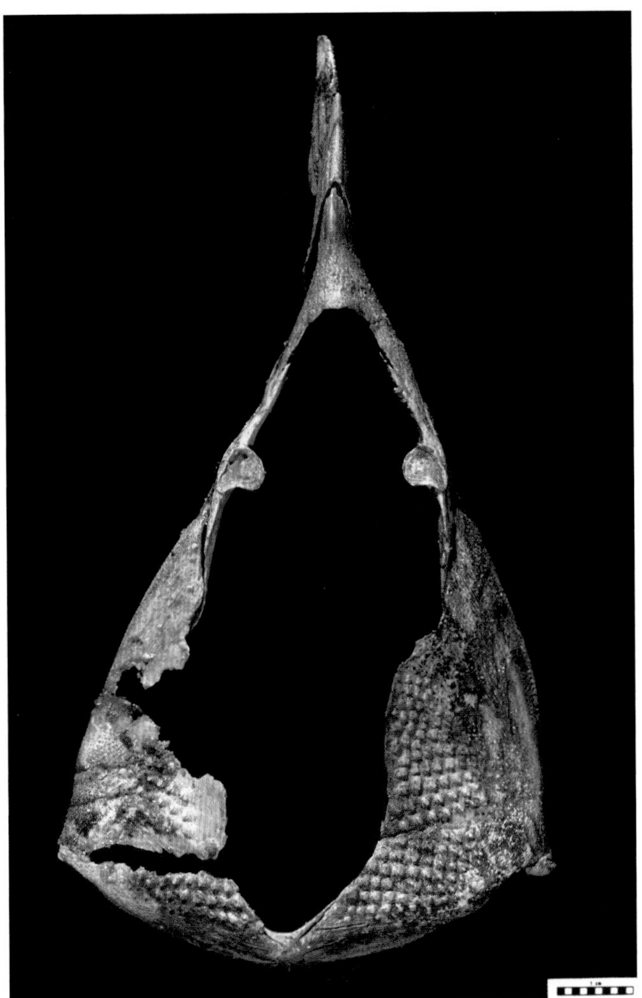

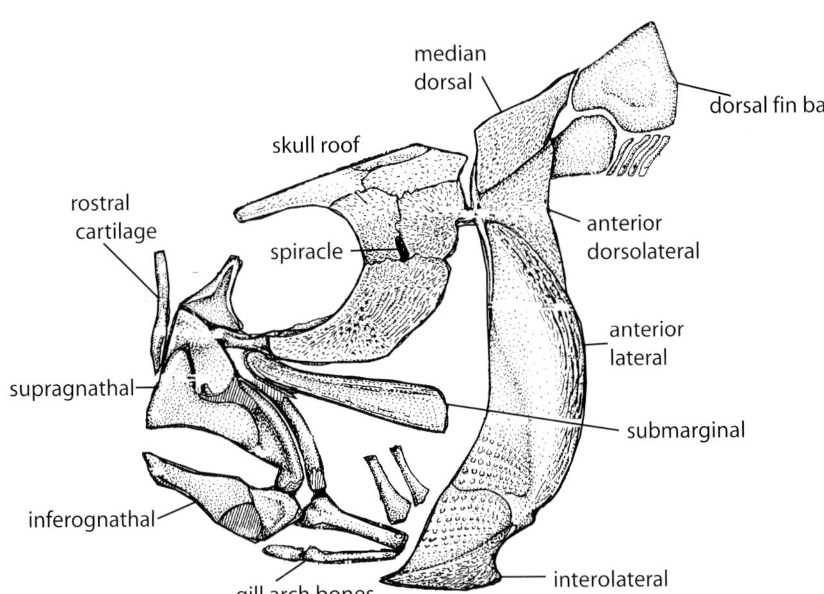

(*Top Left*) The trunk armor of *Campbellodus* in front view, showing the well-developed stellate tubercles on the postbranchial lamina, which lined the inside of the gill chamber.

(*Top Right*) A female of the Middle Devonian ptyctodontid *Rhamphodopsis* from Scotland. The pelvic fins lack claspers that would have been found in a male.

(*Left*) A restoration of the dermal armor and visceral skeleton of the ptyctodontid *Austroptyctodus* from Gogo, Australia.

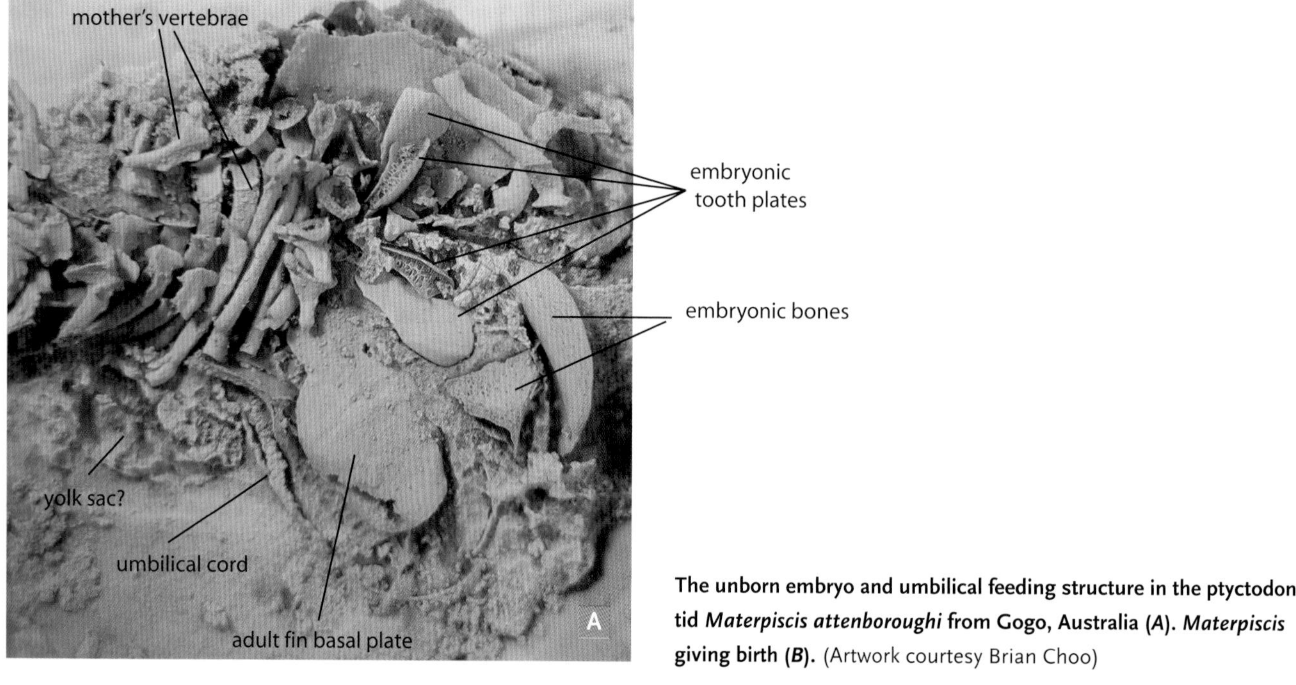

The unborn embryo and umbilical feeding structure in the ptyctodontid *Materpiscis attenboroughi* from Gogo, Australia (*A*). *Materpiscis* giving birth (*B*). (Artwork courtesy Brian Choo)

vived until the Late Devonian. Unusual petalichthyid-like fishes called "quasipetalichthyids" are known from China, which indicates a minor local radiation of the group took place on the isolated South China continental block.

Some of the best-preserved skulls of petalichthyids come from the Early Devonian limestones near Taemas, New South Wales, Australia. Several forms have been described from this region, including *Notopetalichthys*, *Shearsbyaspis*, and *Wijdeaspis*, the latter also known from Europe and Russia. Probably the best fossils of petalichthyids, preserved as whole fishes, come from the Early Devonian Black Shales of Germany. *Lunaspis* is the best-known example, and its remains have also been identified from China and Australia. Petalichthyids were probably bottom-dwelling fishes that swam

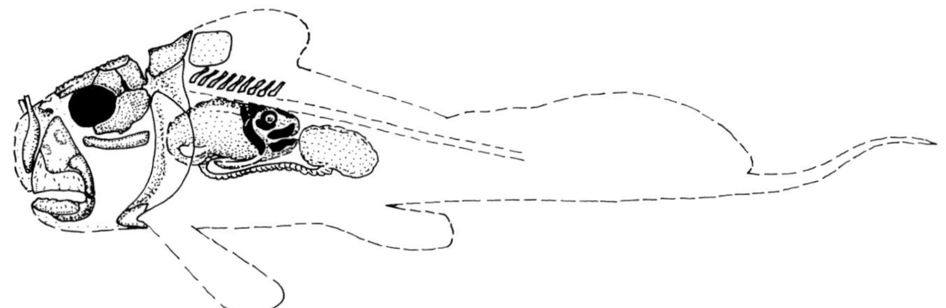

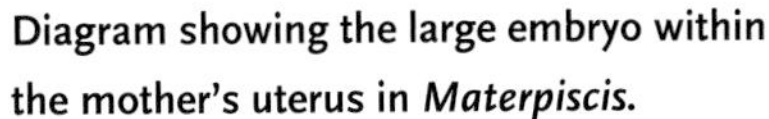

Diagram showing the large embryo within the mother's uterus in *Materpiscis.*

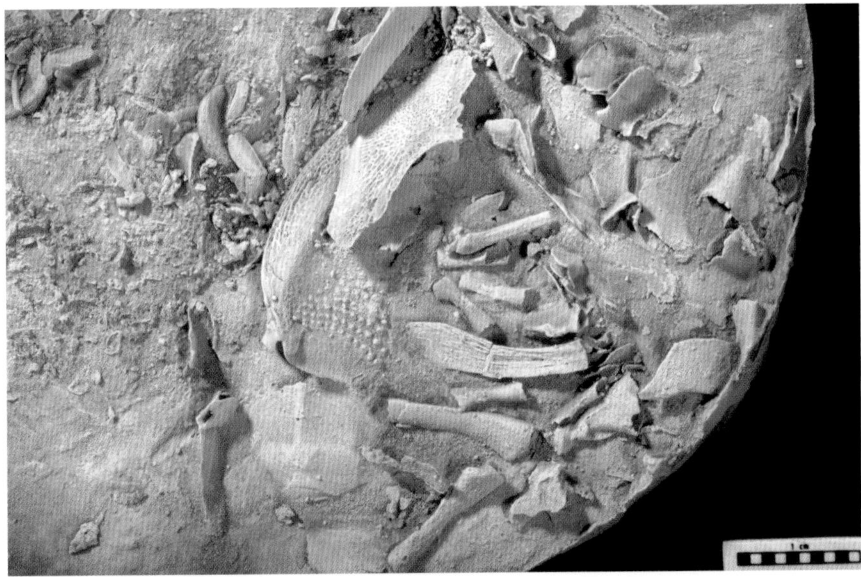

A female *Austroptyctodus* from Gogo, Australia, with three embryos inside her. The embryos appear as little clusters of fragile bones just behind the main trunk shield plates.

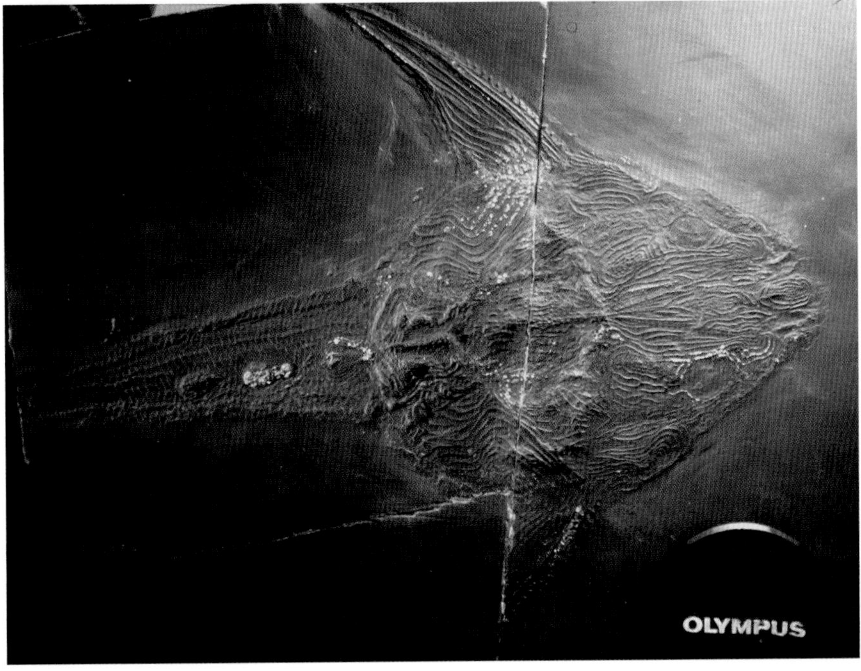

Lunaspis, one of the best known petalichthyid fishes, from the Early Devonian Hunsrück-Schiefer (Hunsrück Slate) of the Rhineland, Germany.

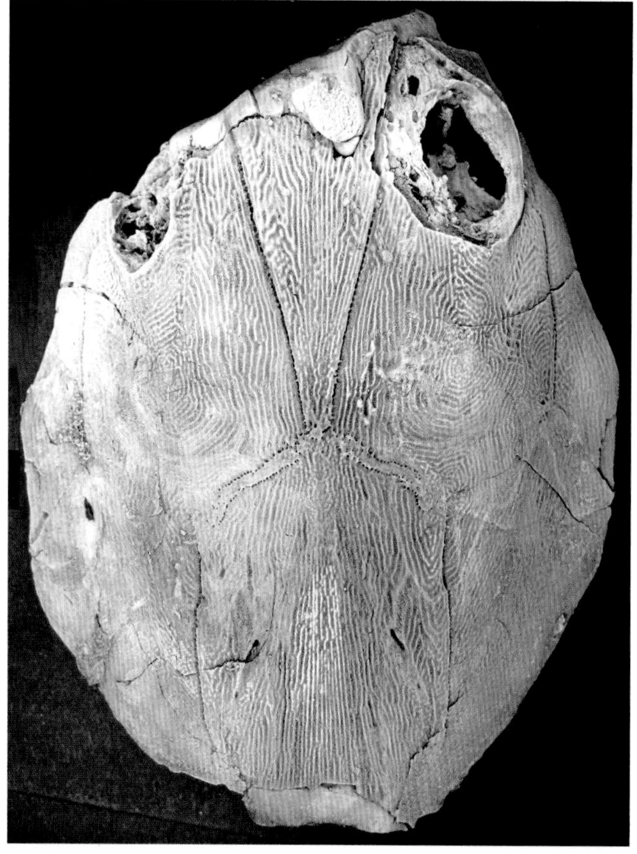

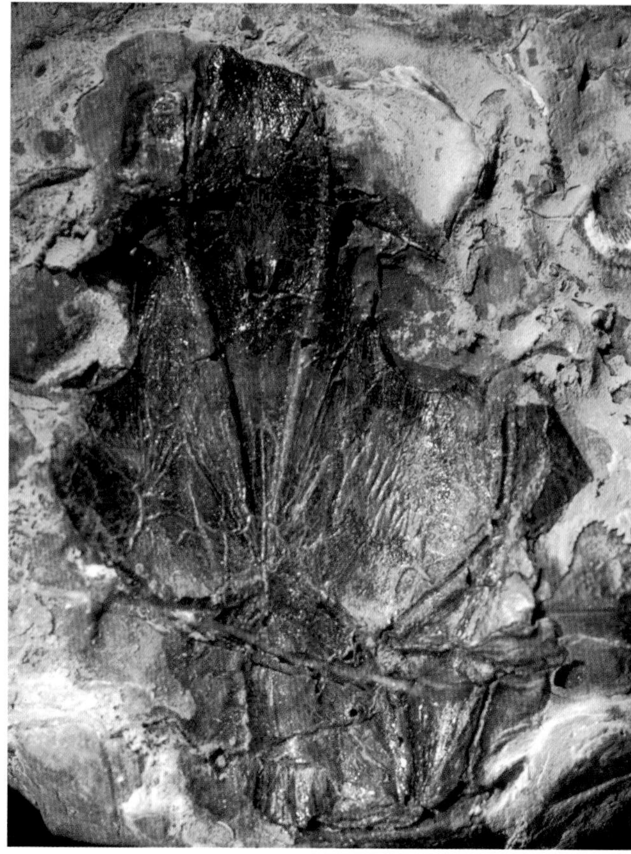

(*Top Left*) A recently discovered petalichthyid skull from the Early Devonian of Taemas, New South Wales, seen in dorsal view. (Courtesy Gavin Young, Australian National University)

(*Top Right*) The petalichthyid *Shearsbyaspis*, from the Early Devonian of Taemas, Australia, seen in visceral view.

(*Left*) A cast of the skull of *Quasipetalichthys* from the Middle Devonian of China, highlighting variations on the standard petalichthyid skull pattern.

slowly searching for prey on the seafloor. Unfortunately, no preserved mouth parts of the known petalichthyids have been found, so we can only guess at their eating habits.

Arthrodira: The Great Placoderm Radiation

The Arthrodira (sometimes called Euarthrodira) were the only group of placoderms to have two pairs of upper jaw tooth plates (called superognathals). The skull had a regular pattern of bones, featuring eyes located to the sides of the head, and a separate cheek unit that

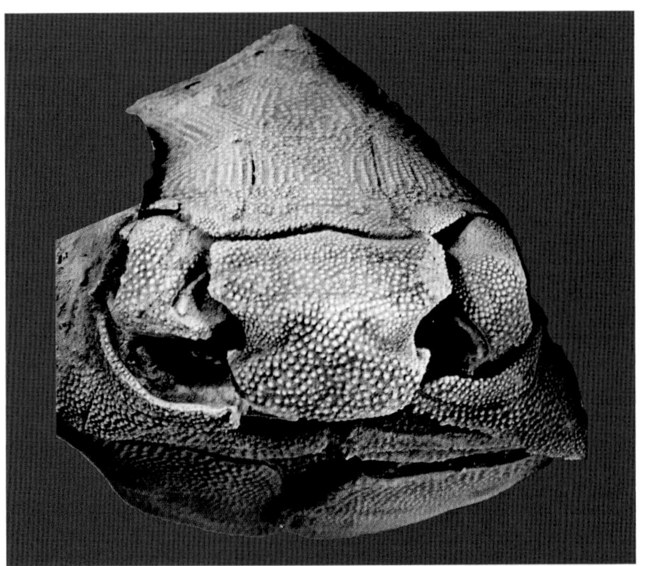

hinged along the side of the skull roof. The head and trunk shields were connected by a ball-and-socket-type joint in all advanced arthrodires; as in the most basal members (grade group Actinolepidoidei; Dupret et al. 2009), they had a sliding neck joint.

The arthrodires generally had fusiform sharklike bodies with a single dorsal fin, broad fleshy paired

(*Left*) An arthrodire, *Buchanosteus*, from the Early Devonian of Taemas, New South Wales, in anterior view. (Courtesy Gavin Young, Australia National University)

(*Bottom*) The braincase of *Buchanosteus* seen in ventral view. Note the small tuberculated parasphenoid bone adhered to the endocranial floor near the front of the skull.

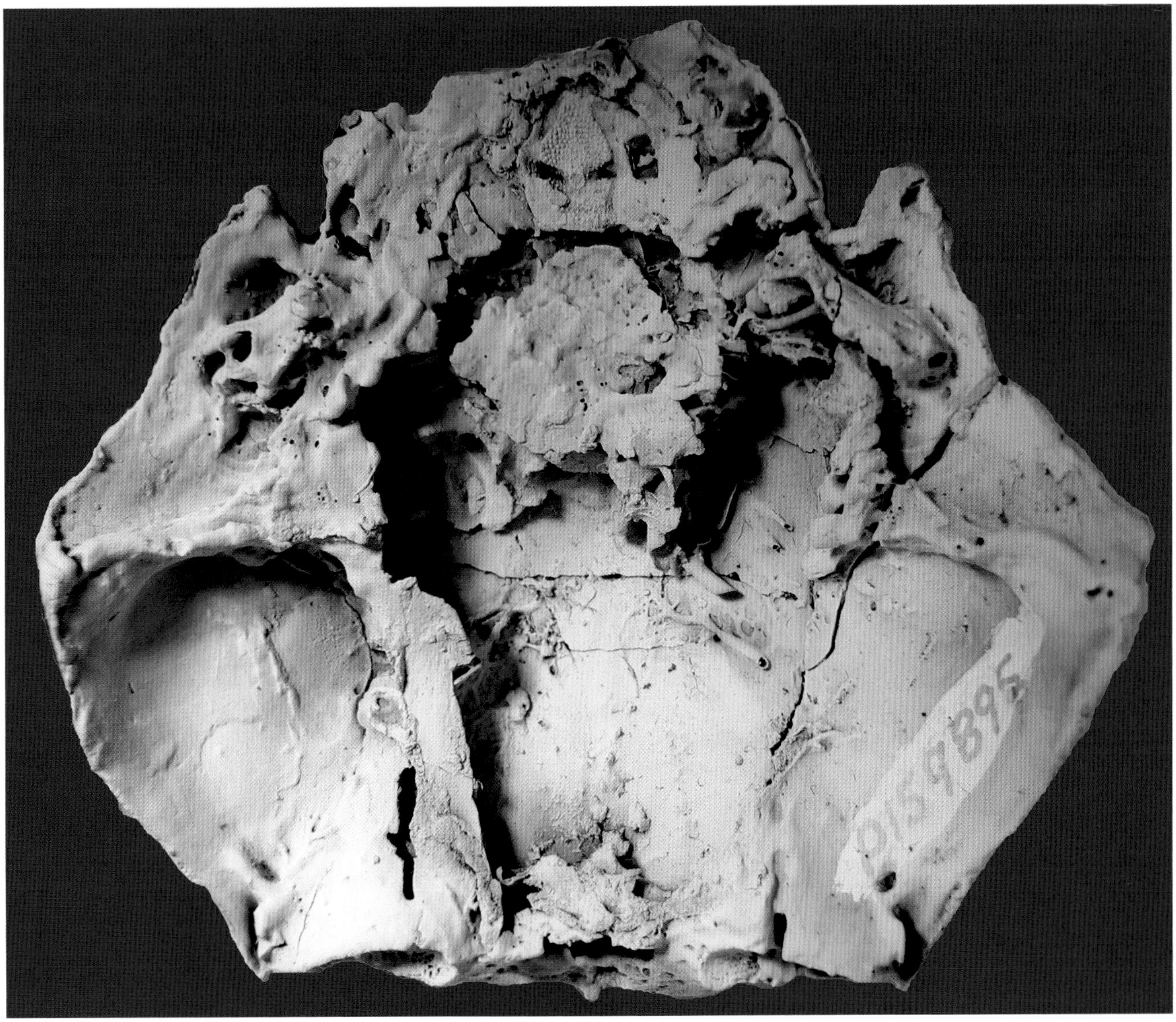

pectoral and pelvic fins, and an anal fin. The tail was primitively covered in non-overlapping scales made up of small polygonal bony platelets often with tuberculated ornament, but the tail was naked in many advanced forms. There were many different families of arthrodires, but in general they are divided into the primitive grade groups, which had long trunk shields with large spinal plates (antarctaspidids, wuttagoonaspidids, actinolepids, and phyllolepids), and the advanced groups (phylctaenids), which had a shortened trunk shield with reduced spinal plates and some with pectoral fins not fully enclosed by the trunk shield without a spinal plate (aspinothoracids). Arthrodires are known throughout the Devonian rocks of the world in both freshwater and marine deposits.

The most primitive arthrodires were the long-shielded antarctaspidids and wuttagoonaspidids, which include a few odd stem group arthrodires such as *Yujiangolepis*, described by Vincent Dupret and colleagues from the Early Devonian of China. Its closest relative is *Antarctaspis* from the Middle Devonian of Antarctica. Both forms have long narrow nuchal plates that contact the pineal plate in the skull. The strange wuttagoonaspidids from Australia and China (*Yiminaspis*) represent the next group up as they and all other arthrodires lack contact between the nuchal and pineal plates. First described in detail by Alex Ritchie in 1973, *Wuttagoonaspis* was a large fish reaching around a meter in size. Its strange head shield had anomalous small bones not seen in other placoderm skull patterns. Its elaborate linear ornamentation led some researchers to suggest an early link between *Wuttagoonaspis* and the flattened phyllolepids. The weak jaws of *Wuttagoonaspis* indicate that it may have been a bottom-feeder, scrounging around in shallow estuarine or marine waters in search of small worms or other prey. Wuttagoonaspidids have now also been recorded from the Early Devonian of China by Vincent Dupret and Zhu Min (2008).

The actinolepids were a group of primitive placoderms known largely from Euramerica, represented by Early Devonian forms like *Aethaspis* from the Bear Tooth Butte locality in Wyoming and described by Robert Denison in the 1950s. The best-known arthrodire of this era is probably *Kujdanowiaspis* from the Early Devonian of Podolia, described in the 1940s by Erik Stensiö in detail from a grinding series. By the arrival of *Actinolepis* in the Early Devonian of Russia and Arctic Canada, forms were beginning to have a shorter, broader nuchal plate, which characterizes most of the advanced arthrodires. *Dicksonosteus* from Spitsbergen, Norway, is known from superbly preserved specimens showing the details of the braincase and its internal anatomy, described in detail by Daniel Goujet (1984).

These Early Devonian actinolepidoid arthrodires lacked a true ball-and-socket neck joint, instead having a simple sliding neck joint, where the head shield sat on a flat platform of bone protruding from the trunk shield. The phlyctaeniid arthrodires were the first of

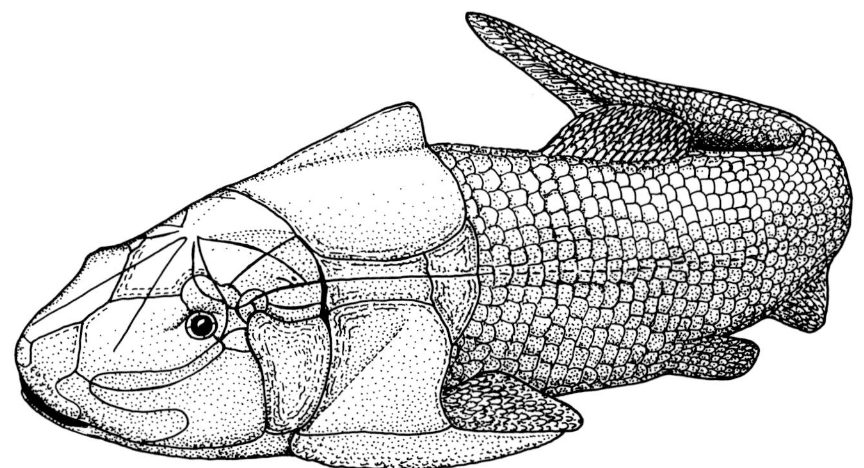

A reconstruction of the basal arthrodire *Wuttagoonaspis* from the Early-Middle Devonian of Australia.

(*Left*) Head shield in dorsal view of *Aethaspis*, a primitive arthrodire from the Early Devonian of Wyoming.

(*Bottom*) Daniel Goujet searching for placoderm fossils in the Early Devonian Wood Bay Group in Spitsbergen, Norway. (Courtesy Daniel Goujet, Musée Nationale d'Histoire Naturelle)

their kind to evolve the primitive ball-and-socket joint, enabling greater vertical mobility of the head shield. Consequently they were better able to open their mouths and catch prey and also had improved respiratory ability. Some of these, such as the Early Devonian *Sigaspis*, *Lehmanosteus*, and *Heintzosteus*, described by Goujet from specimens found at Spitsbergen have very widely splayed spinal plates and look very streamlined. Like

Sigaspis, a primitive long-shielded arthrodire from the Early Devonian of Spitsbergen, described by Daniel Goujet.

phyllolepids (described below), their jaws had many small teeth, which were probably used for gripping soft-bodied worms and such.

The phyllolepids (meaning "leaf scale") were flattened armored fishes, long thought to be jawless, until Erik Stensiö demonstrated in 1936 that they were true placoderms. In 1984 I described one of the first complete phyllolepid specimens from Mount Howitt in southeastern Australia. These had the toothed jawbones preserved, leaving no doubt as to their placoderm affinity. Phyllolepids are readily distinguished by their flat armor made up of a single large plate on the top of the head and trunk rimmed by a regular series of smaller plates. Their most characteristic feature is that each plate had a radiating pattern of raised fine ridges and tubercles.

Only one genus, *Phyllolepis*, was known until recent discoveries from Australia and Antarctica began filling in the gaps of phyllolepid evolution. *Phyllolepis* is known from the Late Devonian (Famennian stage) of Europe and North America, whereas the Australian phyllolepids, *Austrophyllolepis* (meaning "southern leaf scale") and *Placolepis* (meaning "plate scale") appeared earlier in time (Givetian-Frasnian) and seem to be far more primitive than *Phyllolepis*. *Cowralepis*, another recent phyllolepid, was discovered near Canowindra, New South Wales, Australia, and described in 2005 by Alex Ritchie. Like *Austrophyllolepis*, it had a median ventral plate but differs in the shape of its head-shield plate arrangement. Recent study of the gill arches and feeding mechanism of *Cowralepis* by Robert Carr and colleagues (2010) showed that it was an ambush predator capable of gulping prey whole aided by a powerful buccal pump mechanism, similar to the way angel sharks (*Squatina*) catch their prey.

Austrophyllolepis is known from two species from Mount Howitt, eastern Victoria, and its fragmentary remains have also been found in Antarctica. These specimens have revealed more about the anatomy of phyllolepids than have those from any other site, as they show the outline of the whole fish, along with preserved

(*Left*) This *Phyllolepis woodwardi* from Dura Den, Scotland, was the only articulated specimen of this unusual placoderm group known for almost a century.

(*Bottom*) *Austrophyllolepis*, from the Middle Devonian of Mount Howitt, Australia, was the first phyllolepid described with complete jaws and tail preserved. Here are the jaws and parasphenoid from a latex cast.

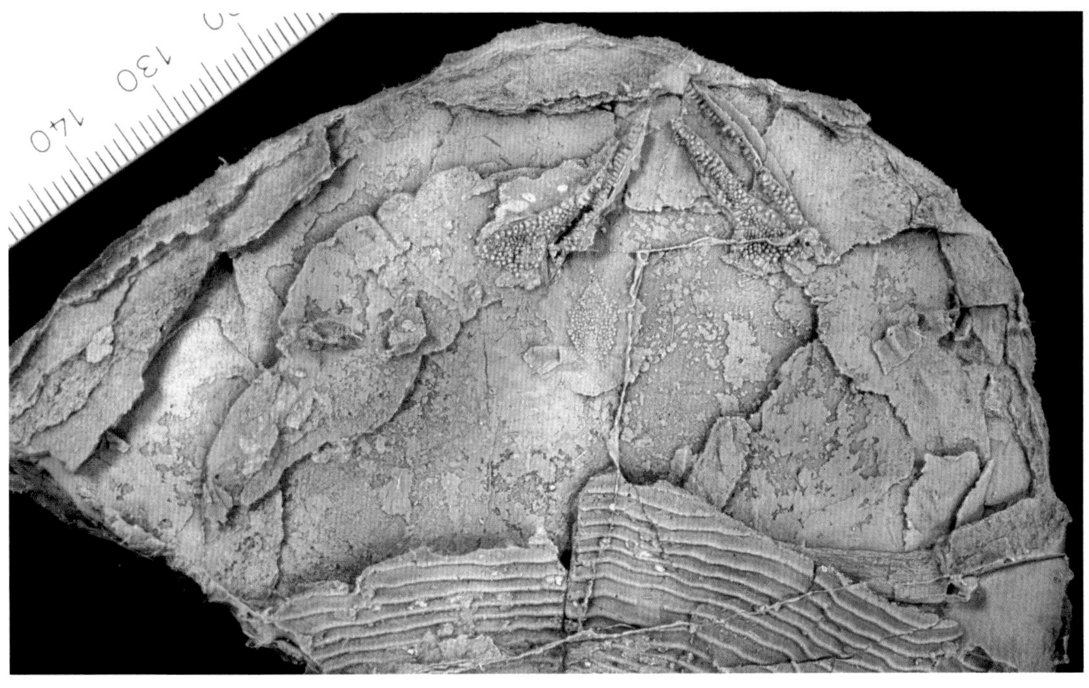

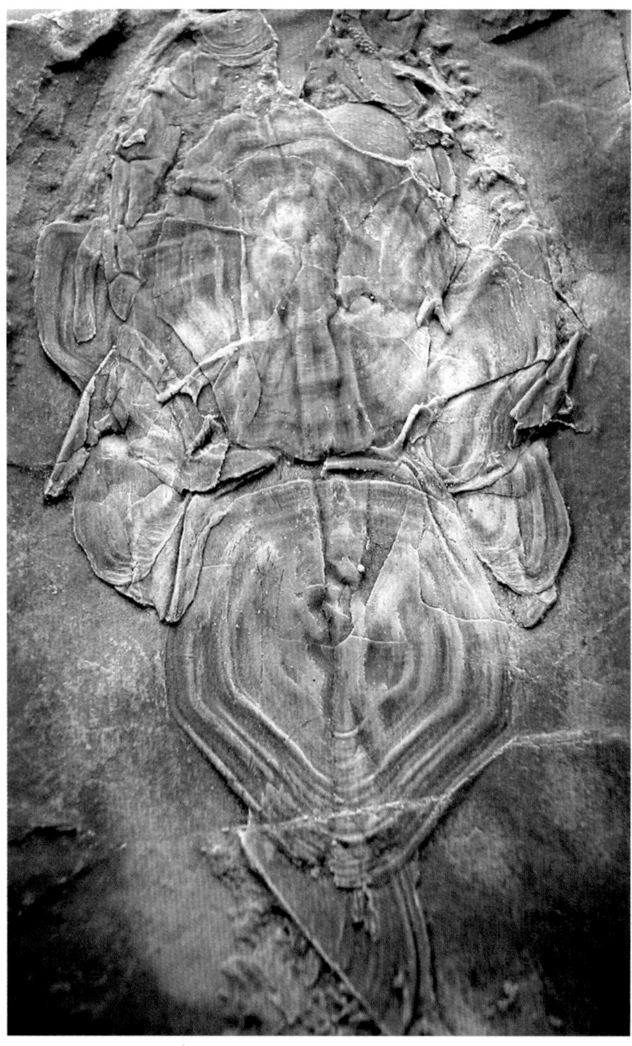

Cowralepis, from the Merriganowry site in New South Wales, has the best-preserved branchial skeleton of any arthrodire. This specimen of a cast shows the armor in dorsal view.

impressions of the jaws, pelvic girdles, palate bone (parasphenoid), and parts of the cheek. From study of these, it is possible to build a picture of phyllolepids as flat predators, lying in wait on a muddy lake bed for some unsuspecting fish to swim above. Then, using its unusually long tail, it could thrust upward to catch the prey in its set of gripping jaws. Phyllolepids were probably blind, as they lacked bones circling the eyes and the head shield is not embayed for eyes as in most placoderms. They had especially well-developed radiating patterns of sensory-line canals that may have helped them detect prey swimming above them while they were concealed below the surface under a fine layer of sediment. The largest phyllolepids were about only about 50–60 cm in length.

Recent finds from China indicate that phyllolepids might have originated from forms like *Gavinaspis* from the Early Devonian of Yunnan. Vincent Dupret and Zhu Min (2008) argued that the group originated in China before spreading to East Gondwana (Australia, Antarctica) in the Middle Devonian and finally reaching the northern hemisphere in the Late Devonian. Recent work has shown that the pelvic fins of phyllolepids may have been modified with long lobes for copulation in the males.

The Brachythoraci are the most advanced group of arthrodires. They are characterized by having a head shield with a ball-and-socket joint. The most basal member of this group is *Holonema westolli*, represented by superb 3-D material from the Gogo site in Western Australia, described by Miles (1971). *Holonema* had a long barrel-shaped trunk shield and the head featured unusual concave lower jaw tooth plates with parallel ridges of raised dentin. The discovery of numerous small stones inside its gut led me to suggest that it was possibly an oncolite feeder, scooping up the algal balls from the fore-reef slopes of the ancient tropical reef where it lived (Long 2006).

Other well-known primitive brachythoracids include the groenlandaspids (named after the common genus *Groenlandaspis*), a group with high-crested median dorsal plates that first appeared in the Early Devonian (*Tiaraspis*) and were abundant in the Middle Devonian of eastern Gondwana, becoming widespread throughout the world by the end of the Devonian.

Australia has an excellent fossil record of the first advanced arthrodires, the eubrachythoracids, beginning with superb three-dimensional plates and skulls of Early Devonian age from near Wee Taemas in New South Wales and Buchan, Victoria. Numerous species have been discovered in these rocks, ranging from small forms about 20 cm long to giants with skulls nearly 40 cm long, suggesting a whole body length of up to 3 m.

The most common species from southeastern Aus-

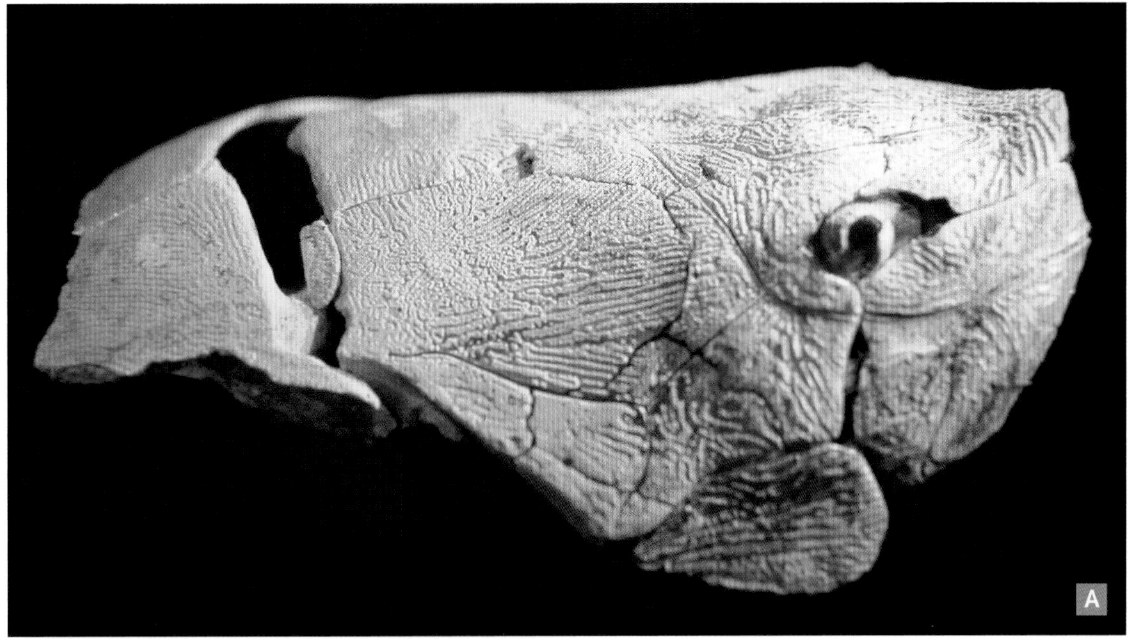

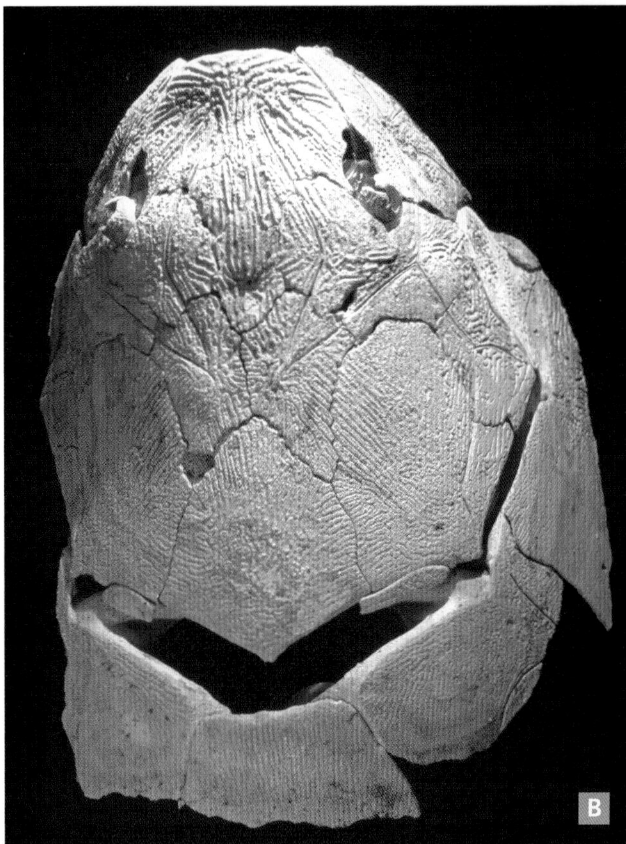

The head of *Holonema westolli* from the Late Devonian Gogo Formation Western Australia in side view (*A*) and top view (*B*).

tralia is *Buchanosteus*, named after the area of Buchan where it was found. The species was first described by Frederick Chapman of the National Museum of Victoria, who misidentified it. Edwin Hills of the University of Melbourne recognized it as an arthrodire and pioneered a method of removing the specimen using acid. Unfortunately he used hydrochloric acid, which also damaged the specimen, but not before he extracted a lot of new information on the structure of the braincase. Although Hills recognized the species as an arthrodire, he cautiously assigned it to the well-known Scottish genus *Coccosteus*. Erik Stensiö of Sweden in 1945 reassigned Hill's specimen to a new genus.

In the late 1970s, Gavin Young, from the Australian Geological Survey Organisation, published a detailed study of the anatomy of *Buchanosteus*, reconstructing much of the braincase and cranial anatomy from new specimens prepared using acetic acid. Using these data, he was able to propose a new theory of arthrodire interrelationships and thus to herald a new era for the study of placoderms. Buchanosteid arthrodires have now been found in China and Russia and are regarded as among the most primitive family of the great radiation of advanced arthrodires having a shortened trunk shield.

Other placoderms found with *Buchanosteus* include

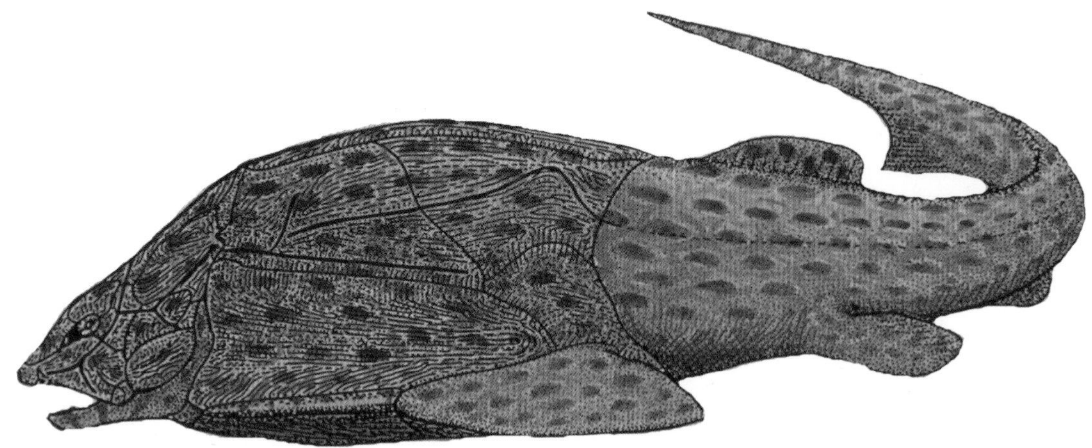

(*Top*) Restoration of *Holonema*. The color scheme based on the assumption that it is a reef-dwelling fish.

(*Left*) The partial skull of *Dhanguura*, a very large homostied arthrodire from the Early Devonian of Wee Jasper, New South Wales, Australia. The restoration shows that the whole skull would have been around 55 to 60 cm long.

(*Left*) The underside of the head shield showing the jaws of *Groenlandaspis*, a commonly found arthrodire in Middle-Late Devonian rocks around the world.

(*Below*) A reconstruction of *Groenlandaspis* based on complete specimens with the tail preserved from Mount Howitt, Australia.

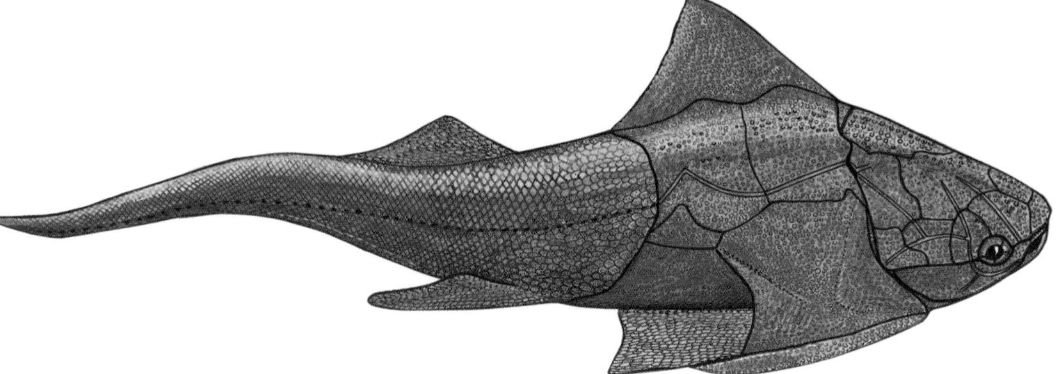

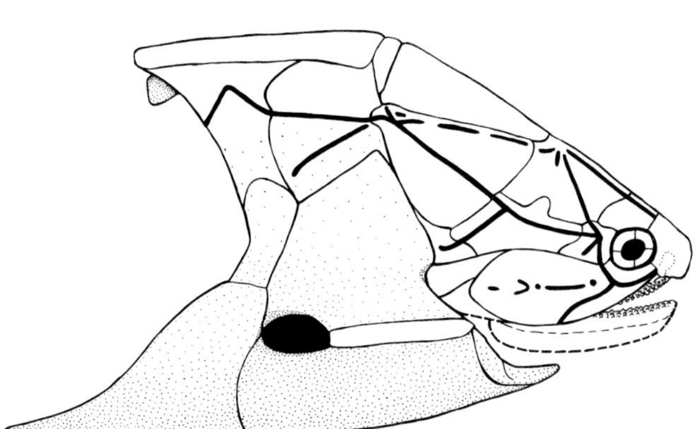

Buchanosteid armor restored in side view. (After a Museum Victoria specimen).

the elaborately ornamented *Errolosteus,* named in honor of Errol White, a paleontologist at the British Museum of Natural History who pioneered studies of Taemas-Wee Jasper placoderms; and *Arenipiscis,* a form with a slender skull roof and delicate sandgrain-like ornamentation (hence, the name meaning "sand fish"). The largest arthrodire skull found at Taemas belongs to a group called the homosteids and was named by Young as *Dhanguura.* It has a large, flat skull about 40 cm long with a small T-shaped rostral plate at the front—a feature linking the generally primitive arthrodires of the Early Devonian with the more advanced forms common in the Middle Devonian. Homosteids were quite large

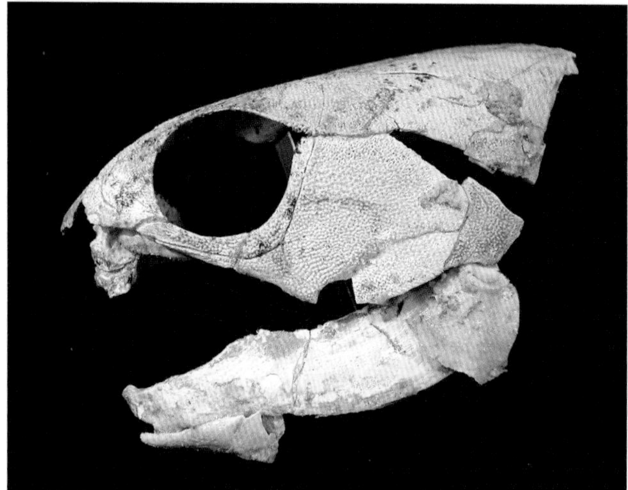

(*Top*) ***Fallacosteus*****, a streamlined, long-snouted arthrodire from the Late Devonian Gogo Formation of Western Australia.**

(*Left*) ***Incisoscutum sarahae*****, a durophagous arthrodire that probably fed on hard-shelled invertebrates on the ancient reef system.**

arthrodires, some reaching 3 m long. They had weak, toothless jaws and may have been early filter-feeders, like whale sharks.

The Late Devonian Gogo fauna of Western Australia includes more than 20 different arthrodires, some belonging in their own unique families. The camuropiscid and incisoscutid arthrodires are small forms that flourished on the ancient Gogo reefs; they are characterized by their elongated, spindle-shaped armor, large eyes, and crushing tooth plates. Extreme elongation of the armor is seen in forms like *Rolfosteus* and *Tubonasus*, which evolved tubular snouts to improve their streamlining. These fishes were only about 30 cm long and may have been active top-water predators, probably chasing the small shrimplike crustaceans that abounded in the warm tropical waters. *Incisoscutum* had the trunk shield incised to free the pectoral fin from being enclosed by bone, probably improving the mobility of its pectoral fins. It was one of the commonest placoderms found at Gogo and also had strong crushing tooth plates, like those of camuropiscids.

The most diverse group of arthrodires found at Gogo are the little predatory plourdosteids. The group is named after *Plourdosteus*, which is well known from the Late Devonian of Canada and Russia. Many different species of plourdosteids occur at Gogo, each distinguished by a unique pattern of skull roof bones or dentition (for example *Torosteus*, *Harrytoombsia*, and *Mcnamaraspis*). The lower jaws feature several well-developed cusps and toothed areas, with numerous teeth along the midline where the jaws meet. In addition they have well-developed bony buttresses that braced the cartilaginous braincase for their powerful biting actions. They were carnivorous fishes, ranging in size from about 30 to 50 cm. The broad, robust shape of their armor suggests that they hunted near the seafloor or within the cavities of the ancient reefs, lunging out after small fishes or crustaceans.

The Gogo fauna includes large predators with huge

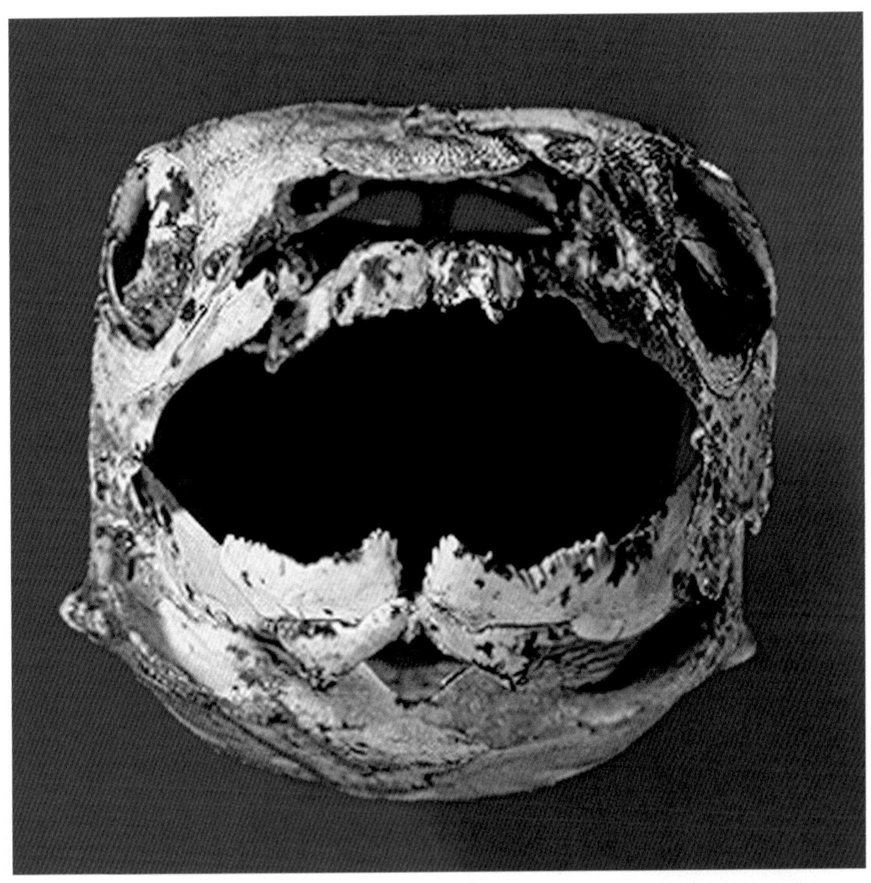

(*Left*) *Compagopiscis*, from the Late Devonian Gogo Formation of Western Australia, showing jaws and teeth. (Courtesy K. Brimmell, Western Australia Museum)

(*Bottom*) This mother arthrodire, *Incisoscutum*, has an embryo inside her. The embryonic bones are miniature replicas of the adult plates.

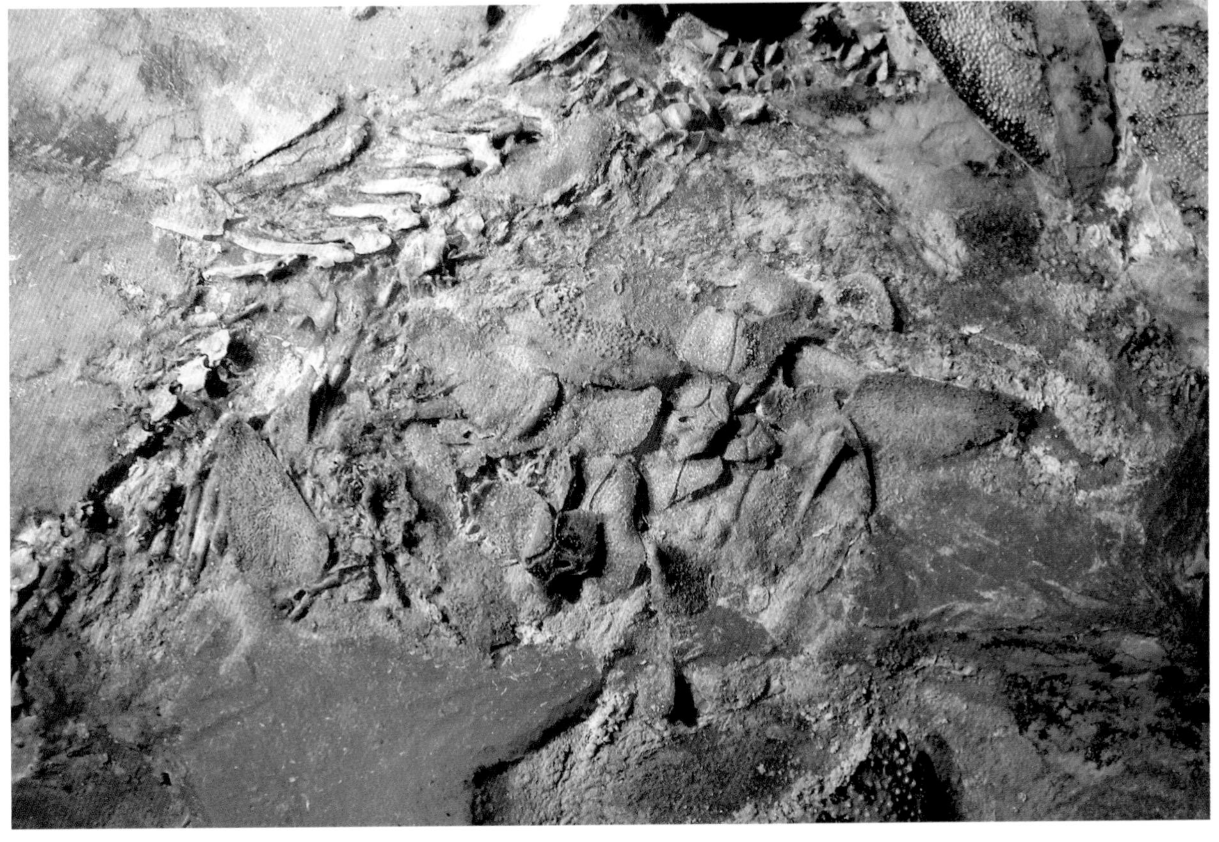

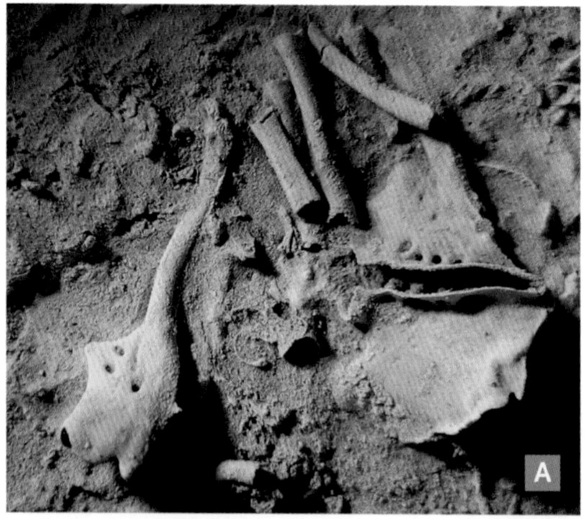

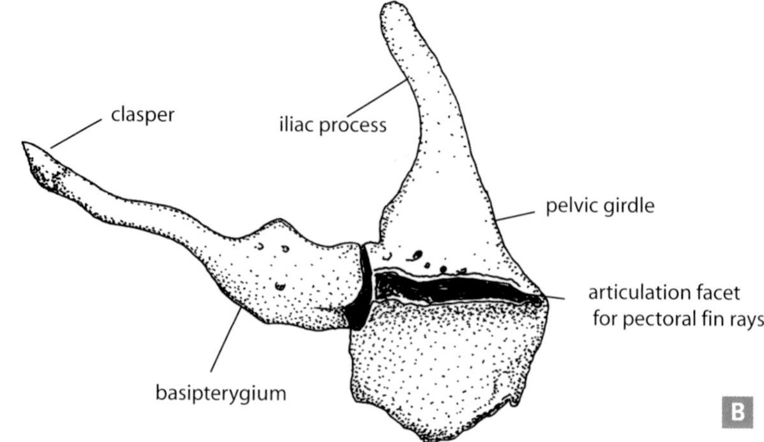

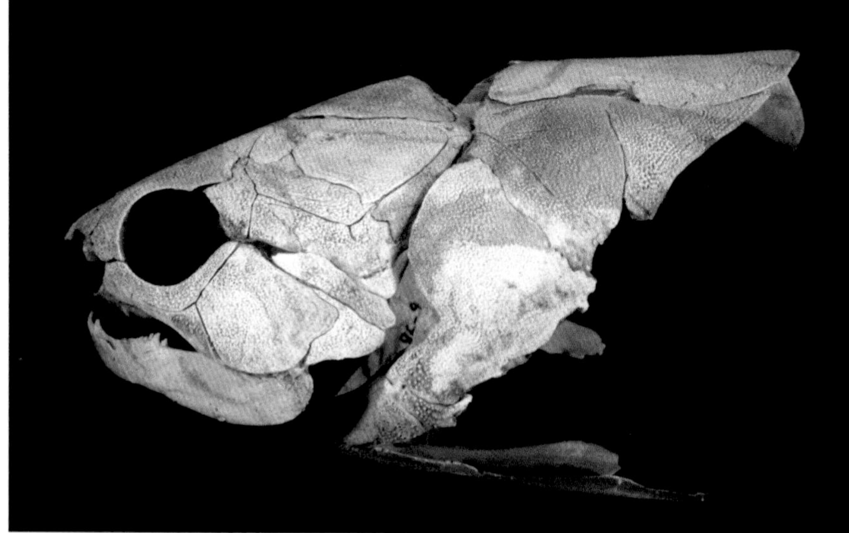

(*Top*) The male clasper and pelvic girdle of *Incisoscutum ritchiei* (*A*); the clasper restored attached to pelvic girdle (*B*).

(*Left*) The only placoderm to become the center of a political campaign, *Mcnamaraspis kaprios*, from Gogo, was declared the official state fossil emblem of Western Australia in 1995.

dagger-like cusps on their jaws—such as *Eastmanosteus calliaspis,* which grew to about 3 m long—as well as many small predatory forms armed with sharp toothlike structures on their jaws. *Eastmanosteus* shows us the basic pattern adopted by the largest placoderms—the dinichthyids (meaning "terrible fish"). In recent analyses, the Gogo species was determined to be a primitive sister species to the larger more ferocious forms. Large skulls and bony plates of these monsters have been excavated from the Cleveland Shales of Ohio and New York State and from limestones exposed in the northern Sahara Desert of Morocco. The largest of these had skull roofs more than a meter long and suggest total lengths of 6–8 m. Most of these species, like *Dunkleosteus* and *Gorgonichthys,* had trenchant, pointed cusps on the lower and upper jaw tooth plates. Biomechanical studies by Phil Anderson have shown that *Dunkleosteus* had one of the most powerful bites of any creature, able to exert forces of up to 5300 newtons at the back of the jaws. *Titanichthys,* from Morocco, was at least 7 m long but had weak jaws that lacked the sharp cusps of the dinichthyids and may have been a giant filter-feeder like the modern-day whale shark. Elga Mark-Kurik has already suggested for an earlier group of large placoderms, the homosteids. Studies of the giant placoderms from the Cleveland Shale suggest that they may have preyed on other arthrodires or been carrion feeders. Perhaps they tried to catch the cladoselachian sharks that lived with

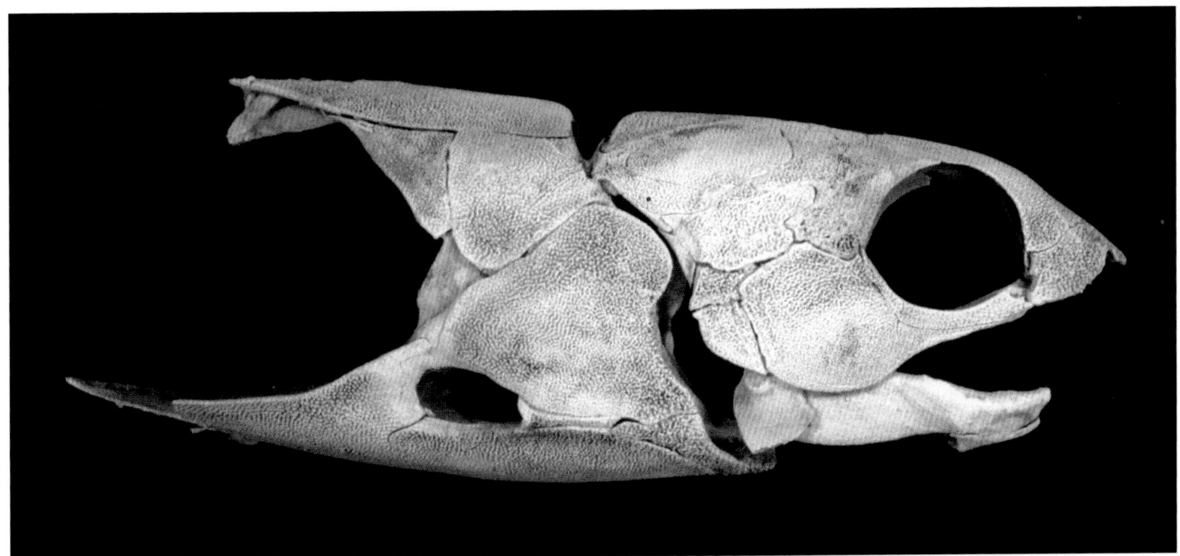

them, but it seems hard to imagine the heavily armored predators ever catching a streamlined little shark.

Despite the size and other measures that would have allowed these and other placoderms to compete with the ascendant groups of sharks and bony fishes, the placoderms mysteriously became extinct at the end of the Devonian, some 355 million years ago. Nonetheless, they dominated the vertebrate life in the waterways for nearly 60 million years and must be regarded as one of the most successful groups of vertebrates to have ever lived.

(*Top*) Another streamlined arthrodire from Gogo, *Latocamurus* has strong jaws with crushing tooth plates for feeding on hard-shelled prey. Armor length is 10 cm.

(*Bottom*) *Eastmanosteus* is known from several species found around the world. This species (*with eye and jaw at right*) reached an estimated 2.5 m in length and is the largest placoderm known from the Gogo Formation. Armor length is approximately 25 cm.

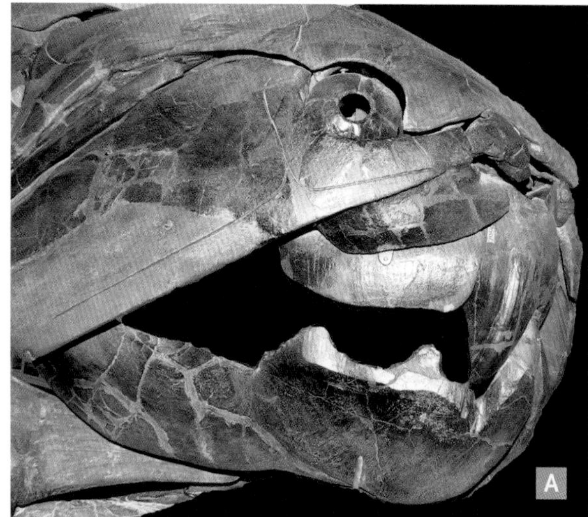

(*Top*) The giant placoderm *Dunkleosteus* grew to 4–5 m in length and is known from several fine specimens from the Upper Devonian Cleveland Shale of Ohio. This specimen (A) mounted is at the Cleveland Museum of Natural History. A restored model (*B*) of *Dunkleosteus* by Esben Horn of Denmark shows it to be a fearsome-looking fish. (Courtesy Esben Horn)

(*Middle*) The lower jaw, or inferogathal, of *Diplognatus*, a large predatory Late Devonian arthrodire from the Cleveland Shales of Ohio.

(*Bottom*) *Oxyosteus*, a tubular snouted arthrodire from the Upper Devonian Wildungen site of Germany.

CHAPTER 5

Sharks and Their Cartilaginous Kin

Killers from the dawn of time

Sharks are one of nature's great success stories. Since their first appearance at least 420 million years ago, they have changed very little, merely improving their ability to hunt and gather food by evolving more effective feeding structures and more streamlined body shapes. The basic body plan of the shark underwent two successful major modifications: one by the start of the Carboniferous Period, when the holocephalomorphans (chimaeras and rabbitfishes) first appeared, and one in the Jurassic Period, when the flattened rays evolved. The acme of chondrichthyan evolution came in the Middle Paleozoic, immediately following the demise of placoderms. Many bizarre forms of Paleozoic sharks have been discovered, some with coiled whorls of serrated teeth, others with huge bony structures on their dorsal fins. Today, sharks, rays, and holocephalans (the group containing living chimaerids) are known from more than 970 species, including the largest fish on Earth, the massive filter-feeding whale shark, which grows up to 15 m in length. Huge predators such as the great white shark, today reach lengths in excess of 7 m. As recently as 2 million years ago, voracious predatory sharks similar to the great white grew up to 15 m long—the most gruesome underwater death machines that ever lived.

Sharks are perhaps the most feared yet the most intriguing of all fishes. Despite the grizzly images we might hold of efficient killing machines that relentlessly attack humans, humans actually consume shark meat at a rate

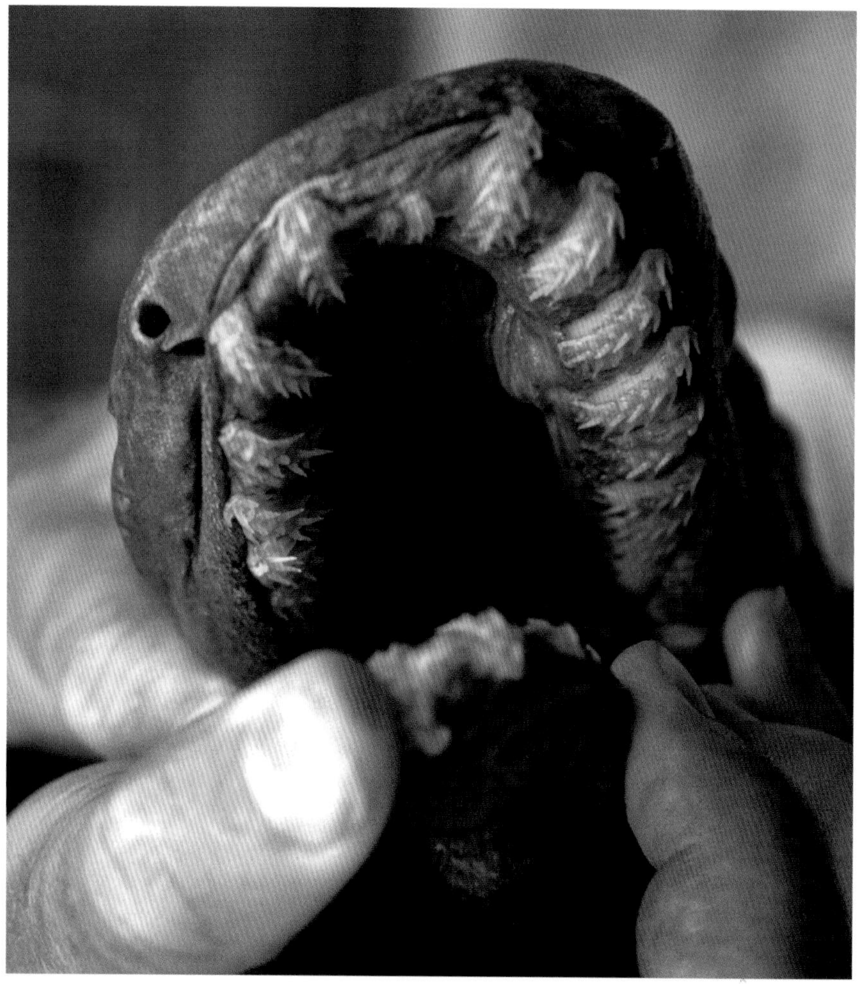

The jaws of a primitive shark, *Chlamydoselache*, showing the rows of teeth that are replaced through life by new teeth growing from behind. Some sharks shed up to 20,000 teeth in their lifetimes, explaining why fossils of their teeth and scales are so abundant.

many millions of times more by volume than sharks consume people: we kill around 73 million sharks each year, and sharks kill about 20 to 30 people. Only a handful of the known species of living sharks and rays have any record of attacking people or would even be capable of inflicting harm. Those that are known to be man-eaters, such as the great white shark, *Carcharodon carcharias*, usually feed on seals or large fishes and may occasionally take a human if the person enters its hunting grounds. However, when we look at sharks from the perspective of their incredibly long history, we see a totally different story, one of a highly successful and efficient predator that has changed little from its first appearance.

Sharks are among the earliest known jawed fishes. The evolution of sharks is known principally by the fossil record of their teeth, spines, and scales, punctuated by a few exceptional specimens showing skulls, and in rare cases whole body preservation. Their scales have been found in rocks dating back to the dawn of the Silurian Period, with possible sharklike scales recorded in Late Ordovician sediments. However, the fact that shark scales occur at this time without the associated fossilized teeth, suggests that these early protosharks may not necessarily have had teeth or even jaws. The mystery remains as to how sharks first evolved. Did they evolve from a heavily scaled agnathan group, like the thelodonts, or from an as-yet-undiscovered ancestral fish group? Perhaps they are closely allied to the early acanthodians (see Chapter 6).The current fossil evidence is incomplete but appears to favor the latter scenario.

The best clues we have to their distant origins are from the peculiar anatomy of oldest well-preserved fossil sharks, which include some enigmatic forms from

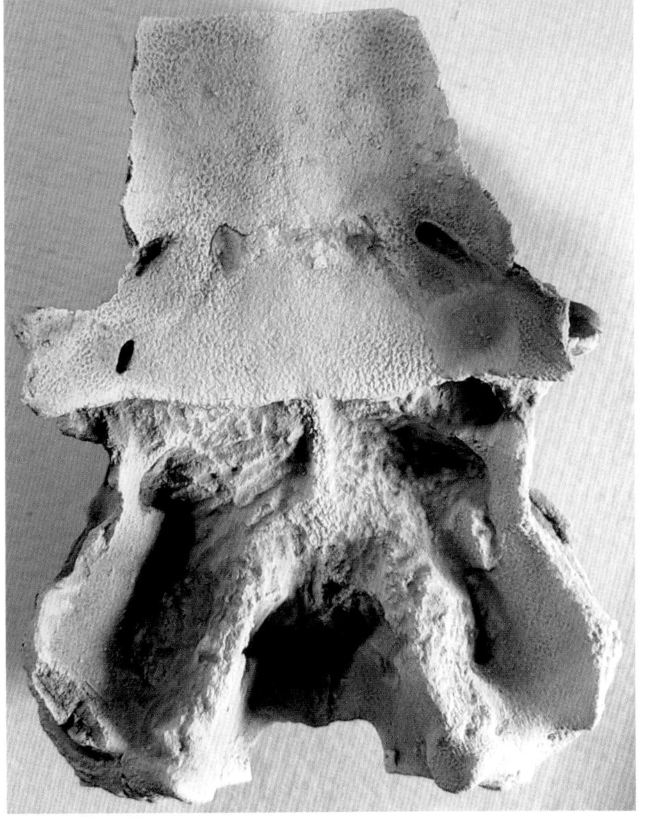

(*Top*) The great white shark, *Carchardon carcharias*, reaches lengths of up to 6.5 m and is one of the oceans top-end predators. This species has been around for about 15 million years, but its prehistoric relatives grew to sizes up to 15 m. (Courtesy iStock)

(*Left*) The skull of an Early Devonian shark from Bolivia, *Pucapumpella*. This is the earliest known three-dimensional preserved shark braincase. (Figure in Maisey 2007)

the Early Devonian MOTH locality in British Columbia (e.g., *Seretolepis, Kathemacanthus*) and a specimen of *Doliodus* from the Emsian of Quebec (Miller et al. 2003). Both these localities have revealed that the oldest sharks have a variety of forms having fin-spines preceding some or all of the fins.

As sharks are a living group of fishes, we have a thorough knowledge of the anatomy, physiology, and lifestyle of many species. Research shows that the shark is far more advanced in many ways than previously thought. Some modern-day sharks, for instance, lay

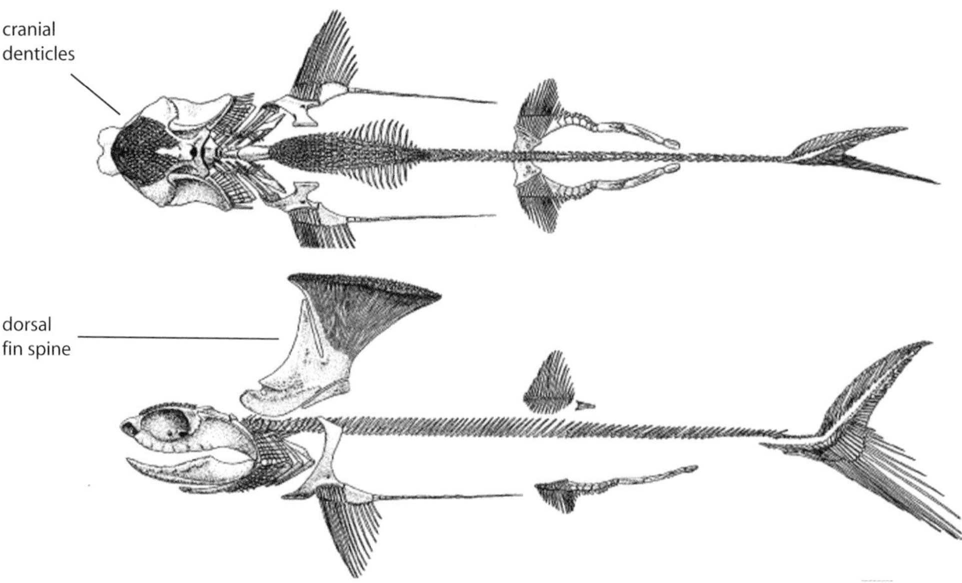

The restoration of the skeleton of *Akmonistion* (*A*), an Early Carboniferous stethacanthid shark. Note the development of the anterior dorsal fin spine into brush of tooth-like denticles. Reconstruction of a Lower Carboniferous marine scene (*B*) from Bearsden, Scotland, showing the shark *Akmonistion* (*center*), with large *Rhizodus* (*upper left, background*). Varieties of early ray-finned fishes (*center left and right*) and acanthodians (*below, center*) are seen alongside a primitive decapod crustacean (*below, right*). (Courtesy Michael Coates)

The head of *Callorhynchus*, an extant chimaerid (holocephalan). This group of chondrichthyans have upper jaws fused to the braincase and crushing dentition of hard tooth plates. They also have gill arches covered by an opercular flap of skin.

eggs and let the young hatch themselves, whereas others give birth to live young and have the equivalent of a mammalian placenta. Some sharks even have quite complex mating behavior and rituals. All sharks possess extremely sensitive snouts that house numerous pores containing ampullae of Lorenzini, special organs that enable them to detect sensitive electric fields so that they can find prey that may even be buried below the surface sands of the seafloor. Sharks' chemosensory organs are so highly refined that they can detect parts per billion of blood in the water from over a kilometer away. Hammerhead sharks, which have the eyes widely spaced on stalks, are now believed to have evolved such bizarre heads because this enables them to spread their ampullae over a wider area to increase their electrosensitivity when foraging for food, such as fish or crabs buried in the sand.

Sharks live in a great variety of habitats, from rivers and nearshore reefs to open oceans. They have been found at all depths, from surface waters to deep abyssal plains, and some species, such as the Zambesi river shark, may travel hundreds of kilometers inland. Some, such as the great white shark, are tachymetabolic, maintaining slightly higher ambient body temperatures than the surrounding seawater, a physiological advance that enables them to move their muscles with great speed and regulate their body temperature irrespective of the changes in the water around them.

Sharks and rays and their cousins, the chimaerids, are all classed within the Chondrichthyes (meaning "cartilage fish"), as they have a dominantly cartilaginous skeleton and only retain few bony tissues within their skeletons. The mineralized spines, teeth, and scales that they possess are invariably made up of dentinous tissues (semidentin or orthodentin) with remnants of acellular perichondral bone present in some of the early forms (e.g., *Akmonistion*), suggesting that sharks once had the potential to develop bone but eventually lost this trait in the course of their evolution (Coates et al. 2002). They have a fascinating evolution that is more akin to the long-term fine-tuning of a perfect machine rather than periodic leaps into new models. From their first appearance, sharks seem to have got it right.

The Origin of Sharks

The origin of sharks is still shrouded in mystery. Some scientists regard sharks as the most primitive of all the jawed fishes, whereas others see them as highly specialized forms that did not require the complex bony ossifications of other fish groups. Sharks appear to be

Basic Structure of Primitive Chondrichthyans

Sharks, rays, and rabbitfishes belong to the class Chondrichthyes (meaning "cartilage fishes"), so-named because they lack an internally ossified bony skeleton, having instead a special type of globular calcified cartilage forming the braincase, jaws, gill arches, vertebrae, and fin supports. The only hard bony tissues are developed in their defensive fin-spines, teeth, and scales. Although many biologists once thought of the cartilaginous condition of sharks as being primitive, as a precursor to the evolution of true bone, it is now regarded as a highly specialized condition that enables sharks to function more efficiently. The cartilage forming their internal skeleton may be strengthened with needles of the mineral calcite to give a high degree of strength while not adding unnecessary weight. The lack of a swim-bladder inside the shark means that buoyancy in the water column is achieved by other means, such as having a large oil-filled liver in association with the much-lightened skeleton. Their broad winglike pectoral fins give sharks their hydrodynamic lift within the water as long as they keep moving.

Sharks have simple jaws consisting of the primary upper and lower jaw cartilages (Meckel's cartilage and palatoquadrate), armed with rows of teeth that grow throughout life and replace damaged or shed teeth from the front rows. The teeth of sharks and rays are in general composed of a thin layer of enameloid (or multilayered enameloid in modern groups) over an orthodentine crown with a trabecular dentin base, sometimes pierced by canals. In food-crushing species, they may be flattened combs of pleuromic dentin and trabecular dentin that forms a grinding pavement. Chondrichthyans have placoid scales, which have similar histology as the teeth, and these are set into the skin but do not generally overlap each other. The scales may be simple bladelike structures or be complex with several generations of growth on each base.

One of the features that distinguish chondrichthyans from other fishes is that they reproduce by internal fertilization. The males have intromittent organs, called claspers, attached to the pelvic fins. They insert these into the cloaca of the female to fertilize her eggs. Even the fossil forms show the presence of claspers, and many well-preserved sharks from the Early Carboniferous Bear Gulch Limestone of Montana show distinct shape and fin differences between the sexes. Males not only have pelvic claspers but also may possess elaborate dorsal fin-spines with brushlike structures possibly used in mating (as in *Falcatus*) or cranial claspers to assist in mating behavior (as in holocephalans).

closely related to the now-extinct placoderms (see previous chapter), and it is possible that both these groups may have arisen from a scale-covered jawless ancestor before the Early Silurian. The presence of sharklike scales of this age and earlier (Late Ordovician), and their striking similarity to the scales of the jawless thelodonts, has led some researchers to suggest that thelodonts and sharks could be close relatives. The discovery of the remarkable fork-tailed thelodonts from Canada supports this view, although it casts no light on the origin of sharks' distinctive jaws and teeth.

Sharklike scales from the Early Silurian of Mongolia, such as those of *Mongolepis* and *Polymerolepis*, resemble the simple placoid scales of modern sharks, but their internal structure differs in having complex odontodes growing from the bases to form long thin tubes of a dentinous tissue devoid of little tubes (canicules) termed *lamelline* (Karatajute-Talimaa 1995). It has been suggested by earlier paleontologists that fishes having such scales should be placed in a group called the Praechondrichthyes. Late Ordovician scales with sharklike features are also known from the Harding Sandstone of North America (Smith and Sansom 1998) and central Australia (e.g., *Areyongia*; Young 1997) but again differ in fundamental histological features, which precludes their being real sharks.

The first undoubted chondrichthyans are known from scales that date back to the Lower Silurian (Llandoverian). Forms like *Elegestolepis* and *Ellesmereia* show the presence of true dentinous material making up the scale. The scales, however, did not grow, being shed and replaced by larger ones as the fish grew. True growing scales evolved in sharks by the end of the Early Devonian, represented by *Ohiolepis* and other species. The

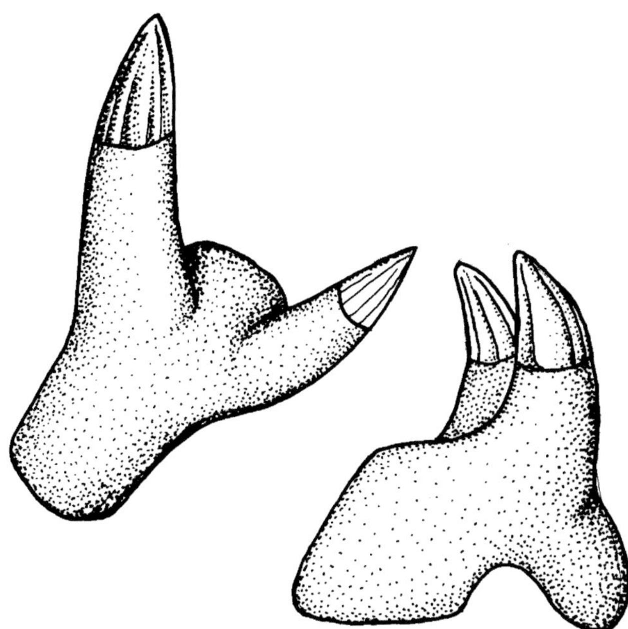

These two-pronged teeth are from the Early Devonian shark *Leonodus*, from near Leon, Spain (size about 2–4 mm).

oldest known shark's teeth date back to the beginning of the Early Devonian. The shark that these tiny teeth (under 4 mm) belonged to is named *Leonodus* (after Léon, Spain) and has been found at several sites in Europe. Other Early Devonian records of fossil shark teeth are from Saudi Arabia.

The MOTH locality in Canada has revealed complete articulated fish, such as *Seretolepis*, which had sharklike scales and a variety of fin spines present both as paired and median elements (Hanke and Wilson 2004). *Doliodus problemticus* from the Early Devonian of Canada further demonstrates that early sharks had paired pectoral fin-spines, as in placoderms, acanthodians, and early osteichthyans such as *Guiyu*.

The oldest articulated sharks include *Doliodus problematicus* from the Emsian (late Early Devonian) of New Brunswick, Canada, and *Ptomacanthus* from the early Devonian of Great Britain. *Doliodus*, first described from isolated teeth as an acanthodian, is now known to have possessed paired pectoral fin-spines. Its braincase had a precerebral fontanelle, lacked a ventral otic fissure but had a posteriorly located endolymphatic fontanelle. In these features, it showed basic shark characteristics not seen in other jawed vertebrates. Isolated teeth of *Doliodus* have been found in older sediments near the Emsian-Pragian boundary. *Doliodus* lived in an estuarine-to-freshwater habitat, but most other Early Devonian shark remains have been found in marine deposits. *Ptomacanthus* was for many years classified as an acanthodian, until recently Martin Brazeau (2009) redescribed it and analyzed its braincase, showing it to be either the most primitive known chondrichthyan or stem gnathostome more closely allied to sharks. In

A shark's tooth of an undescribed new species from the Frasnian Gogo Formation, Western Australia, in normal view (*A*) and a micro-CT scan (*B*) (height 3 mm). (Courtesy Ken Walker, Museum Victoria)

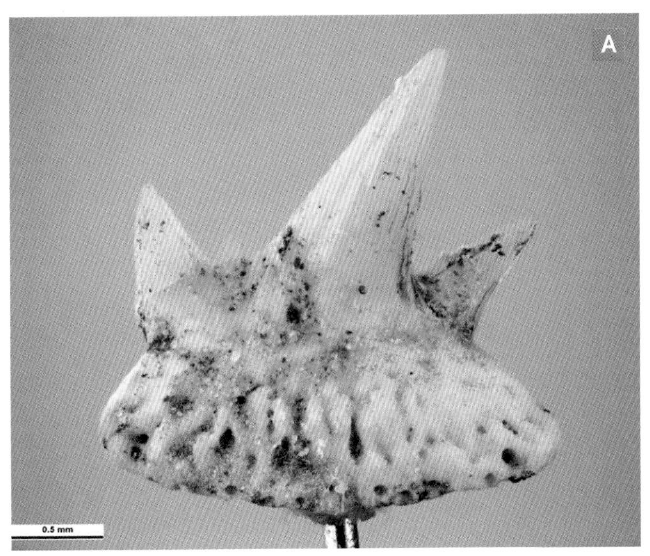

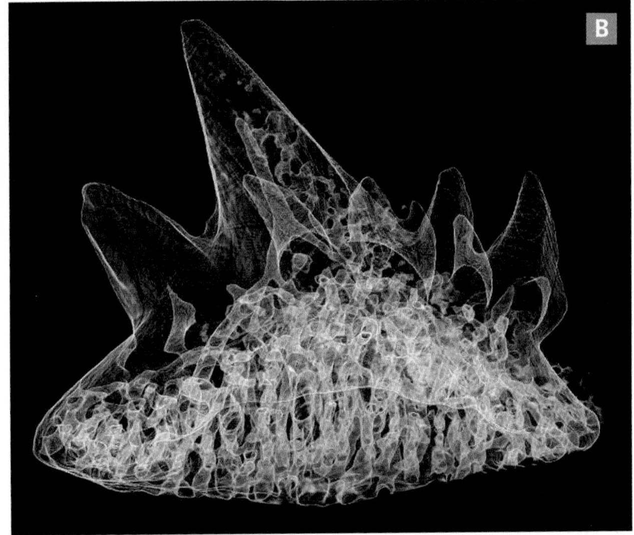

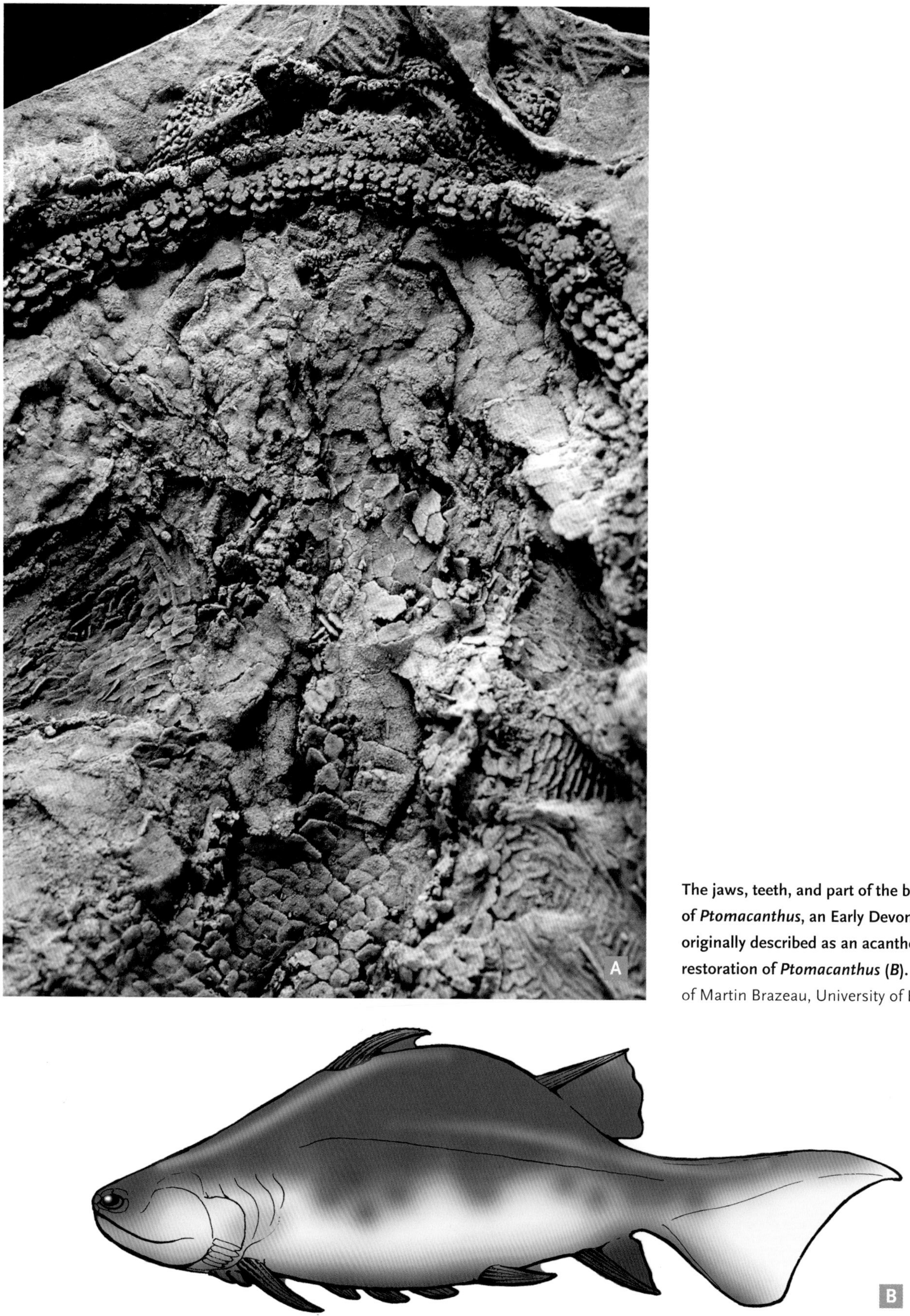

The jaws, teeth, and part of the braincase of ***Ptomacanthus***, an Early Devonian shark originally described as an acanthodian (***A***); restoration of ***Ptomacanthus*** (***B***). (Courtesy of Martin Brazeau, University of Berlin).

his cladistic analysis of the acanthodian fishes, Brazeau suggested that sharks could have emerged from one of several distinct groups of early acanthodians and that this larger group could have come from placoderm-like ancestors.

Other Early Devonian shark fossils include an isolated braincase from the Emsian of South Africa (Maisey and Anderson 2001; Maisey et al. 2009). The specimen shows that primitive shark braincases also had some features previously thought to be found only in basal osteichthyan fishes (such as a ventral otic fissure, posterior dorsal fontanelle, and a palatobasal jaw articulation).

The earliest diverse assemblage of shark's teeth from one formation is from the Middle Devonian Aztec Siltstone of Antarctica, where at least five taxa occur, some like *Portalodus bradshawae,* having large bicuspid teeth almost 2 cm in height (Long and Young 1995). The teeth of others, such as *Mcmurdodus,* have complex crowns with many cusps.

This connection that sharks had undergone an early radiation in the Middle Devonian of Antarctica dawned on me after a grueling field trip collecting fossil fish remains in Antarctica's South Victoria Land in late 1991 to early 1992. I was searching the Lashly Range where, 20 years before, Gavin Young and Alex Ritchie had found the world's oldest partially articulated shark remains. Young had discovered a little shark, about 40 cm long, which he named *Antarctilamna* ("lamnid shark from Antarctica"). Impressions of the braincase of *Antarctilamna,* as well as fin-spines and teeth have also been found in Australia, Venezuela, and Saudi Arabia. *Antarctilamna* had a robust fin-spine preceding the large dorsal fin, and its teeth had two large splayed cusps with smaller median cusps in between. It may have had paired pectoral fin-spines as well (suggested by Miller et al. 2003). The structure of its teeth shows that they had a thin layer of enameloid covering an orthodentin crown over a trabecular dentin root (Hampe and Long 1999). Another peculiar little tooth from the Aztec Siltstone of Antarctica was named *Mcmurdodus* by Errol White of the Natural History Museum, London, in 1968. Recently this type of tooth was found in central Australia in rocks of late Early Devonian age. The teeth are up to 5 mm or so wide and have several sharp, flat cusps arranged almost symmetrically along a broad root. The significance of *Mcmurdodus* arises from the suggestion made by Sue Turner and Gavin Young that its multilayered enameloid crown indicates that it is the earliest known member of the neoselachian group to which all modern sharks belong. The root structure of *Mcmurdodus* has penetrating canals, another feature of modern shark teeth.

Nearly all of these Early-Middle Devonian occurrences suggest that the oldest true chondrichthyans—those definitely having teeth of characteristic form and tissue

The tooth of *Mcmurdodus*, from the Middle Devonian of central Australia, shows the complexity of some early shark dentitions (width 4 mm).

Search for shark fossils in the Lashly Range, south Victoria Land, Antarctica. In 1970 the remains of the then oldest partially articulated shark, *Antarctilamna*, were found here by Australian scientists Gavin Young and Alex Ritchie. *Doliodus* from the Early Devonian of Canada is now the oldest articulated shark fossil.

types—came from Gondwana and that the first great radiation of sharks may have taken place there also. This is supported by the fact the oldest definite shark's teeth are *Leonodus* from the Lochkovian-Pragian of Spain, then part of western Gondwana (Armorica). These teeth predate those of *Doliodus* from the Emsian of Canada by at least 10 million years.

Middle Devonian sharks are well known from many isolated teeth and scale genera as the group rapidly spread around the world. An articulated skull found in the Givetian sediments of Germany, named *Gladbachus adentatus*, was at first thought to represent a primitive toothless shark, but a recent study by Michael Coates of the University of Chicago has shown that it possessed many tiny teeth. By the late part of the Middle Devonian, many distinct groups of sharks had evolved each with characteristic tooth types. *Omalodus* from North America represents the earliest member of the phoebo-

Portalodus **was one of the largest sharks from the Middle Devonian Aztec Siltstone, Antarctica. This tooth is almost 2 cm high and may have come from shark close to 2 m long.**

dontid sharks. Xenacanthid sharks also appeared at this time, represented by *Dittodus*, as did typical cladoselachian forms.

By the Late Devonian, sharks had become truly cosmopolitan with more than 80 species known largely from teeth and scales, plus about a dozen taxa known from complete or partial remains. In the United States, the Late Devonian Black Shales of Ohio and Pennsylvania have yielded many fine specimens of complete fossil sharks and isolated shark remains, belonging to approximately two dozen different species, but most commonly found are the species of the well-known genus *Cladoselache*. Some specimens may have been up to 2 m long. *Cladoselache*, with its fusiform body, large winglike pectoral fins, and two triangular dorsal fins, each preceded by a short, stout fin-spine, looked quite like many modern sharks. The superb preservation of some Cleveland Shale sharks shows the bands of muscles and even the shark's last meal in some cases. *Cladoselache* had five long gill slits, and its powerful jaws have many rows of small multicuspid teeth. This type of tooth, in which there is a central main cusp and a several smaller cusps on either side, is generally called the "cladodont" type (after *Cladoselache*) and was adapted for swallowing prey whole, rather than gouging or grasping prey, as the method used by many modern sharks.

Studies by Mike Williams of the Cleveland Museum of Natural History showed that, in 53 fossil sharks with prey preserved in their gut regions, about 64% had their last meal of small ray-finned fishes, about 28% fed on the shrimplike *Concavicaris*, 9% fed on conodont animals, and in one specimen the remains of another shark were found. The orientation of the prey showed that the sharks were catching their prey tail first, then swallowing them whole. Another Late Devonian Cleveland Shale shark, *Diademodus*, looked similar to *Cladoselache*

The head of *Antarctilamna*, as preserved in the holotype specimen (*A*), from Antarctica. The gill slits are determined on each side of the head, and isolated teeth can be seen circled on the specimen. The braincase of *Antarctilamna* (*B*), from the Middle Devonian Bunga beds, New South Wales, Australia. A restoration of what *Antarctilamna* might have looked like (*C*).

but had distinctive teeth with several equal-sized cusps on each root. Sharks were definitely on the rise as the Devonian drew near its end. They must have been efficient swimmers to avoid being eaten by the giant predatory placoderms, such as *Dunkleosteus*, that also inhabited these seas.

Most of the known Devonian sharks are represented solely by their teeth. The great variety of tooth types shows that they had began adopting a diverse range of feeding strategies. Whereas primitive sharks tended to have two main cusps on each teeth (for example *Doliodus, Leonodus, Antarctilamna, Portalodus*), Late Devonian sharks predominantly had multiple cusps on each tooth. One extreme example of this is the unusual teeth of *Siamodus* (meaning "tooth from Siam"), that I found in Thailand, which has up to eight equal-sized cusps on a strongly arched root (Long 1990).

Other common Late Devonian sharks, known almost exclusively from teeth, are the *Phoebodus* group. The teeth have well-developed roots, which may protrude prominently, and the crown has three main cusps, which may or may not be flanked by smaller interme-

(*Top*) An outcrop of the Cleveland Shale, Ohio, where some of the best-preserved Late Devonian sharks have come from.

(*Bottom*) *Cladoselache* from the Late Devonian Cleveland Shale of Ohio is known from many hundreds of superb specimens, making it one of the best-known early sharks.

diate cusps. The many species of *Phoebodus* have been reported from around the world by Michal Ginter of Poland in rocks of Middle-Late Devonian age, and their occurrences are used to date rocks. The later forms (Carboniferous and Permian age) that had been attributed to *Phoebodus* are currently under revision and may actually belong to different genera. Unusual phoebodontids include *Thrinacodus* and *Jalodus. Thrinacodus* was first described from Australia by Sue Turner and is now recognized from around the globe. The teeth of this species actually resemble little grappling hooks. *Jalodus* teeth occur in deep-water facies, whereas *Phoebodus* teeth represent open shelf environments. The shallow marine platforms are dominated by the broad-toothed *Protacrodus* (Ginter 2001).

By the close of the Devonian, sharks had become firmly established in both marine and freshwater habitats around the world. One of the most unusual Late Devonian sharks was *Plesioselachus* from the Famennian of South Africa. It had a very high dorsal spine and a short, deep body shape reminiscent of the later holocephalans. Unfortunately the specimen is not well-enough preserved to place it confidently with any major group of chondrichthyans (Anderson et al. 1999).

These unusual teeth belong to small (c. 50 cm) sharks, *Ageleodus* and *Ctenoptychius*. No doubt such specialized dentition accompanied an unusual method of feeding.

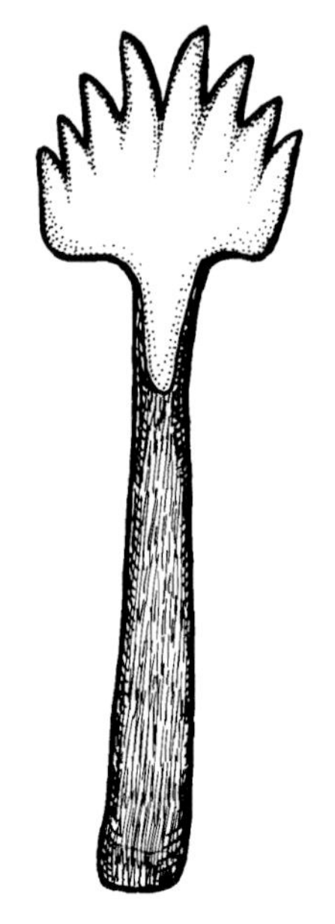

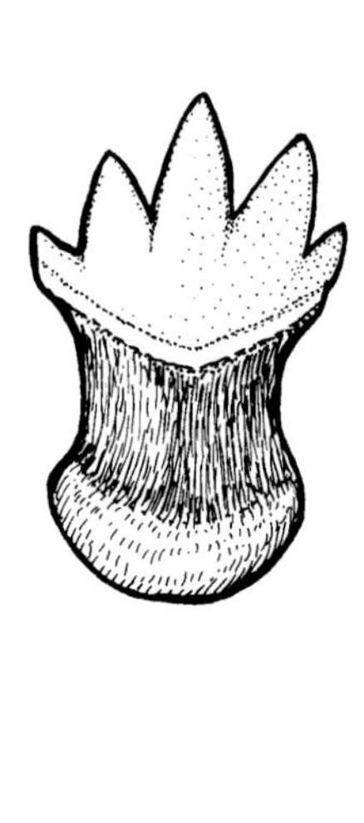

Tooth Tissues of Primitive Chondrichthyans

As most chondrichthyans are known principally from their teeth and scales in the fossil record, the nature of their tooth tissues is of prime importance in their identification. Several kinds of tissue have evolved over millions of years to make teeth harder and more resilient.

Vitreodentin: The enameloid outer layer of the tooth (Zangerl 1981), but this is missing in some groups (e.g., xanacanthids). The main cusp is made up of a type of dentin over a pulp cavity and base made of spongy trabecular dentin.

Orthodentin: Dentin characterized by distinct growth lines (Owen's lines) parallel to tooth surface. Deposited around the pulp cavity.

Pallial dentin: Develops during the initial phase of dentin formation, seen in cross-section as a zone of less density. May form a zone of hypermineralized dentin.

Trabecular dentin: Spongy dentin with large spaces interspersed within it, sometimes the cavities (luminae) may form parallel to semiparallel alignment in cross-section.

Pleromin: Hypermineralized dentin matrix (= pleromic hard tissue), a term first used for scales of psammosteid agnathans by Tarlo (1964).

The Late Paleozoic Chondrichthyan Radiation

By the Early Carboniferous, following the dramatic extinctions of the armored placoderm fishes and several other major fish groups, the chondrichthyans underwent another major radiation. Many new families and higher groups appeared, including the first holocephalans, which are discussed separately below. Rays are also specialized forms of sharks, but they did not arise until the Mesozoic Era.

Paleozoic sharks showed a great variety, much more so than that seen in living sharks. The stethacanthid sharks lived from the Late Devonian through to the Permian and had a characteristic bony brushlike structure adorning the main dorsal fin. *Stethacanthus* and *Akmonistion* had a massive dorsal brush with many sharp denticles, while others, such as *Damocles* and *Falcatus,* had narrow cylindrical dorsal fin spines. A superb whole fossil specimen of *Akmonistion* was found from the Bearsden deposit near Glasgow by fossil collector Stan Wood, of Edinburgh, and was described by Michael Coates and Sandy Sequoia (2001). It is now in the Hunterian Museum. Stethacanthids are more commonly known throughout the Late Palaeozoic by their teeth, which have characteristic broad roots with many smaller cusps flanking a principal median cusp. Teeth of Early Carboniferous forms from North America measure up to 7 cm in width, with cusps nearly 4 cm high. This gives an estimated mouth width of nearly 1 m if a regular number of tooth rows (about 12) are assumed to be present, perhaps giving the fish an estimated body length of 6 m or more. Such gigantic hunters prowled the seas as the largest vertebrates in the marine realm of their day, while their predatory counterparts in inland waters were the 6- to 7-m-long rhizodontiform fishes.

Other early sharks of the Carboniferous and Permian Periods include the ctenacanth group, typified by the genus *Ctenacanthus* (from the Greek, *ctenos,* "comb," *acanthos,* "spine"). These sharks had fin-spines elaborated ornamented by many fine rows of nodes, giving them a comblike appearance. Complete body fossils of *Goodrichichthys* and *Ctenacanthus* are known from the

***Falcatus*, a well-preserved Early Carboniferous shark from the Bear Gulch Limestones of Montana. This one is a male.** (Courtesy Richard Lund, Carnegie Museum)

Early Carboniferous sites near Edinburgh, Scotland. These were generally small sharks, under 50 cm long.

Xenacanth sharks are another successful group in the Late Paleozoic and early part of the Mesozoic. Xenacanths have characteristic teeth with two main cusps and a well-developed button of bone on the root (termed the *lingual torus*). *Xenacanthus*—well known from complete fossils of the Permian-Triassic Periods in Western Europe and from well-preserved skulls and spines from the redbeds of Texas—had a large serrated defensive spine protruding from its neck, and the tail was straight, not heterocercal as in most sharks. Xenacanths were predominantly freshwater predators that invaded river systems from the seas, as their teeth are also well known from marine deposits. Nearly complete xenacanthid shark remains have also been discovered at the Triassic Somersby fish site, near Gosford in New South Wales, Australia.

Edestoid sharks were among the most freakish looking of any fish. Some of the relatively complete body fossils of these sharks indicate a streamlined, fast swimming lifestyle (e.g., *Fadenia*). Others, such as *Helicoprion* and *Edestus*, evolved complex tooth whorls that coiled about on themselves and were probably overhung from the lower jaws. There is still much speculation as to how these whorls were used in life. It has been sug-

***Ctenacanthus*, an Early Carboniferous shark with robust fin spines. This specimen from the Cleveland Museum of Natural History shows the shoulder girdle and dorsal fin spine.**

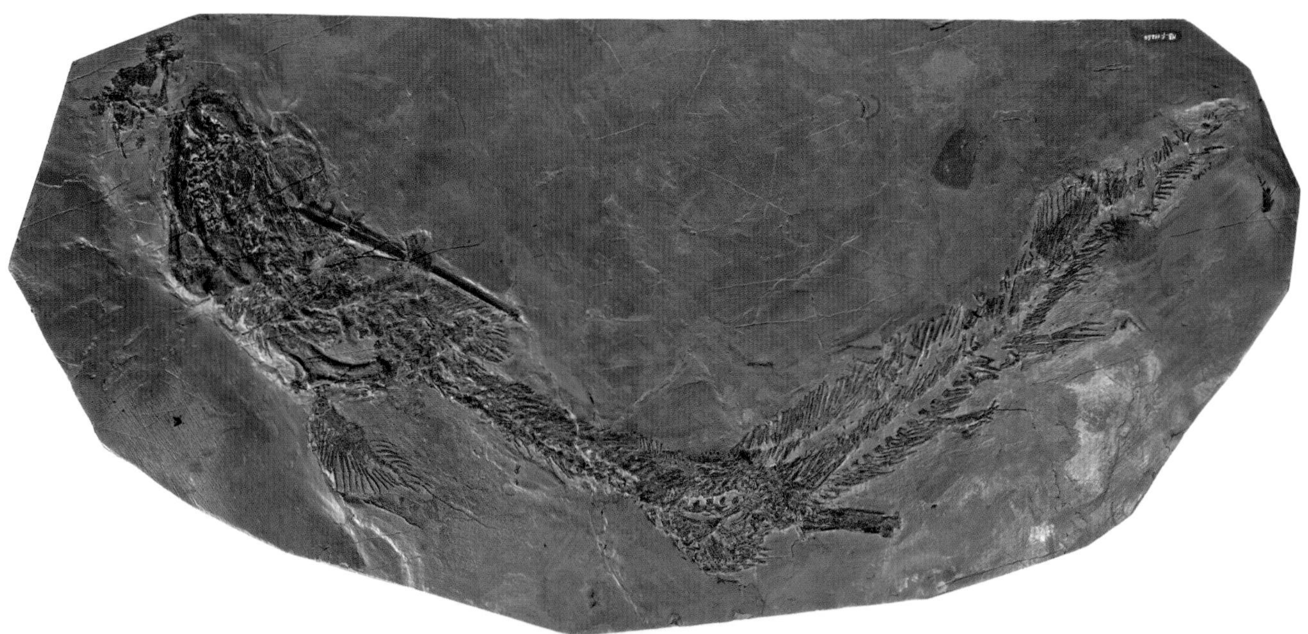

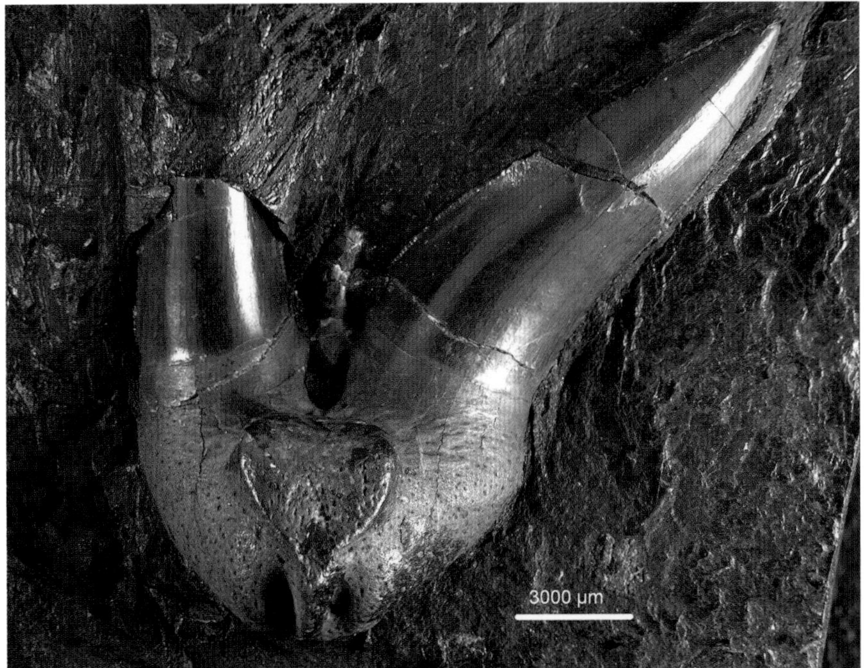

(*Top*) A beautifully preserved fossil shark, *Xenacanthus meisenheimensis*, from the Early Permian of the Saar-Nahe Basin, southwest Germany. (Courtesy, Carola Radke, Museum für Naturkunde, Berlin)

(*Left*) A double-pronged tooth of the xenacanth *Orthacanthus kounoviensis* from the Carboniferous of the Czech Republic and Germany. (Courtesy Oliver Hampe, Berlin)

gested that they may have mimicked ammonites, coiled shellfish that were abundant at this time, and thus could have attracted prey to the shark if used in a particular fashion. It seems more likely that these sharks used the jagged tooth whorls when charging into a school of fish or ammonites and thrashing about to snag prey on the projecting array of teeth. Similar feeding methods are used today by the saw sharks. Despite the mystery of how they utilized the tooth whorls, the group was highly successful, spreading around the world during the middle part of the Permian; *Helicoprion* tooth whorls have been found in Russia, North America, Japan, and Australia.

Orodontid sharks had elongated bodies, lacked spines, and had characteristic broad, crenulated crushing teeth. *Orodus* teeth are commonly found in Carbon-

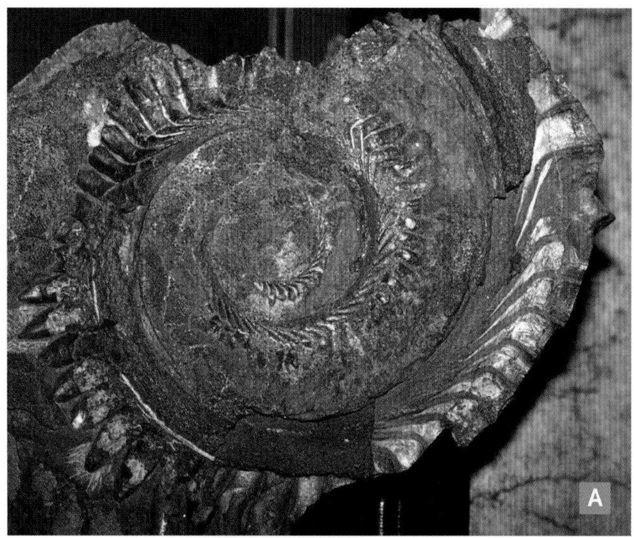

This strange whorl of teeth (*A*) belong to *Helicoprion*, a Permian shark that had these deadly coils at the front of the lower jaw. The sharks may have used these coils to snag squidlike creatures or other fish. This specimen is from the Permian of Russia Restoration of *Helicoprion (B)* by Oleg Lebedev, Borissiak Paleontological Institute, Moscow.

iferous deposits around the world. The petalodonts were a diverse group of flattened raylike sharks that lived in the Late Paleozoic. They are mostly known from teeth, except for a few well-preserved specimens. *Belanstea montana* from the Bear Gulch Limestone of Montana, in the United States had a deep-body form with large feathery pelvic and dorsal fins. Similarly *Janassa* had broad, rounded pectoral fins like a ray. The identification of most petalodonts is largely through the teeth. Their peculiar teeth are strongly compressed and indicate a diet of hard-shelled invertebrates. Where articulated dentitions are known, they have only a few teeth in each jaw. Characteristic teeth, such as those of *Petalodus*, are broad, flat, and slicing with collars of folded

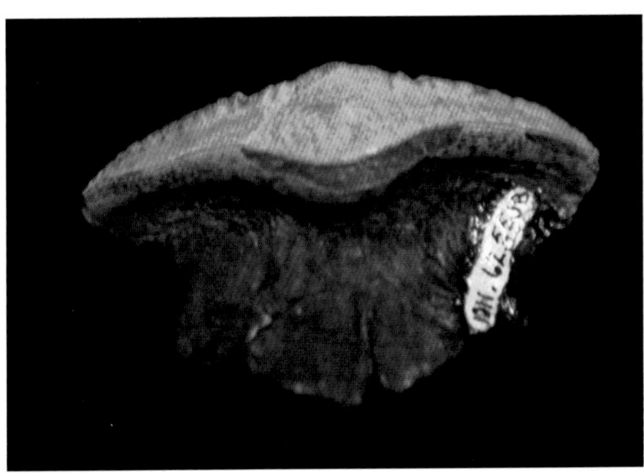

The strange tooth of a petalodont shark from the Lower Carboniferous (size c. 6 mm wide).

enamel at their base. *Ctenoptychius* and *Ageleodus* are two typical genera, whose teeth have from 4 to 30 small pointed cusps on the flat, broad root (Downs and Daeschler 2001). At the other extreme, we see *Megactenopetalus,* which has a single large multicuspid upper tooth and one large triangular shearing lower tooth forming the entire dentition (Hansen 1978).

Humble Beginnings of Modern Chondrichthyan Faunas

The great radiation of sharks continued throughout the Mesozoic, and by the end of that era most of the modern shark families had appeared. One particular group was prominent in the early half of the Mesozoic, the hybodontids. These first appear as early as the Carboniferous Period, represented by a beautiful complete fossil of *Diploselache* from the Carboniferous of Scotland.

Hamiltonichthys from Kansas is another good example of an early hybodontid shark and is identified by its well-developed enlarged head spines, stout dorsal fin-spines, broad multicuspid teeth, and peculiar scale shapes. The group is best represented by the many species of *Hybodus,* a blunt-headed shark up to 2.5 m long that inhabited the seas of Europe, Africa, Asia, and North America throughout the Mesozoic Era. The teeth of many hybodonts had numerous cusps arranged on a broad root, but some had almost crushing dentitions of flat plated teeth (such as *Acrodus* and *Asteracanthus*). Smaller gripping teeth were set in the front of the mouth. *Triodus* is a Cretaceous hybodont known from a perfect three-dimensional cranium showing the jaws and braincase.

The oldest euselachians, or true selachians, were once thought to be traced back to *Palaeospinax* from the Lower Jurassic (Maisey 1977), but this group is now thought to belong within the basal galeomorphs (Klug 2009).

The seven-gilled sharks (order Hexanchiformes) appeared in the Mesozoic Era. The living frilled sharks and cowsharks, *Hexanchus, Heptranchias,* and *Notorhynchus,* are primitive forms with only one dorsal fin and six to seven pairs of gill slits. Their teeth have many flat cusps along a broad root, with marked differences between the upper jaw and lower jaw dentitions. They often inhabit deep waters and feed chiefly on fish. Fossil

Reconstruction of *Hybodus*, a common genus of Mesozoic shark that grew to lengths of about 2 m. (After the work of John Maisey)

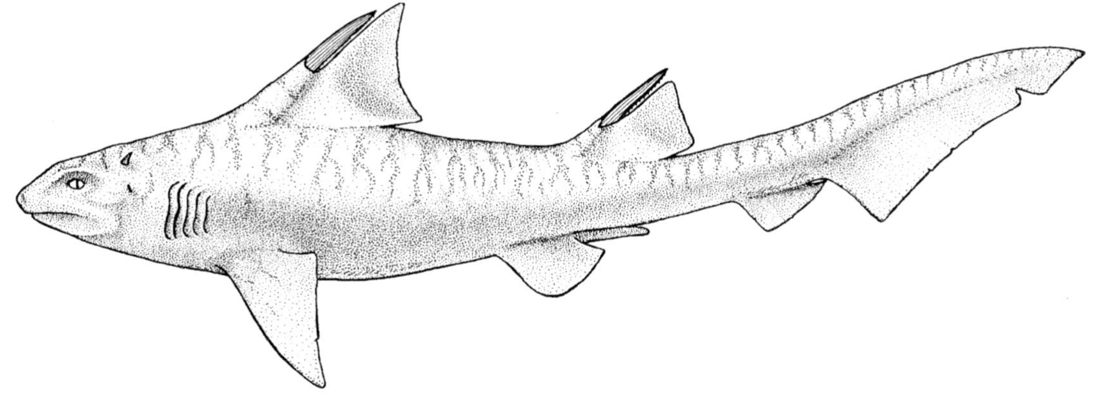

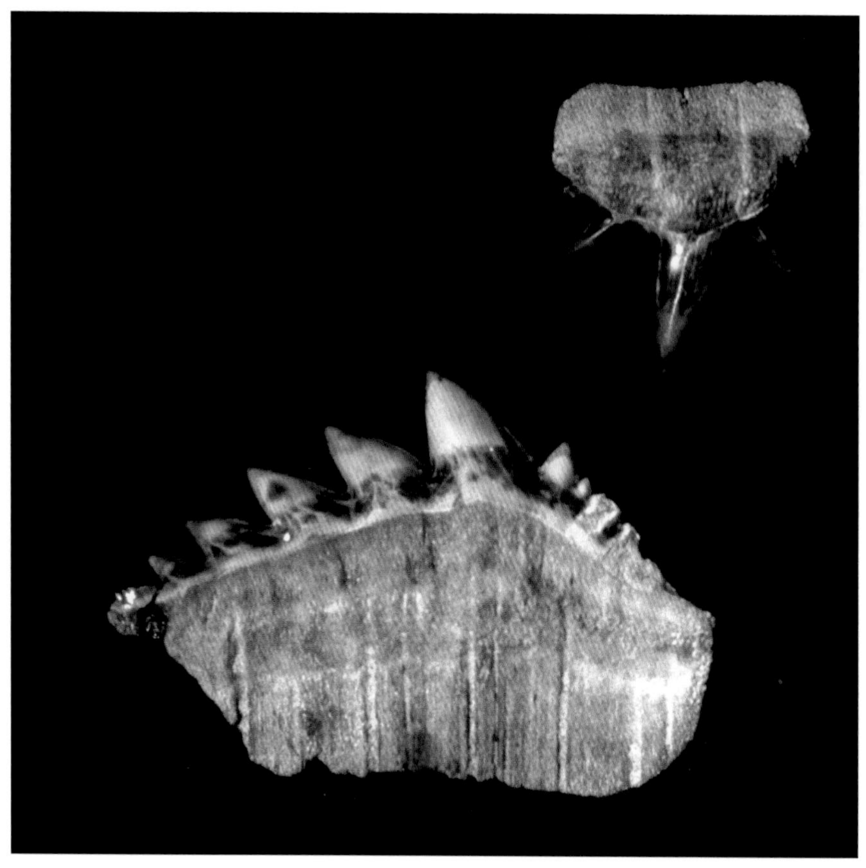

The teeth of a fossilized six-gilled shark *Notidanus*. Note the different shapes for upper and lower teeth. (Courtesy David Ward)

teeth of the group show that forms such as *Hexanchus* have been around since the Early Jurassic.

During the Mesozoic Era, many groups of sharks evolved. Some of the commonest fossil teeth found in Cretaceous deposits around the world include the narrow curved teeth of the bullsharks (*Carcharias*) and the broader toothed lamnids (*Cretolamna, Cretoxyrhina*). There are far too many to list here. As far as shark evolution is concerned, most of the action was over by this time: many of the living families had already appeared. Beside teeth, other commonly found remains of sharks of this period include their vertebrae, preserved because of the high degree of calcification in the cartilage. Great lamnid sharks appear in the Cretaceous, known from teeth and large vertebrae that indicate forms, such as *Cardabiodon* from Western Australia described by Mikael Siverson, grew to 5 to 6 m long and were cruising the seas at the same time as the great reptilian predators, the ichthyosaurs, mosasaurs, and plesiosaurs. The megamouth shark, *Megachasma*, another large living lamnid, also is first known from teeth in the Cretaceous.

During the Tertiary Period, the remaining living genera of sharks evolved, including rare forms like the deep-water goblin shark, *Mitsukurina*, and can be traced back through their fossil teeth to their first appearance. The largest predatory sharks of all time evolved by the start of the Miocene, about 23 million years ago. Once thought to be ancestors of the great white shark, the evolution of these giant killers is now hotly debated. The largest species of fossil mako, *Isurus hastalis*, commonly found in Miocene and Pliocene deposits around the world, may have reached sizes of 6 to 8 m and, if it had developed serrations on its teeth, it could have given rise to the great white shark (Ehret et al. 2009). Many fossil species of *Isurus* are known from Miocene and younger rocks around the world, including the teeth of the extant species *Isurus oxyrinchus*.

The extinct species *C. megalodon*, *C. angustidens*, and *C. auriculatus* all fall into the genus *Carcharocles*. How-

The living whale shark *Rhinodon typicus* is one of the two largest fish alive today, reaching sizes of 15 m, along with its filter-feeding cousin, the basking shark. (Courtesy iStock)

Teeth of the giant lamniform shark *Carcharocles megalodon* are up to 18 cm high, suggesting a maximum size for the fish of about 15 m. This was perhaps the largest predatory shark ever to have lived. In addition to its gigantic teeth, its fossilized vertebrae have also been found. It is not regarded as an ancestor of today's great white shark, *Carcharadon carcharias*, but instead is believed to have come from an extinct lamnid lineage closer to modern mako sharks. (Courtesy David Ward)

ever, advocates of the hypothesis that the great white shark is a descendent species from the giant *C. megalodon* and its kin hold that these extinct species should all be placed in the same genus, *Carcharodon*. These researchers argue that older occurrences of *Carcharodon carcharias* predate the intermediate serrated teeth and that the molecular clock method of tracing the rate of change of DNA places a divergence time for *Carcharodon* and *Isurus* about 43 million years ago. There are also transitional teeth showing the link from *Otodus obliquus* into serrated identical forms of *Carcharocles* species, as shown to me by David Ward of Orpington, United Kingdom.

The teeth of the largest predatory shark to have ever lived, *Carcharocles megalodon,* were up to 17 cm high. Although early estimates grossly exaggerated the maxi-

This rare complete fossil carcharhiniform shark, *Galeorhinus cuvierii*, is a relative of the living tope shark *Galeorhinus galeus*. The specimen is from the Eocene age and comes from the famous Monte Balca site in Italy. (Courtesy Lorenzo Sorbini, Verona)

mum size of this shark to 25 m or more, it is now possible to constrain the estimates by what we know of how sharks grow and how their teeth increase in size proportionately to overall body length. The revised maximum size of *Carcharocles megalodon* is about 13 to 15 m, still almost twice the size of the largest great white shark ever caught (just under 7 m). The large triangular serrated teeth of this monster are found in rocks of Miocene and Pliocene age all around the world. *Carcharocles megalodon* evolved from an earlier large species, *Carcharocles angustidens*, of Oligocene age. *Carcharocles* and its descendents are believed to be from the lamnid lineage of *Otodus obliquus*, a large form that terrorized the seas during the Late Palaeocene / Early Eocene. *Otodus* teeth reach sizes of 8 cm or more and indicate that this early large lamnid may have grown to 6 m.

It is probably no coincidence that the largest *Carcharocles* species evolved the same time as the large filter-feeding baleen whales first appeared. It appears that an arms race of sorts ensued with larger species of *Carcharocles* evolving in step with the first baleen whales by the start of the Miocene. Reports of giant great white sharks up to 10 m long in recent times have lead some scientists to suggest that *Carcharocles* might still be alive out there somewhere, although I regard most fishermen's estimates of "the one that got away" with extreme caution. If the giant *Carcharocles megalodon* was still alive today, its teeth would almost certainly have been found in fairly recent marine deposits, but no fossils have been found in strata dating more recently than about 2.5 million years ago. One can only speculate as to why these mighty killers died out. The first drastic climatic cooling that was to herald the Pleistocene ice ages began in the late Pliocene about 2.6 million years ago. Perhaps they could not adjust to the climatic changes as well as their possible prey, the warm-blooded marine mammals, did. We know from recent discoveries of fossil baleen whales of this age in Antarctica that the great whales began inhabiting Antarctic waters at this time. Was it just for feeding, or also to escape from the colder-blooded predators that could not cope with the near-freezing Antarctic waters?

Holocephalans: An Early Radiation?

The holocephalomorphans include the living chimaerids (holocephalans) and a host of fantastic extinct forms, such as the iniopterygians. These are cartilaginous fishes, like sharks, that have the upper jaws fused to the braincase and have the gills covered by a large soft operculum. Recent phylogenetic analyses point to an early radiation of the holocephalans stemming from sharklike ancestors similar to *Falcatus* or *Damocles* (Coates and Sequeira 2001). Lack of data from other key forms makes the cladogram less robust, so alter-

A female *Echinochimera* from the early Carboniferous Bear Gulch Limestones of Montana, USA. (Courtesy Richard Lund, Carnegie)

natively they could be the sister group to selachians (sharks).

Holocephalans are a small subclass known today from about 34 species, such as the elephant shark, chimaerids, and rabbitfish. Like sharks, they have a cartilaginous internal skeleton and reproduce using internal fertilization, the males bearing clasping organs near the pelvic fin and a median clasper on the face. However, unlike other chondrichthyans, the holocephalans have an operculum covering the gill arches and their upper jaw is fused to the braincase, providing a powerful bite for their crushing plate dentition. They have long bodies with whiplike tails and are often found at great depths in the abyssal plains. They swim by fluttering movements of the pectoral fins and slow sideways movements of the tail.

Before the dramatic discoveries made by Dick Lund of Adelphi University in the Bear Gulch Limestones of Montana, the early history of holocephalans was virtually unknown. Early theories put out by Tor Ørvig of the Swedish Museum had them evolving from the ptyctodontid placoderms. These theories are not widely accepted by the paleontological community, as ptyctodontids have been known for some time to have been a valid monophyletic group of placoderms (Goujet and Young 1995). While it was once thought that holocephalans and sharks diverged from a common ancestral form, it is now clear from the fossil record that the holocephalans are an early split off from the mainstream of Devonian sharks, whose basal ancestors include stethacanthids like *Akmonistion* (Coates and Sequiera 2001).

The Bear Gulch Limestone of Montana has yielded an extraordinary diversity of complete holocephalan fossils; more than 40 new types have been found in the past three decades, and still more are found almost every field season. Some like *Harpagofutator* was a simple eel-like form with the males having elaborate spined appendages trailing off the face presumably for use in mating. A most significant find was *Delphyodontos*, based on two just-born fetuses about 5 cm long, which suggested that these fishes were giving birth to live young and that there may have been interuterine competition as far back as 340 million years ago (Lund 1980).

The cochliodontiforms are a diverse group known primarily from teeth and spines. They feature strongly

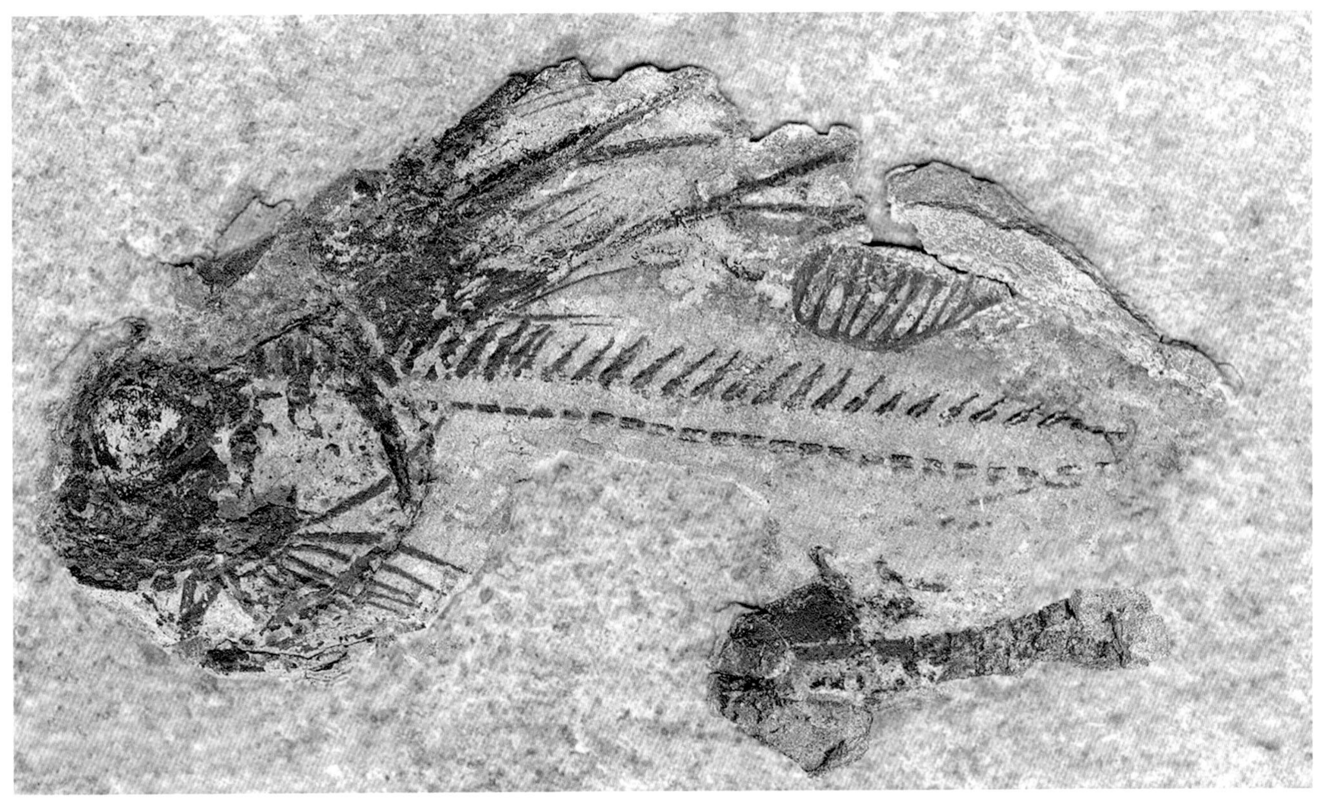

The iniopterygian *Rainerichthyes zangerli* from the Early Carboniferous Bear Gulch Limestone of Montana. Iniopterygians were bizarre Late Paleozoic holocephalans with high dorsal fins and robust sharp fin spines. (Courtesy Richard Lund, Carnegie Museum).

convex and often knobbly shaped crushing tooth plates. Forms such as *Cochliodus*, *Deltodus*, and *Sandalodus* are well-known from Carboniferous deposit throughout Europe and North America. Species of *Psammodus* are believed to have had up to 12 similar bricklike tooth plates that fit tightly together to make crushing pavements. Copodontids such as *Copodus* are thought to have had only two crushing tooth plates in each jaw.

The menaspiform fishes include well-preserved examples of deep-headed fishes with bony plates on the head and well-developed head spines that project strongly outward in some forms (e.g., *Menaspis armata*). *Deltoptychius* is a fish known from the Carboniferous deposits of Bearsden in Scotland. The body is covered by scales and the head enclosed by armor made of layers of dentin with thickly ornamented surface sculpturing. The fish grew to about 45 cm in length and had a short tapering tail. The back was protected by rows of stout spiny scales.

Chimaerids were bottom-dwelling fishes that fed chiefly on shellfish and crustaceans. They first appeared in the Carboniferous Period and emerged as a successful group in the Mesozoic Era, reaching a peak of diversity in the early Cretaceous. *Echinochimaera meltoni*, from the Carboniferous of Montana was an early chimaeriform having a typical body much like that of any modern chimaerid. The tail was long and whiplike, with an anal fin close to it. The head was large with big eyes. There is a large bony spine preceding each dorsal fin. The males have a brushlike structure on the first dorsal spine and several feathery spines over the eyes. Females have a single eye spine and a broader first dorsal fin.

Squaloraja was a most unusual raylike fish that lived in the Jurassic Period in Europe. It had a long triangular rostrum extending out in front of its face supported by a rostral cartilage more than twice the length of its braincase. It had two pairs of upper jaw tooth plates and one pair forming the lower dentition. Other peculiar chimaerids with elongated rostra include the myriacanthids such as *Acanthorhina* from the Jurassic of Germany.

More typically fossil chimaerids are well-known from tooth plates found in Mesozoic and Cenozoic marine deposits, such as *Edaphodon* and *Ischyodus*, both known from many species. Some of these belong to fishes of large size, reaching 3 m or more in length. The typical dentition seen in these fishes includes large paired mandibular tooth plates forming the lower jaw, with palatine and vomerine plates on each side of the upper jaw. The extant chimaerids are principally small deep-water forms. In some countries such as New Zealand, species are commercially fished and sold as "flake." The fossil record of living chimaerid genera such as *Callorhynchus* and *Chimaera* show they are relatively unchanged since their appearance in the Mesozoic Era.

Iniopterygians were a small group of bizarre-looking little fishes, known from five genera only, that all lived in the Late Carboniferous of North America. Once thought to be an entirely separate group of chondrichthyans, they are now regarded as a highly specialized holocephalans. They all had stout pectoral fin-spines projecting high up on the shoulder girdle, some with much larger pectoral spines (e.g., *Promexyele*). In some forms the upper and lower jaws are free (e.g., *Polysentor*), whereas in most the upper jaws are fused to the braincase. Symphysial tooth whorls are present at the front of the lower jaws. The dentition consists of simple pointed cones or wider teeth with smaller lateral cusplets. A recent remarkable study used a synchrotron particle accelerator to generate 3-D images of a 300-million-year-old iniopterygian skull inside a nodule of rock and found the brain actually fossilized inside the skull (Pradel et al. 2009).

The Rays Shine Through

The first fossil stingrays (batomorphs) date from the Jurassic Period as more or less flattened sharks, which had enlarged their pectoral fins and reduced the tail. Many types of sharks specialized to adapt to a bottom-feeding lifestyle. Living species of angel shark (genus *Squatina*) show well this intermediate shape between the active swimming shark and the flattened ray. The main difference between sharks and rays is that rays swim by wavelike motions of their wings or pectoral fins, whereas sharks are propelled solely by their tails. Other features seen in rays that distinguish them from typical sharks are that they have a downward-facing mouth with rows of flat crushing tooth combs and a well-developed tail-spine (actually a modified dorsal fin-spine).

A lower jaw (mandibular) tooth plate of the extinct chimaerid fish *Ischyodus* from the Eocene of England. (Courtesy David Ward)

Despite having a conservative body plan, rays evolved to make use of numerous feeding niches. Some giants such as the manta ray are filter-feeders like their shark counterparts, the whale shark and basking shark. Others such as the electric rays have developed the ability to project powerful electric fields to stun the fishes on which they feed or to deter attackers. Many rays simply feed on crustaceans and shellfish by cruising along the seafloor detecting food with their highly developed electrosensory systems.

The oldest fossil rays include *Spathiobatis,* an almost guitar-shaped form known from complete specimens found in France. *Jurabatos,* from the Early Jurassic of Germany, is represented by a single isolated tooth. By the Middle Jurassic there were several genera of rays. *Spathobatis,* known from the Early-Late Jurassic, is known from well-preserved body fossils of Late Jurassic age in France. *Belmnobatis,* also from France, and *Asterodermus* from Germany, are other well-preserved rays of similar age. All of these early rays belong to the family Rhinobatidae, which includes several living forms, such as the shovel-nose ray and the guitarfish. These all have relatively long, sharklike bodies.

The Myliobatidae, or eagle rays, made their first appearance in the Late Cretaceous, rapidly became widespread throughout the Tertiary and are still thriving today. The eagle rays have complex pavement-type dentitions consisting of several rows of broad, flat crushing plates flanked by many smaller polygonal units. They use these to grind up food such as crabs and shells. Fossil rays, such as *Myliobatis,* and the many other known families of rays are commonly found in Tertiary marine deposits around the world, represented by these distinctive crushing plates as well as occasional fossil stingers (modified dorsal fin-spines) and scales.

The rays began invading freshwater river and lake habitats early in the Tertiary, as seen by the well-known complete body fossils of *Heliobatis* from the famous Early Eocene Green River Shales of Wyoming (Carvhelo et. al 2004). *Heliobatis* probably invaded the lake systems when the sea levels rose high enough for saltwater to

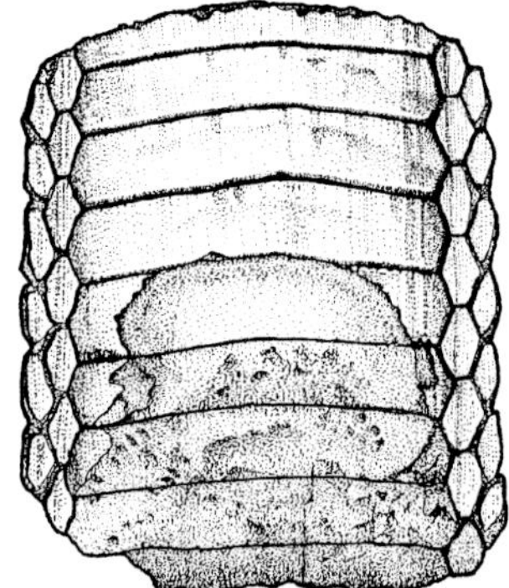

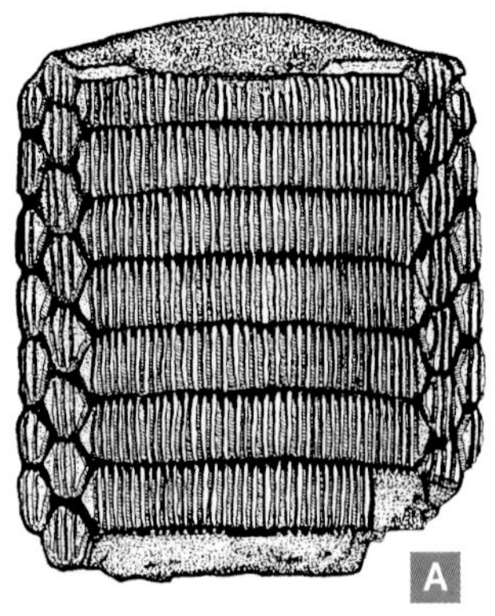

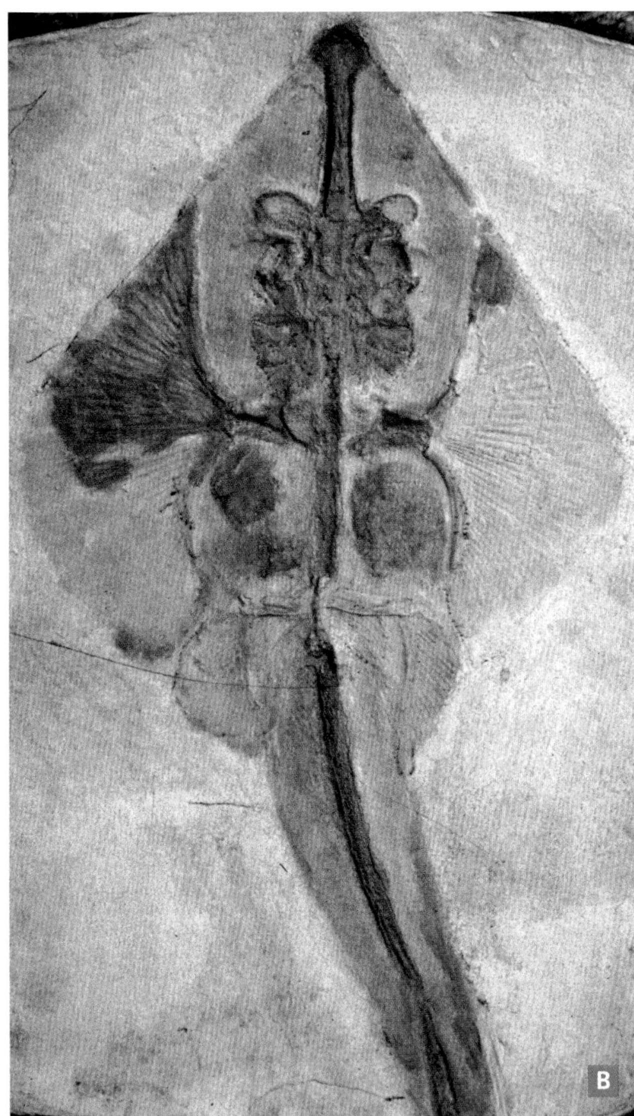

The flat crushing tooth plates of extinct species of eagle ray, *Myliobatis*, are found as fossils (*A*) in marine deposits back through the past 40 million years or so. Modern stingrays (*B*) resembled the fossil forms right from their first appearance, such as this Jurassic rhinobatid from Solnhofen, Germany. Most of the fossil rays have closely related living relatives. (Courtesy David Ward)

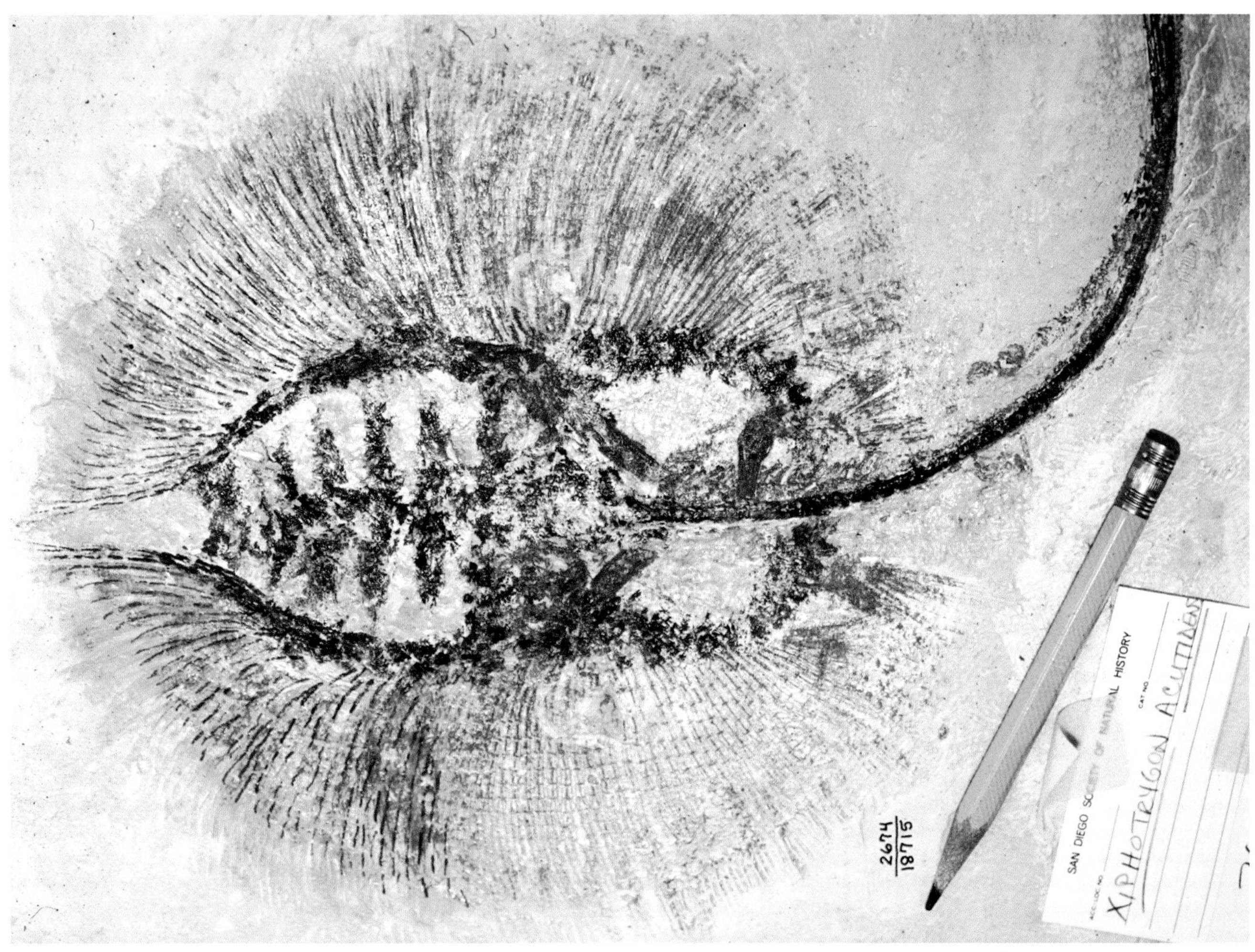

This superb fossil stingray *Xipohotrygon* comes from the Eocene Green River Shales of Wyoming.

flood the lake system, and so the Green River stingrays are not considered to be primarily freshwater fishes.

Sharks and their kin have thus been evolving and radiating into a wide variety of shapes and forms since their first appearance some 420 million years ago. They survived four major mass extinctions and today form an integral part of nearly all oceanic food chains, where they sit at both ends of the spectrum, from top-end predator to giant filter-feeders. Their ability to develop a flattened body shape as rays was their most recent great success, with ray species now outnumbering sharks and chimaerids combined. The humble chimaerids appeared to survive without much change since the Carboniferous Period, mainly due to their adaptation to life in deep-water habitats. Nonetheless we rely on all these sharks and their relatives as food in nearly all parts of the world. Our challenge ahead is to protect and nurture the modern chondrichthyan fauna in the face of an increasing demand for their meat, especially the unsustainable practice of catching sharks just for their fins alone.

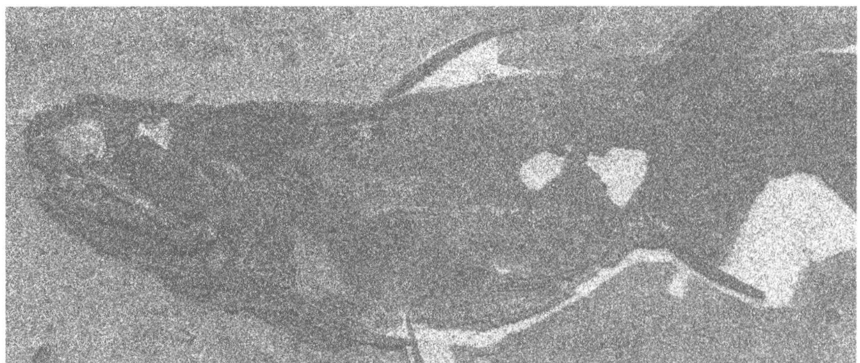

CHAPTER 6

Spiny-Jawed Fishes

The odd mixture of early jawed fishes such as acanthodians and their kin

The acanthodians were an unusual group of mostly small fishes that appeared in the Silurian Period and reached a peak of diversity in the Devonian. They were the only group of fishes that had ornamented bony spines in front of all fins and tiny scales that mostly had bulbous, swollen bases. There were three main groups of acanthodians: the Climatiiformes, the Ischnacanthiformes, and the Acanthodiformes, each distinguished by the presence or absence of bony armor bracing the pectoral fins, the presence of one or two dorsal fins, and the structure of their teeth and scales (Denison 1979).

The earliest forms may have lacked fin spines, as in the Silurian *Yealepis*, or had only median fin spines, as in the Early Devonian *Paucicanthus*. The climatiiform acanthodians, predominant in Early Devonian faunas, had elaborate bony shoulder girdle armor and numerous sharp spines (Miles 1973b). The ischnacanthiforms were predators with teeth firmly fused to the jawbones. The longest surviving group of acanthodians were the acanthodiforms. These were fast-swimming filter-feeders that lacked teeth in the jaws and had gill rakers for straining food through the gill chamber. The demise of the acanthodian fishes at the close of the Permian Period is seen as another victim of the most dramatic extinction event in the history of the planet in which more than 90% of marine species are believed to have been rendered extinct.

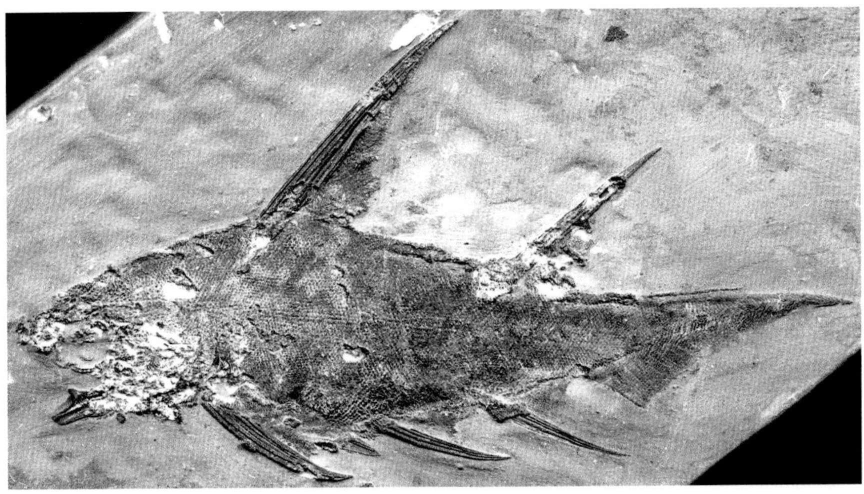

***Broachmonadnes*, an Early Devonian acanthodian with numerous defensive spines from the MOTH locality of Northwest Territories, Canada.** (Courtesy Mark Wilson)

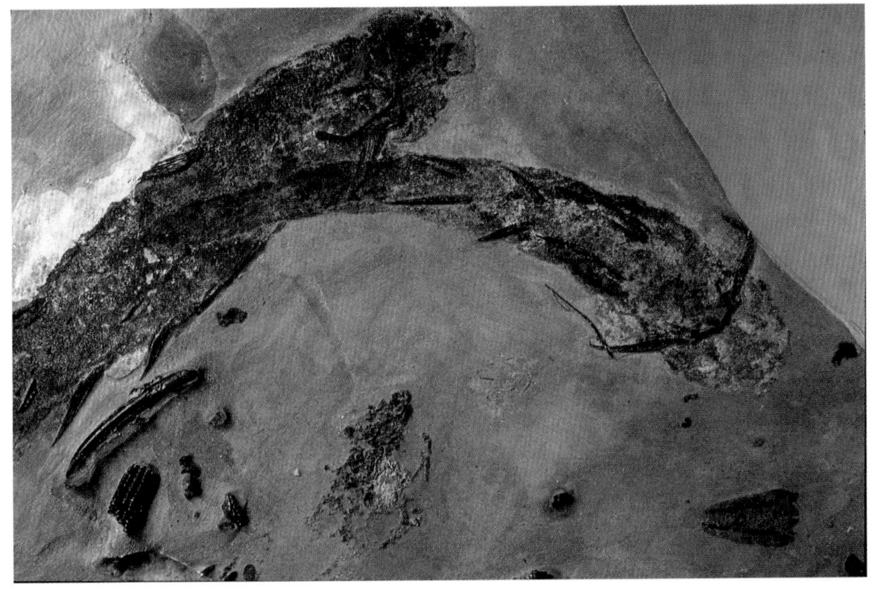

The Early Devonian acanthodian *Luposyrus* showing the shoulder girdle armor underneath the fish (ventral view). (Courtesy Mark Wilson)

Acanthodians were first described by French paleontologist Louis Agassiz in 1844, who named *Acanthodes*, *Cheiracanthus*, and *Diplacanthus*. The group takes its name from the Greek *akanthos*, meaning "spine," as the group's most distinguishing features are the presence of deeply inserted paired fin spines in front of the pectoral and pelvic fins and a single fin spine preceding the dorsal and anal fins. Many books dubbed them "spiny sharks," and increasing new fossil evidence supports their being very closely related to early sharks. It is their distinctive spines, and the characteristic shape of the acanthodian scale, that makes the group widely recognized from every continent in sedimentary rocks of Silurian-Permian age, even though more recent discoveries are showing that fin spines were quite widespread in nearly all early fish groups. Although the oldest acanthodians date back to the beginning of the Silurian, the earliest near-complete remains are stem gnathostomes from the Late Silurian of Australia (*Yealepis*) and the oldest complete acanthodian fossils from the earliest Devonian rocks of Canada (e.g., *Paucicanthus*) and Europe (e.g., *Climatius*).

Acanthodians remain as one of the most enigmatic of all ancient fish groups, about which we have the least amount of anatomical knowledge and few real clues as to their affinities with other types of fishes. Unlike the superb Gogo or Taemas fossil fishes, we lack acid-prepared, three-dimensional skulls of acanthod-

Mark Wilson (*left*) and Gavin Hanke (*right*), at the Early Devonian MOTH site in Canada, where some of the best-preserved early acanthodian fossils have been found. (Courtesy Mark Wilson, University of Alberta)

ians (although we do have some acid-prepared isolated bones from them). Instead, our knowledge of the group is known largely based on the shape of the body and fins, the preservation of some external head structures in early forms, and the well-ossified head and gill-arch skeleton in one of the last and probably most specialized species, *Acanthodes bronni* (Miles 1973a).

In recent years, there has been a renewal of studies on acanthodian fossils because of spectacular finds of near-complete body fossils from the Northwest Territory in Canada made by University of Alberta's Mark Wilson and Gavin Hanke of the Royal British Columbia Museum, Canada. Although acanthodians are rare as complete fishes, their minute scales are useful in determining the relative age of Palaeozoic sedimentary rocks and have been studied intensely for that purpose. However, the study of their anatomy—for solving their evolutionary origins and relationships—has only blossomed in the past 20 years with new discoveries from Canada, Europe, and Australia. Several of the newest species of acanthodians, described in the past 5 years or so are from the Early Devonian sites in Canada, where several new unusual forms have been found exquisitely preserved. These fossils have been crucial to readdressing the question of the group's origins.

The Origins and Affinities of Acanthodians

The arrangement of scales in many rows per vertebral segment (known as *micromeric squamation*) that typifies acanthodians is also seen in fossil sharks, some placoderms, and in some basal actinopterygians, such as *Cheirolepis* (see Chapter 8). None of the potential ancestral groups of jawless fishes had similar micromeric squamation, although thelodonts came the closest in this respect. Acanthodians appear to share more specialized characteristics with osteichthyans than with other

Basic Structure of an Acanthodian

The acanthodians were mostly slender, fusiform fishes with long heterocercal tails and short, blunt heads, although a few did develop deep bodies. The mouth is nearly always large, except in a few specialized forms (such as diplacanthoids), and the head is usually invested with many small dermal platelets called tesserae. There are two dorsal fins in all acanthodians except for the Acanthodiformes, which have only one dorsal fin situated close to the tail. The eyes have sclerotic bones of varying number encircling them, and there are two pairs of nostrils (excurrent and incurrent nares) at the front of the head. The jaws are ossified as a single lower jaw cartilage that may be supported by a strip of external, ornamented bone (the mandibular splint), but externally ornamented-tooth-bearing bones are not present. There are five gill slits opening at the side of the head and these may have dermal branchiostegal plates preceding them in some species, a feature seen otherwise only in the bony fishes.

The gill arch bones and braincase are known only in *Acanthodes*, one of the most specialized of all acanthodians. The gill arches feature a series of basibranchial bones, large hyophyals and ceratohyals, large epihyal and epibranchials, and short, rearward-facing pharyngobranchials. The upper jaw (palatoquadrate) may have a simple articulation with the braincase or have a complex double articulation (in some acanthodiforms).

The braincase, known only in *Acanthodes,* is incompletely ossified, made up of four bones that were held together by cartilage. The large dorsal ossification covered most of the top of the head and protected the brain, and a smaller occipital ossification was present at the rear of the head, serving as a site for trunk muscle attachment. The underneath of the braincase had a large anterior basal ossification pierced by a canal for the hypophysis—a space where the pituitary gland is housed and the internal carotid arteries converge from outside the braincase, and a smaller rear section below the occipital ossification. In general, the shape and proportions the braincase of *Acanthodes* resemble that of a primitive ray-finned fish. In some acanthodians the inner ear canals are preserved, indicating that three semicircular canals were present and furthermore that otoliths, or ossified ear stones, were present in the inner ear of some species (for example *Carycinacanthus*). These gave

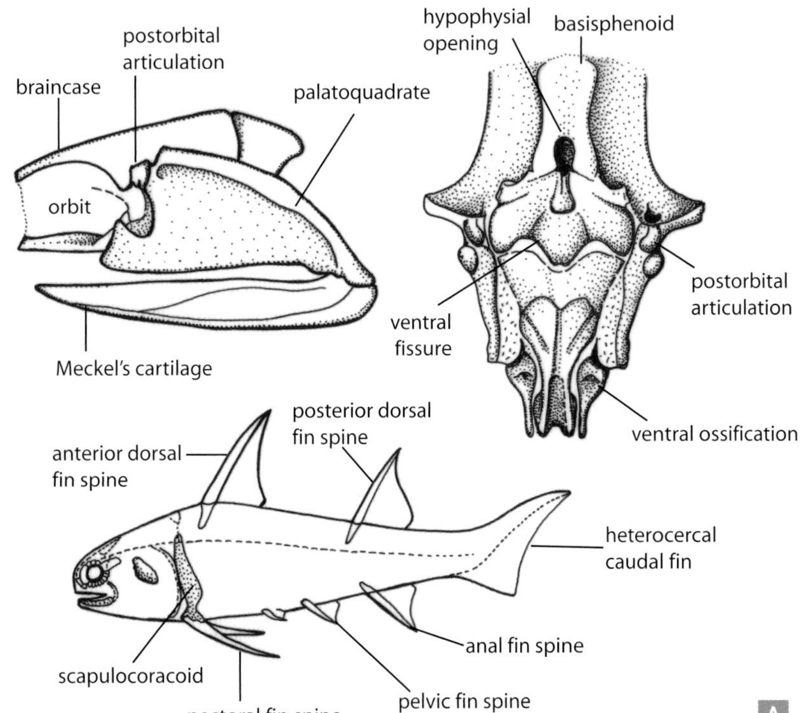

Basic anatomy of acanthodian fishes (*A*). Scanning electron micrographs of a scale of *Nostolepis,* a climatiid acanthodian from the Middle Devonian of Antarctica (*B*). Scale bars are 0.1 mm. Note the ribbed crown, constricted neck, and tumid base. (Courtesy Carole Burrow, Queensland Museum)

Basic Structure of an Acanthodian, cont'd.

the fishes an improved sense of balance and direction, necessary for fast turns and maneuvers.

The body has a shagreen of many tiny close-fitting imbricating scales, and these have a swollen base that lacks a pulp cavity. The scales grew by concentric addition of new layers like an onion. Two main scale types are recognized by their histology: an *Acanthodes* type, which has the crown made of true dentin and a thick acellular bone base; and a *Nostolepis* type scale that has a dentin crown penetrated by vascular canals and a base of cellular bone. The crowns of acanthodian scales are generally quite flat or weakly domed and are separated from the base by a well-defined "neck." Ischnacanthid acanthodians have well-developed jawbones with teeth and may develop complex dentitions with tooth "fields" present. Climatiiform acanthodians had spiny individual teeth and complex tooth whorls at the front of the lower jaws, whereas acanthodiforms were toothless filter-feeders with well-developed gill-rakers for filtering food.

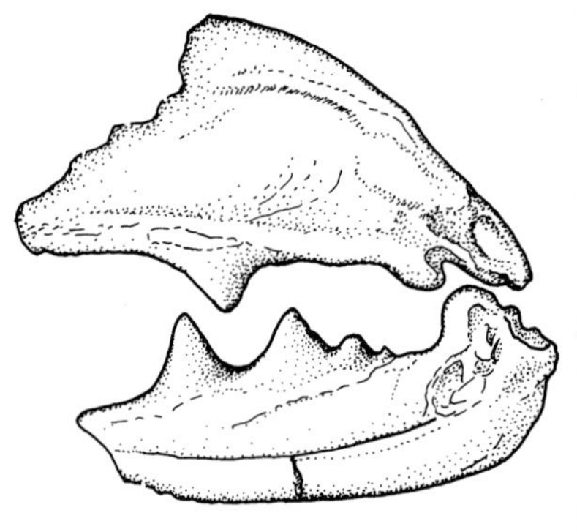

The jaws of *Acanthodopsis*, a large predatory acanthodiform from the Carboniferous Period.

groups—such as the presence of a similar-shaped braincase (*Acanthodes*), the presence of branchiostegal rays, the nature of the hemibranchs (gill filaments) along the gill arch, and the possible presence of endochondral bone in the braincase. This last is controversial but advocated here. Their sharklike features would seem to be mostly primitive characteristics seen in many early jawed fishes and not necessarily suggestive of a close affinity with any one gnathostome group, despite similarities with early sharks. This viewpoint, however, is not held by all, as Erik Jarvik, of Stockholm, argued that acanthodians are actually close relatives of the primitive seven-gilled sharks. The most recently published analyses of gnathostome interrelationships tend to favor acanthodians as being either the sister group to osteichthyans, or as a grade group with some derived members closer to osteichthyans and others embedded deeply within chondrichthyan origins (Brazeau 2009).

The oldest acanthodian fossils, from the Early Silurian of China, are broad fin spines of fishes named *Sinacanthus* (meaning "Chinese spine"). Early acanthodian scales showed a simple tissue structure but nonetheless still grew continuously throughout life, a condition not seen in fossil sharks but one that has a strong link with the higher jawed fishes such as placoderms and osteichthyans. The structure of the head and the shape of the body in the earliest complete acanthodian fossils tell us little about which other group of fishes they may have evolved from or collaterally with.

The oldest possible acanthodian known from partially complete remains is *Yealepis* from the Upper Silurian of Victoria, described by Carole Burrow and Gavin Young (1999). The specimen shows typical scales of unmistakable acanthodian morphology close to that of *Nostolepis* species, although its poor preservation and lack of spines precluded Burrow and Young from confidently placing it within the acanthodian group.

Paucicanthus from the earliest Devonian (Lochkovian) of Canada, described by Gavin Hanke (2002), was another enigmatic form. It lacked spines on the paired pectoral and pelvic fins but had median fin spines that in all morphological aspects were typical acanthodian structures. This suggests that the early members of the group may have secondarily lost fin spines early in their

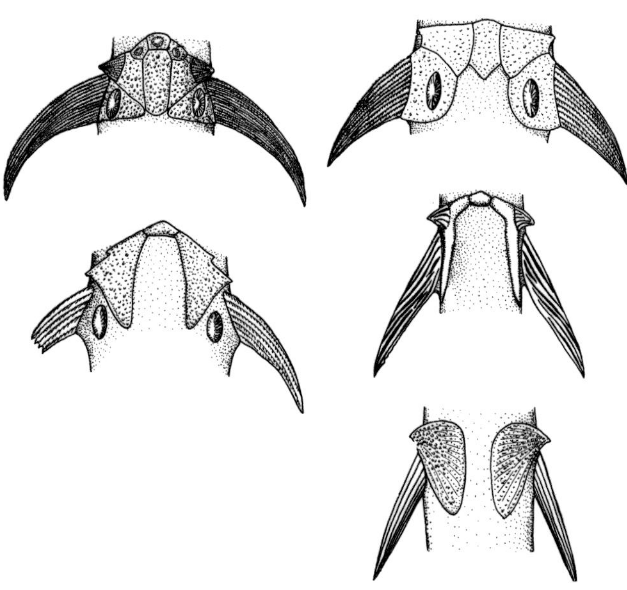

Variations in the ventral shoulder girdle bones surrounding the pectoral fin spines in some Early Devonian acanthodians. (See Long 1986b)

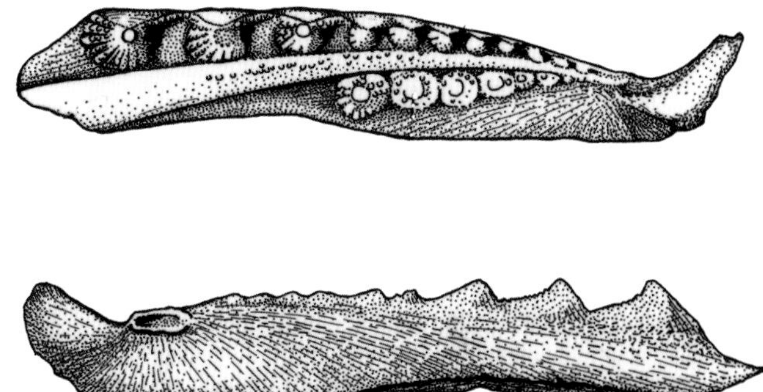

The jaw of *Taemasacanthus* an Early Devonian ischnacanthid from New South Wales, Australia. Ischnacnathids have teeth fused to the jawbones.

evolution, as paired pectoral spines were a widespread feature among many groups of basal gnathostomes (as in sharks like *Doliodus*, all basal placoderms, and some basal osteichthyans like *Guiyu* and *Psarolepis*).

The Climatiiformes: Armored Acanthodians

The oldest group of spiny acanthodians known from complete remains were the Climatiiformes, named after a genus (*Climatius*) found in the Early Devonian Old Red Sandstone of Britain. They are characterized by elaborate bony armor around the entire shoulder girdle. In primitive forms there may be numerous additional intermediate fin spines in paired rows along the belly, between the pectoral and pelvic fins. The scales and some isolated fin spines of *Climatius* and its relatives have been recognized from sites in Europe and Russia, and climatiid-like scales have been found almost everywhere in sequences of similar age. Such heavily armored fishes were not able to move their pectoral fins, which were rigidly fixed to the shoulder girdle armor, and so the fish probably hydroplaned its way along the seafloor searching for prey. The many sharp spines projecting at all angles from its body were no doubt a deterrent to would-be attackers.

Other heavily armored early climatiids include *Parexus*, which had one very large spine, and *Euthacanthus*, which had five pairs of intermediate fin spines along its belly surface. Many of these fishes come from sites in Britain that were freshwater or marginal marine deposits. The earliest acanthodians were principally marine fishes, but later in the Devonian they became predominantly brackish-to-freshwater forms.

After the first radiation of heavily armored climatiids had waned (by the Middle Devonian), the group continued on in a less spectacular form. The diplacanthids were a deep-bodied group of climatiiforms, which had greatly reduced their pectoral fin armor and unusual cheek plates covering the side of the head. The best-known form, *Diplacanthus* (meaning "paired spines") has been found in Middle-Late Devonian rocks in Europe and North America. *Diplacanthus* retained some primitive features, such as intermediate fin spines and fairly rigid pectoral fins. *Rhadinacanthus* looked similar to *Diplacanthus* but had shorter anterior spines in its dorsal fin than in its posterior dorsal fin. *Milesacanthus* was a Middle Devonian diplacanthid known from the Aztec Siltstone of Antarctica, which shows the group's widespread distribution (Burrow and Young 2004). Another diplacanthid known only from East Gondwanan regions is *Culmacanthus*, a deep-bodied fish that had fewer plates around the pectoral fins and lacked intermediate fin spines entirely (Long 1983). Its characteristically ornamented cheek plates were large and have

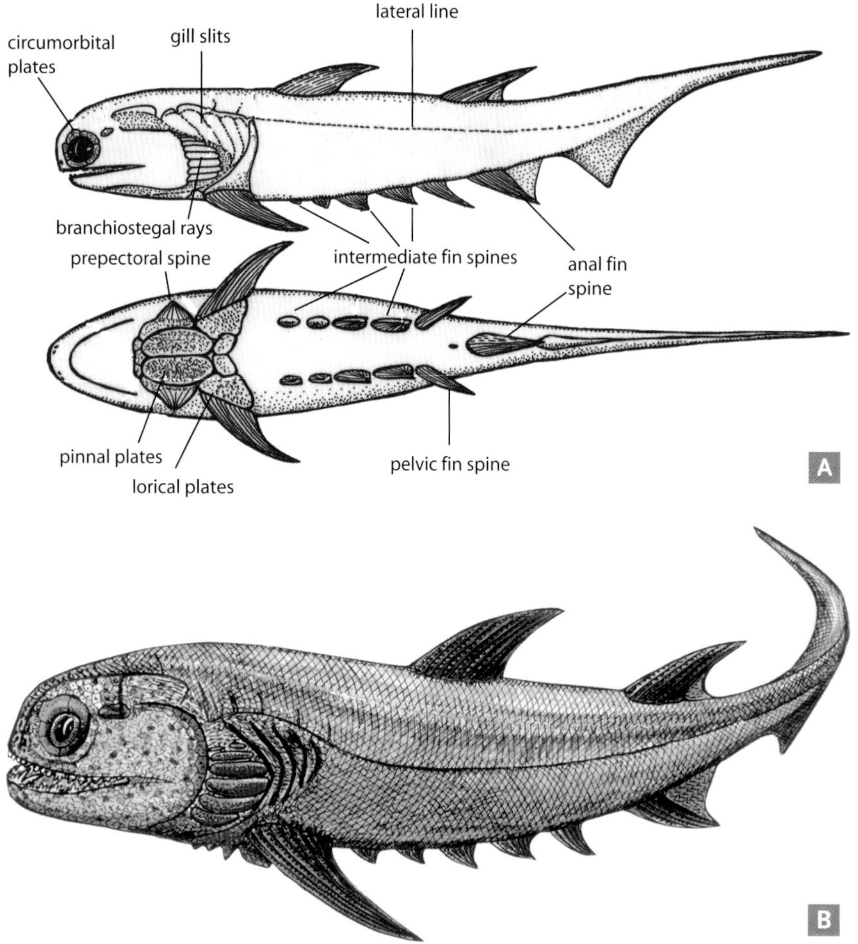

Basic features of a generalized acanthodian, *Climatius* (*A*). Reconstruction of *Climatius* from the Early Devonian of England (B).

been found at a number of sites throughout Australia and Antarctica. The genus is now known from three species, based on the shapes of the cheek plates. The diplacanthids all had small mouths that were devoid of teeth, suggesting that were algal-grazers or detrital-feeders.

The last climatiiforms were the gyracanthids, large acanthodian fishes up to about 1.5 m long, which first appeared in the Middle Devonian of Antarctica (East Gondwana) and radiated out to invade the northern hemisphere countries during the Carboniferous Period. The characteristically large spines with chevron-type radiating ridges are easily recognized in Carboniferous fluviatile deposits around the world. The only articulated specimens showing what these fishes were really like belong to *Gyracanthides murrayi*, found near Mansfield in southeastern Australia. In this species, the shoulder girdle has been freed from the rigid bony plates locking the fins, so that the large pectoral spines were actually capable of a slight degree of movement. These spines were enormous, nearly half the length of the fish, and probably acted as a defense against the large predatory rhizodont fishes that shared its habitat. The remains of a closely related form, *Gyracanthus*, are commonly found as isolated spines and shoulder bones in the British and American coal measures of similar age.

Ishnacanthiformes: The Predators

The special feature that distinguishes the Ischnacanthiformes (from the Greek meaning "thin spines") is that they had robust dermal gnathal, or jaw, bones with rows of teeth fused firmly to them. These special gnathal bones formed the only biting margin of the jaws, the rest being largely cartilaginous or, in some

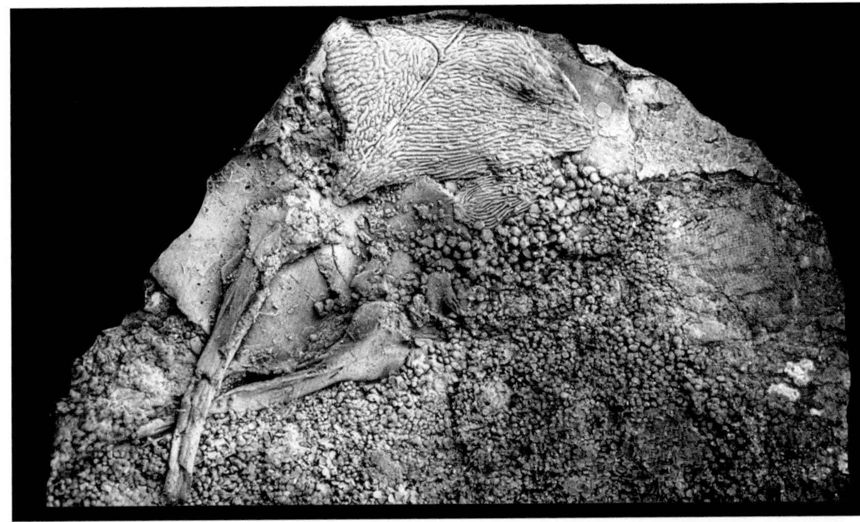

The cheek plates, scales, and spines of *Culmacanthus*, a Middle Devonian acanthodian from Mount Howitt, Victoria, Australia (latex cast).

The deep body and robust fin spines of *Culmacanthus*, a Middle Devonian acanthodian from Australia (latex cast).

species, with an additional bony splint on the outside of the lower jaw. Although ischnacanthid scales are well known from several Middle-Late Silurian sites around the world, the earliest complete fossils of the group are of Early Devonian age, from Canada and Britain. The superbly preserved specimens from the Delorme Formation of Canada's Northwest Territories show that the first ischnacanthids lacked the bony armor around the shoulder girdle that characterized their cousins, the climatiiforms.

One of the best-known forms, *Ischnacanthus,* was a slender-bodied predator with long, narrow fin spines. The distinctive gnathal bones showed a series of large triangular, curved teeth with smaller cusps separating the large teeth. The head had no large bony plates, only some slightly larger bones in the region of the jaw joint. One of the oldest and most primitive ischnacanthids is *Uraniacanthus,* from England, which had intermediate fin spines present between the pectoral and pelvic fins.

The largest of all the acanthodians was the ischna-

Milesacanthus is the only relatively complete acanthodian fish known from Antarctica. It was found by Gavin Young in the Middle Devonian Aztec Siltstone, Transantarctic Mountains in 1971.

canthid *Xylacanthus*, which, like many ischnacanthids, is known only from its jawbones. *Xylacanthus* jaws have been found in the Late Silurian of Canada and the Early Devonian red beds of Spitsbergen, indicating a maximum estimated length of about 2 m, which makes this fish among the largest predators in the shallow marine environments it inhabited. Ischnacanthid jaw elements have been described from many other Devonian marine deposits, some having highly complex dentitions. *Rockycampacanthus* and *Taemasacanthus*, from the Early Devonian limestones near Buchan and Taemas, in southeastern Australia, were small fishes whose jawbones had a second row of teeth in addition to the main cutting row, as well as having isolated scattered teeth and denticles on the jaws. Overall the ischnacanthids are a poorly known group. In the few cases in which they are well preserved as whole body fossils, they appear to be very conservative throughout their evolution. One of the last surviving ischnacanthids was *Grenfellacanthus* from the Famennnian Hunter Siltstone of New South Wales, a large fish that is estimated to have been extinct by the close of the Devonian.

The Acanthodiformes: Filter-Feeders and Survivors

The Acanthodiformes were the most successful group of acanthodian fishes and are easily characterized by their single dorsal fin, lack of teeth, and well-developed

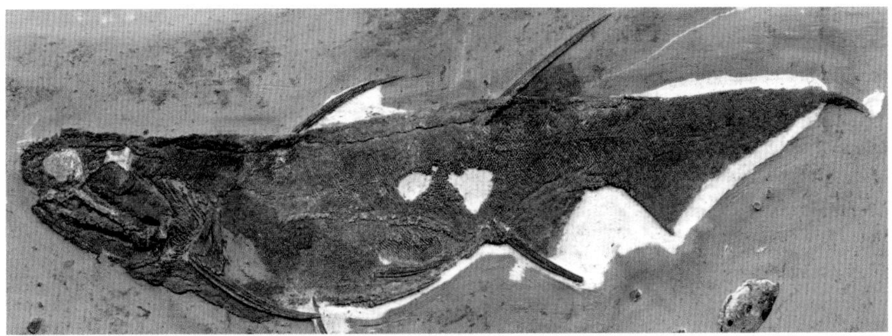

Ischnacanthus **from the Early Devonian MOTH locality in Canada.** (Courtesy Mark Wilson, University of Alberta)

A fin spine of *Gyracanthides*, a large acanthodian known from the Devonian-Carboniferous deposits of Gondwana regions. This specimen is from the Early Carboniferous of Mansfield, Australia.

A close-up view of some of the teeth of the predatory ischnacanthid *Grenfellacanthus* from the Late Devonian of Grenfell, Australia.

gill rakers. These features indicate a free-swimming, filter-feeding lifestyle. A major trend throughout their evolution was the freeing up of the shoulder girdle to have several ossified components, allowing greater movement in the pectoral fins. Perhaps such maneuverability was developed in order to chase the schools of plankton or small crustaceans on which they fed or simply evolved as a more efficient way of escaping their attackers.

The acanthodiforms first appeared in the Early Devonian and were represented by two primitive mesacanthids. These are the only acanthodiforms to have intermediate fin spines present between the pectoral and pelvic fins. *Mesacanthus* from Scotland is known from well-preserved complete specimens from Angus, Scotland. *Paucicanthus vanelsti* from Canada differs in having expanded skull-roof plates and paired nasal bones (Hanke 2002). By the Middle and Late Devonian, the acanthodiforms were flourishing and had achieved a standard, conservative body plan much like that of the ischnacanthids. *Cheiracanthus* was a widespread Middle Devonian form, whose characteristic scales are recognized from sites of this age around the world, although the only complete specimens are known from the Old Red Sandstone of Scotland.

Late Devonian forms such as *Howittacanthus*, from Mount Howitt, southeastern Australia, and *Triazeugacanthus*, from the Escuminac Formation of Quebec, Canada, vary only in minor ways from the later forms

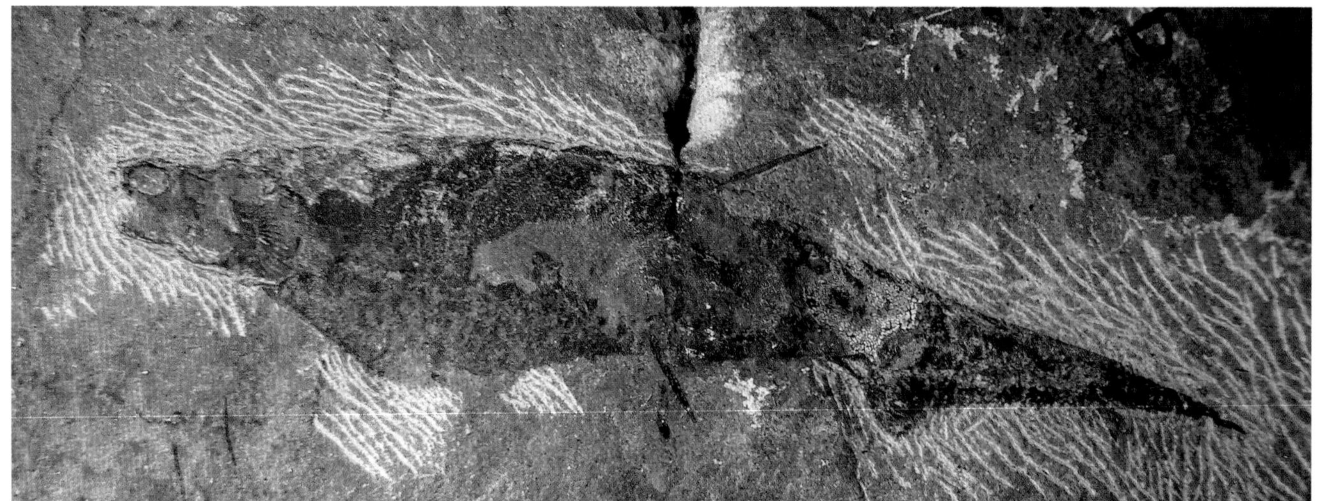

Homalacanthus, a Middle Devonian acanthodiform fish from Bergisch-Gladbach site in Germany. Acanthodiforms had lost the anterior dorsal fin spine and were largely filter-feeding fishes that lacked teeth.

A specimen of *Howittacanthus* from the Middle Devonian Mount Howitt site, Australia.

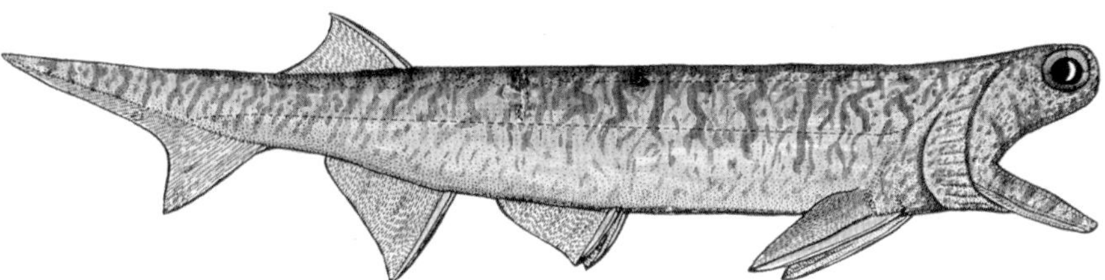

Reconstruction of *Howittacanthus* as a filter-feeding fish.

such as *Acanthodes*, by differences in size of pelvic fins and numbers of ribs on the fin spines. In all advanced acanthodiforms, the scales have flat, unornamented crowns and rather shallow bases and the fin spines tend to have only one large median rib. Advanced acanthodiforms, such as *Acanthodes* and *Utahcanthus* from North America, also had three kinds of otoliths, or ear-stones, one for each chamber of the membranous labyrinth. The presence of well-formed otoliths may have given these fishes a greater degree of balance or a faster sense of direction when swimming rapidly.

Acanthodes was a cosmopolitan genus that lived from the Early Carboniferous to the Middle Permian and is one of the only acanthodians in which the braincase and gill arches are preserved as impressions in material of *Acanthodes bronni* from the Lebach site, Germany. Its numerous gill rakers and wide mouth suggest that it was an efficient filter-feeder. *Acanthodes* is the only known Permian acanthodian, representing the last survivor of this once-flourishing class of fishes.

Acanthodopsis, known from jaws with simple triangular teeth, flourished in the Carboniferous coal swamps of Britain and Australia and may have grown to up to 1 m long. However as this form lacked separate gnathal bones with the distinctive, complex tooth morphology seen in other ischnacanthids, *Acanthodopsis* may be an

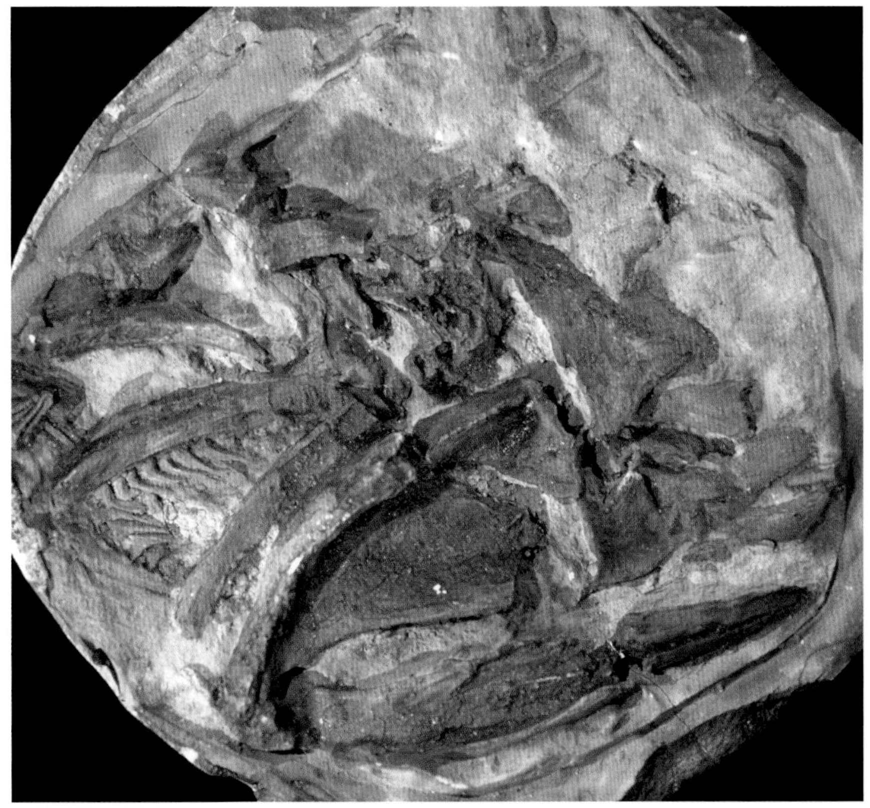

A cast of the jaws and gill arches showing gill-rakers for filtering food, from *Acanthodes bronni*, from the Permian Period of Lebach, Germany.

Two specimens of *Acanthodes* from the Permian of Germany. This genus was the last surviving member of the acanthodian lineage, which became extinct by the end of the Paleozoic Era.

aberrant acanthodiform, turned from a primarily filter-feeding lifestyle to predation. Carole Burrows (2004) of Australia described material of *Acanthodopsis russelli* as three-dimensional jaws and placed the fish within the order Acanthodiformes based on the complex double jaw articulation.

The acanthodians most likely died out because of the rapidly increasing numbers of ray-finned fishes and sharks at this time, which were expanding into a great range of habitats and feeding niches. Today, we can utilize the minute scales of acanthodians from Silurian and Devonian deposits to date the age of sediments and make correlations with other faunas. Acanthodian fossils are becoming increasingly more important in this regard, and there is much research to be done to refine correlation schemes based on their scales. There is still much to be learned about this interesting grade group of spiny fishes. Recently discovered 3-D material of them from the Late Devonian Gogo sites in Australia should help provide a wealth of new information about their anatomy and help shed light on their mysterious affinities to the other early jawed fishes.

CHAPTER 7

An Epiphany of Evolution

The Osteichthyes, fishes having well-ossified internal skeletons

The osteichthyans (meaning "bony fishes") today form the largest and most diverse group of vertebrates and are represented chiefly by the ray-finned fishes (subclass Actinopterygii) and also by the lobe-finned fishes (Sarcopterygii). Their origins date back to the Late Silurian, some 420 million years ago, when they were a rare component of the ancient fish faunas. By the start of the Devonian Period, all the major groups of bony fishes had appeared (Zhu et al. 2009). But before the end of the Devonian, the first land animals had evolved from within the radiation of advanced lobe-finned osteichthyans. Buried within the early evolution of osteichthyan fishes are the key stages to the most complex evolutionary transition in vertebrate history: how a water-breathing fish became a land-living amphibian.

One of the secrets of the success of the bony fishes lies in the presence of an early lung, later modified to become a swim-bladder. The two major groups of bony fishes are the ray-finned (Actinopterygii) and the lobe-finned (Sarcopterygii). The latter includes two main divisions, containing many extinct orders of fishes, the Dipnomorpha (including living lungfishes) and the Tetrapodomorpha (the fish lineage that gave rise to land animals). Aside from their differing physical features, especially their unique patterns of bones forming the skull and shoulder girdles, many of the osteichthyan groups also evolved specific feeding adaptations, some characterized by having unique tooth tissues.

Today when we think of fishes, we most often think of the osteichthyans. They are the group that contains the majority of the approximately 30,000 living species of fish. In contrast to their widespread success in today's waters, the earliest osteichthyans had a real battle to survive among the shoals of primitive Devonian predatory fishes. The first osteichthyans are very poorly known from fossils, represented by scales and mere fragments of broken bones (Botella et al. 2007). The oldest remains, showing what their cranial anatomy was like, are of Middle Silurian age (*Guiyu*, Zhu et al. 1999). During the Early Devonian, the placoderms and acanthodians were increasing in diversity, as sharks and osteichthyans were a noticeably lesser part of the faunal assemblages.

As the Devonian progressed, the osteichthyans diversified into many different groups, mostly lungfishes and lobe-finned predators, and these flourished right to the end of that period, with just a few groups surviving on a bit longer in geological time. Three genera of lungfishes (*Protopterus, Lepidosiren, Neoceratodus*) still survive on the three different southern continents, but only a two species of primitive lobe fin, the coelacanth (*Latimeria*), survived in isolated marine habitats to present times.

In this chapter, the basic anatomy of an osteichthyan fish is presented with an overview of the enigmatic remains of the most primitive fossil members of the group. The evolutionary success of the bony fishes lies not only in their flexible skeletal patterns, which allowed for a range of different feeding styles to evolve (especially in the ray-fins), but also in the nature of their varied

(*Top*) The family of the basal ray-finned fish the gar, *Lepisosteus*, goes back to the Early Cretaceous some 140 million years ago. Gars like this one bridge the gap between heavily armored Paleozoic ray-finned fishes and the more derived thin-scaled forms such as the teleosteans that predominate today. (Courtesy Rudie Kuiter, Museum Victoria)

(*Left*) *Protopterus*, the African lungfish, represents the smaller group of lobe-finned fishes (sarcopterygians), a type of fish that flourished in the Paleozoic era but is today represented by six species of lungfishes and two of coelacanths.

***Latimeria*, the living coelacanth, the most basal member of the extant lobe-finned fishes. This species was discovered in 2007 living off Sulawesi, Indonesia.** (Courtesy SeaPics)

tissue types developed within their skeletons, both as external dermal plates and as a variable range of tooth tissues. The words "bone" and "teeth" are used loosely for a range of toothlike tissues that evolved to give more strength to crushing plates, to provide lightness and strength to principal skeletal components, and to reinforce large stabbing fangs for prey capture.

Unlike the placoderms, sharks, and acanthodians, the osteichthyans achieved one great evolutionary novelty that was to be the harbinger of a later successful invasion of land: a respiratory system that could take advantage of oxygen within air rather than relying on aqueous respiration solely through the gills. The simple internal lung in osteichthyans that supplemented gill breathing in times of oxygen depletion (Clack 2007) would later develop as the prime method of respiration in land animals.

The advantage of the lung is that it can function also as a hydrostatic organ enabling the fish to rise or fall within the water column by regulating its volume of gases. This means that energy is saved by not having to push forward to develop lift with large winglike pectoral fins (as in sharks and placoderms); thus the fins became freed to adapt for improved maneuverability within the water. With this novelty came the great diversity of body shapes achieved by later lineages of ray-finned fishes, such as the deep-bodied palaeoniscoids (for example *Ebenaqua* and *Cleithrolepis*, mentioned in Chapter 8) that modified the lung into a swim-bladder. In many living species of ray-finned fishes, the swim-bladder can also partially function as an accessory breathing organ, enabling some species to exist outside the water for short times. Excellent examples are the mudskippers, small fishes that live among the mangroves and are able to climb trees. The evolutionary transition from a simple lung to an advanced tetrapod lung employing costal, or rib, breathing was therefore not such a complex step but one that involved greater rigidity of the trunk

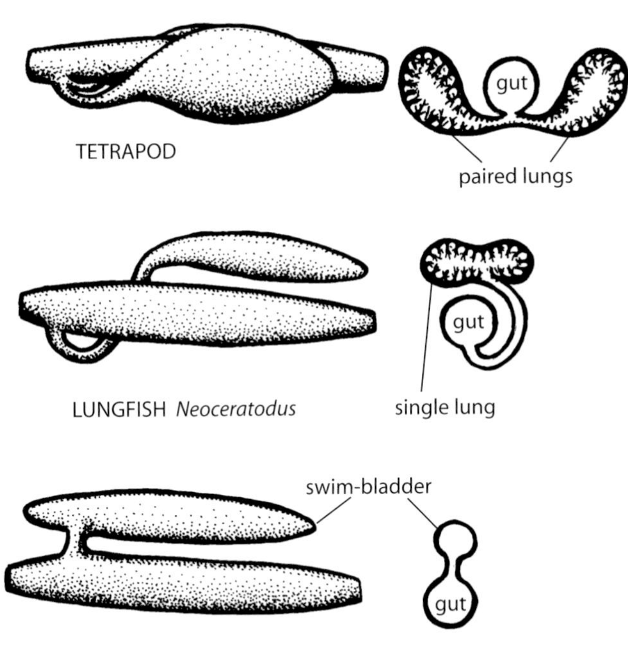

The origin of lungs in higher vertebrates could have come from a primitive condition found in all osteichthyan fishes or as a derived organ from the swim-bladder that first evolved in ray-finned fishes. In all cases, the lung has evolved as an outpocketing from the gut.

and development of the ribcage along with increased musculature control within the animal's trunk region. The skeletons of osteichthyan fishes follow a conservative plan at their earliest origins, and most groups retained a fairly similar pattern throughout their evolution. The notable exception is the ray-finned fishes, which underwent a much great diversification after the Devonian and then just kept on diversifying as new grades of organization appeared (e.g., holostean and teleostean stages), with groups rapidly diversifying right up to the present day. The freeing up of their originally rigid cheek- and jawbones enabled them to adapt to a great variety of specialized feeding mechanisms from those of simple or complex predators, to algal grazers, to tube-mouthed seahorses that suck plankton from the seawater. The variety of teleostean feeding mechanisms is almost endless.

The structure of the most primitive osteichthyan fishes is reasonably well known from the Early Devonian onward, although most of these basal groups are classified as stem osteichthyans. In this chapter, the anatomy of stem osteichthyans is shown, along with the two most basal groups (onychodontiforms and actinistians, or coelacanths), which diverged before the main lineages of dipnomorphans and tetrapodomorphans (these are treated in subsequent chapters).

Lophosteiforms: Enigmatic Ancestors of Bony Fishes

There are only two known genera of lophosteiforms, *Andreolepis* and *Lophosteus*, both of which have been found in the Late Silurian deposits of Oesel, an island in the Baltic Sea, off the coast of Sweden. The lophosteiforms were first described by the famous German paleontologist Walter Gross (Gross 1969, 1971) and are still an enigmatic group known only from isolated scales, teeth, and some rare fragmentary bones. They appear to have been similar in structure to primitive ray-finned fishes, although the teeth lacked a dense hard tissue called *acrodin*, which mineralizes on the tips of teeth in nearly all ray-finned fishes.

The tooth shape is conical with a large central pulp cavity, and the teeth were probably fused into the jaw, lacking a root system. The scales have a similar shape to the rectangular or rhombic scales of early ray-finned fishes but lack the layers of mineralized shiny ganoine that make up the surface ornament in the ray-finned fishes. The shoulder girdle, known from a partial cleithrum, is also of similar form to that in the early ray-fin *Cheirolepis*. These tantalizing pieces give us a sketchy idea that the lophosteiforms looked a bit like the earliest ray-finned fishes but lacked the important tissue types that were to characterize the group and to be a foundation for later major evolutionary changes in their skeletons. Work published in *Nature* by Hector Botella and colleagues (2007) and a recent study by Friedman and Brazaeu (2010) determined that lophosteiforms are stem osteichthyans and that they cannot be aligned more closely to ray-finned than to the lobe-finned fishes.

Another mysterious fish originally thought to be

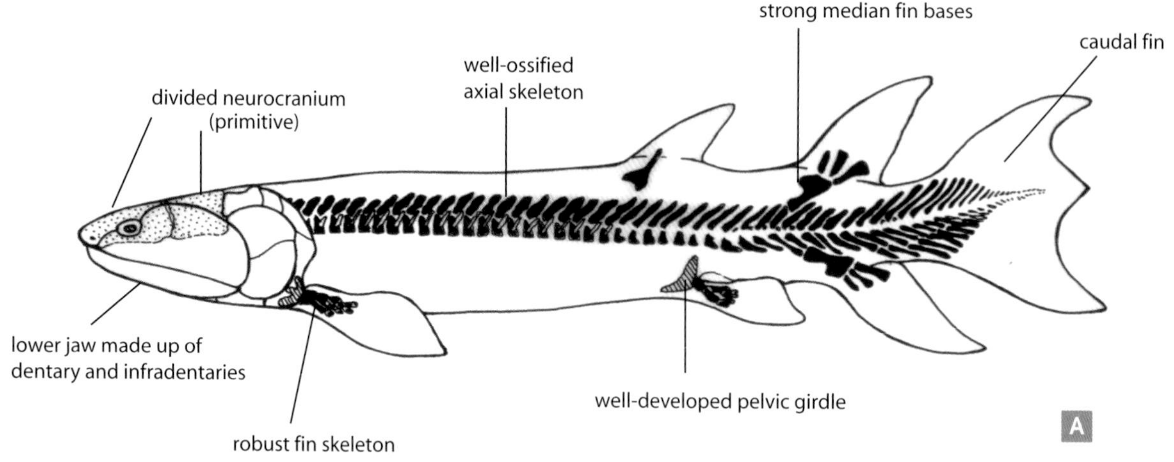

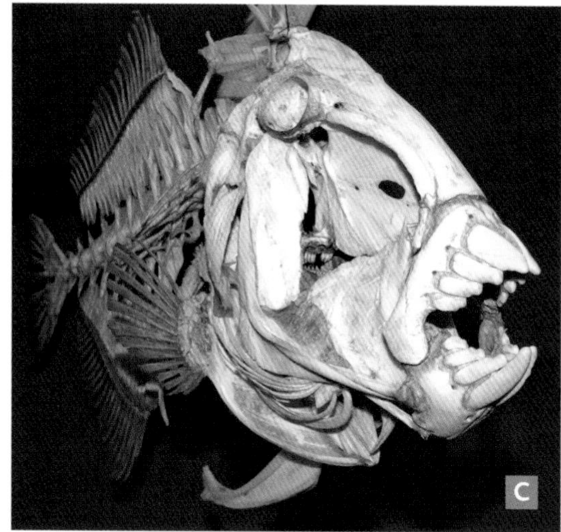

The skeleton of an advanced osteichthyan fish (*A*), the Late Devonian tetropodomorph *Eusthenopteron*, showing major components. A strong internal skeleton characterizes nearly all osteichthyan fishes. The skull (*B*) of *Lycodontis funebris*, a moray eel, highlighting the most successful adaptation of advanced ray-finned fishes: the ability to mobilize bones of the mouth for a wide range of feeding actions. Moray eels actually engage a second set of jaws hidden in the throat to doubly seize their prey. The skeleton of a wrasse (*C*), one of the many hundreds of teleostean fish families alive today. Note the robust teeth situated on protrusable jawbones. Teleosts are the most diverse of all vertebrates, containing most of the approximately 30,000 living species of fishes known. Part of the axial skeleton of the swordfish (*D*), *Xiphias*, showing strongly ossified vertebrae that enable the fish to swim at speeds of 80 kph. Osteichthyans differ from all other fishes by their well-ossified axial skeletons.

Basic Structure of an Osteichthyan

Osteichthyan fishes are characterized by having a well-ossified internal bony skeleton, although the earliest fossil forms show the least degree of ossification of the vertebrae and internal bones. Endochondral bone, formed around a cartilage precursor is present, and marginal teeth are developed on the upper and lower jawbones, specifically the maxilla, premaxilla (upper jaw), and dentary (lower jaw). The premaxilla bears part of the infraorbital sensory-line canal.

The gill arches—bony or cartilaginous arches made up of segments and located on either side of the pharynx that support the gills—are highly evolved in osteichthyans relative to other jawed fishes. Pharyngobranchial elements are present with suprapharyngobranchials on the first two arches, and the hyoid arch has interhyal and hypohyal bones. The first two gill arches articulate to the same ventral median bone (basibranchial), which often bears small toothed bones to help crush up the food. Gular plates cover the underneath of the head, and a subopercular bone is present below the large opercular bone.

The pattern of bones in the shoulder girdle of osteichthyans is unique: a large cleithrum and clavicle make up most of the girdle, with smaller supracleithrum, post-temporal, and postcleithrum or anocleithrum bones. This rigid girdle evolved in parallel to that of the placoderms to give support for the pectoral fin musculature.

The scales of osteichthyan fishes articulate and overlap with one another, more so than for any other fish; and in advanced osteichthyan groups, such as the higher groups of sarcopterygians and later actinopterygians, each rounded scale may have more than 75% of its surface overlapped by neighboring scales. This gives a high degree of rigidity to the trunk that allows for more efficient and more powerful swimming.

The internal anatomy of an osteichthyan is unique among fishes in the presence of a swim-bladder (or lung) that enabled the fish to regulate its depth in the water column (and ultimately to breathe air). The musculature is also highly specialized in the presence of branchial levator, interarcual, and transversi ventrali muscles, which affect the function of the gills and the stomach.

Although these features are found in all members of the primitive osteichthyan groups, the great explosion of actinopterygians has seen some of these features highly modified or secondarily lost in special cases. Furthermore, many of these characteristics apply also to primitive fossil amphibians, again emphasizing that these are really just an advanced lineage of osteichthyans with digits instead of fins. This fascinating transition is explored in detail in the final chapter of the book.

a basal actinopterygian is *Ligulalepis*. The scales were originally described in 1968 by Hans-Peter Schultze. They bear a well-developed dorsal peg, and the histology shows ganoine layers to be present. However a skull and braincase attributed to *Ligulalepis* were recovered from the Early Devonian of Taemas, Australia, that showed it retained a number of primitive osteichthyan features, such as the presence of an eyestalk, a condition lost in most other osteichthyans. The phylogenetic analysis of its anatomy determined it must be regarded as a stem osteichthyan for the present time (Basden et al. 2000).

Early Lobe-Fins or Stem Osteichthyans?

The oldest well-preserved remains of lobe-finned fishes belong to Late Silurian forms from China such as *Guiyu* and *Psarolepis*, which are discussed further in Chapter 10. These two species, along with an Early Devonian form called *Achoania*, bridge a gap between true osteichthyans and other early jawed fishes in displaying a range of very primitive features, such as paired fin spines and broad-based scapulocoracoids, or shoulder bones, supporting the pectoral fins (Zhu and Yu 2009).

Other early lobe-finned fishes are known from Early Devonian species from Arctic Canada, Europe, and China. These fossils consist of skulls belonging to a group that had already achieved its basic body plan and skull pattern—the porolepiforms (see Chapter 11 for details), as well as some peculiar forms that appear to be intermediate in structure between the lungfish and other sarcopterygian groups. A number of enigmatic

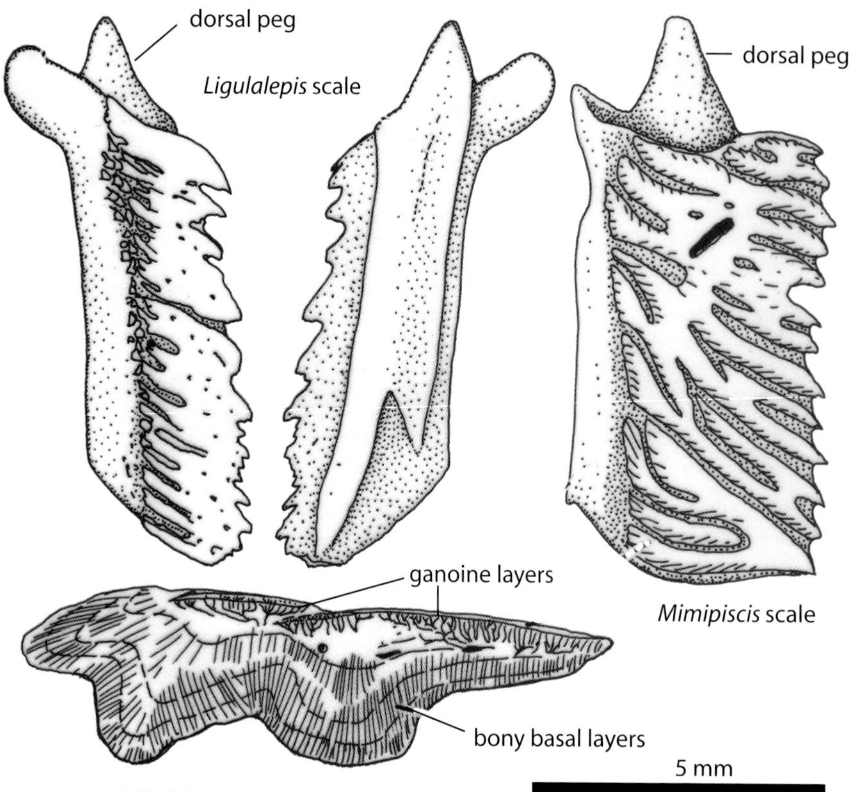

(*Left*) A scale of *Ligulalepis*, a Late Silurian / Early Devonian basal osteichthyan, showing the rectangular shape and well-developed peg and socket that holds the scale within its row.

(*Bottom*) The skull of *Ligulalepis*, from Early Devonian of Taemas, Australia, in dorsal view (*A*) and ventral view (*B*). This is one of the oldest and best-preserved skulls and braincases of any basal osteichthyan found to date. Actual size around 1.2 cm.

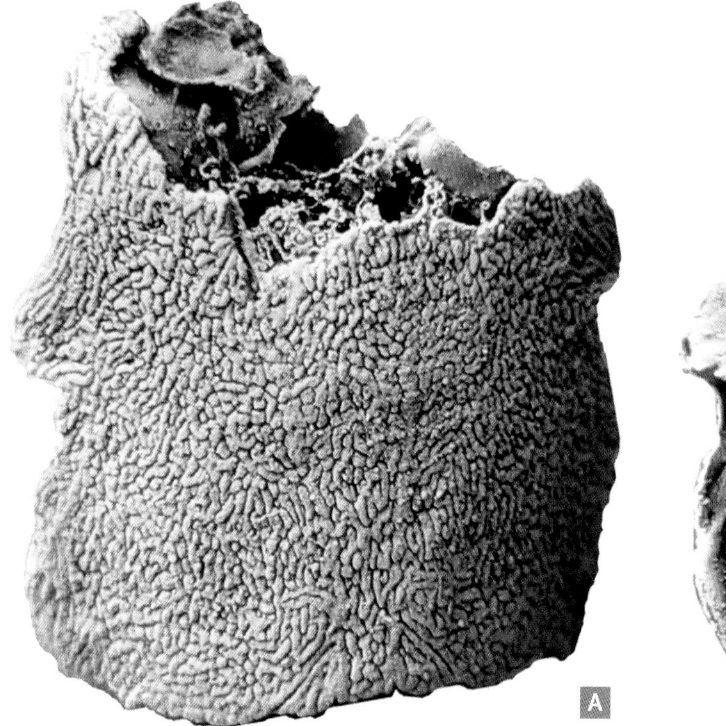

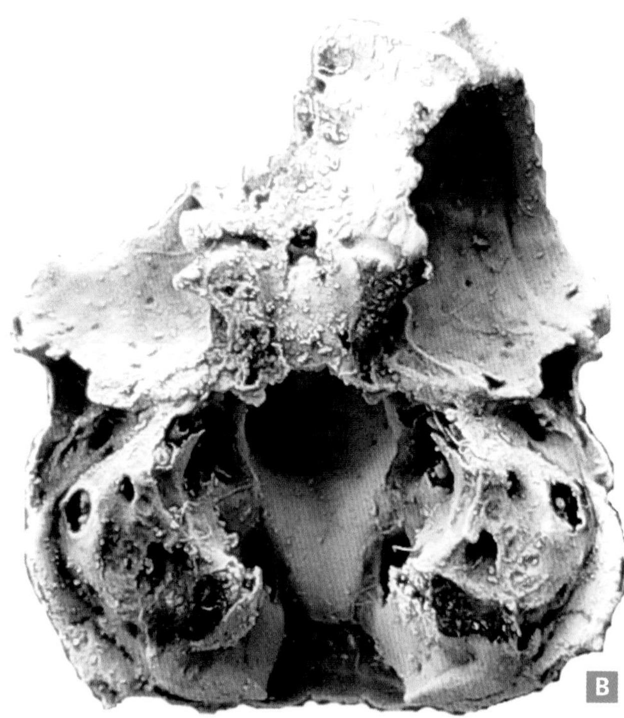

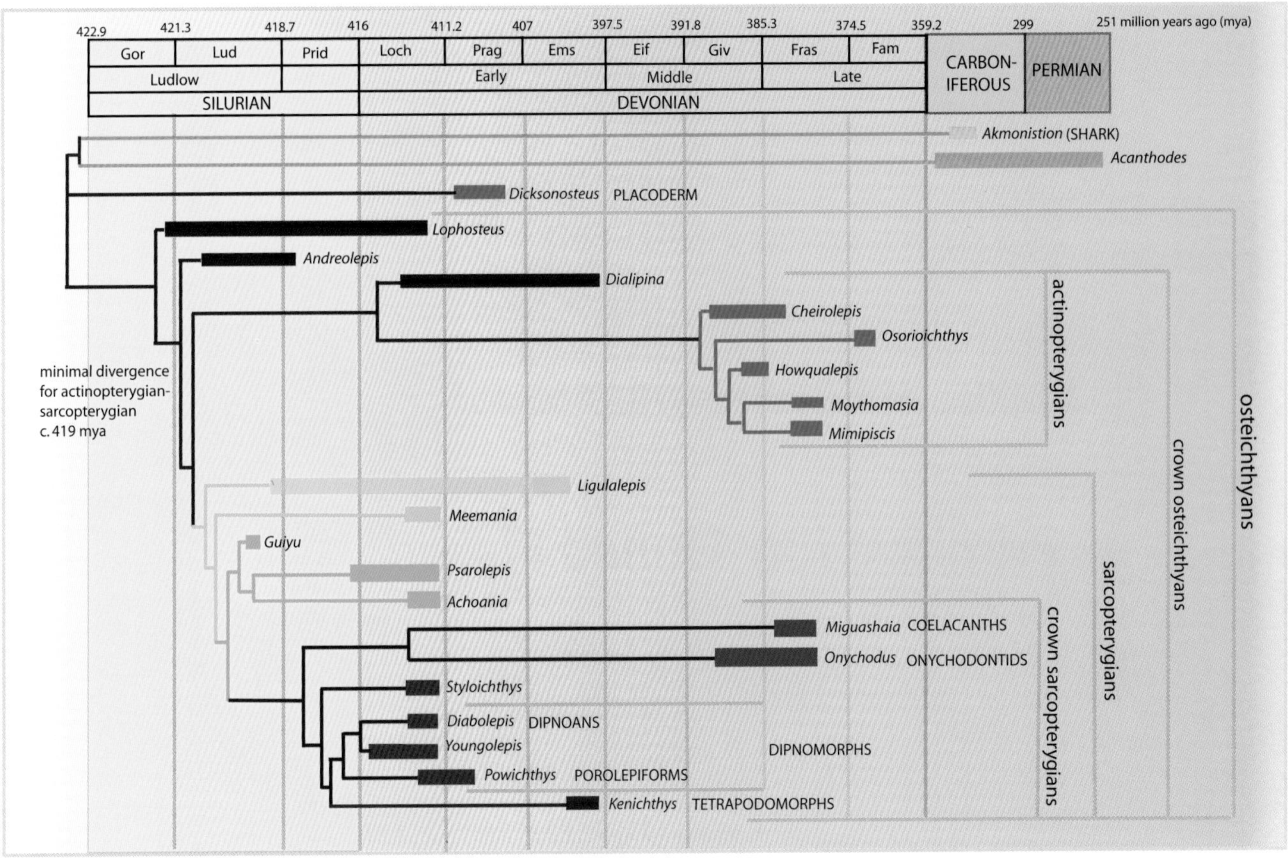

A diagram showing the earliest origin and diversification of the major lineages of osteichthyan fishes. (Modified from Zhu Min et al., 2009).

genera are included in this group of fishes: the possible ancestors of the porolepiforms *Youngolepis* and *Powichthys* and the "proto-lungfish" *Diabolepis* (meaning "devil scale").

These fishes are still poorly known, and their relationships within the osteichthyans have been the subject of much debate. However, a consensus is slowly emerging that they fit into one or other lineage of sarcopterygians, or lobe-finned fishes. Most paleoicthyologists agree that they are closely allied to the porolepiforms and that lungfish and porolepiforms are closely related sister groups (forming the Dipnomorpha).

The current consensus of opinion among fossil fish researchers is that because *Youngolepis* and *Powichthys* possess the characteristic of an enameloid layer that dips into the pores of the cosmine, they are considered to be allied to the porolepiform sarcopterygians. *Diabolepis* is generally regarded as more closely related to the lungfish than to any other group. However, how closely lungfish and porolepiforms are related is still a controversial issue and is one that will be discussed in more detail in Chapter 10. Further discussion of how the osteichthyans evolved into tetrapods that ultimately led to humans can be found in Chapter 13.

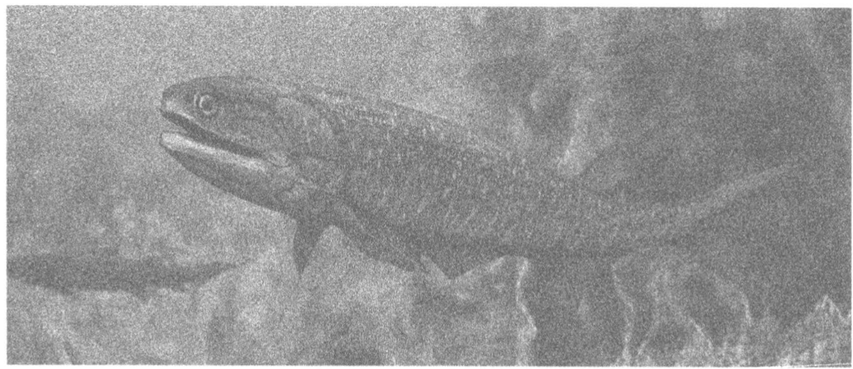

CHAPTER 8

Primitive Ray-Finned Fishes

The rapid rise of the first bony fishes

Of the bony fishes that were discussed in the last chapter, the ray-finned fishes (Actinopterygii) are the largest group, comprising at least 29,000 of the approximately 30,000 living species. The oldest ray-fins appeared in the Late Silurian, some 420 million years ago, known only from scales and fragmentary bones and teeth. The first complete fossils of the group appear some 25 million years later, belonging to *Cheirolepis*, a peculiar fish with tiny scales and fleshy pectoral fin lobes from the Middle Devonian of Scotland and Late Devonian of North America. In the Late Devonian about a dozen other ray-fins appeared, including well-preserved species found in the Gogo Formation of Western Australia. These fossils reveal much information about the anatomy and structure of the first ray-finned fishes, a group generally referred to as palaeoniscoids. In the Late Paleozoic, the group underwent a great burst of diversification, with some 50 or more families appearing in the Carboniferous and Permian Periods. During this time the first of the neopterygians appeared, represented by forms like *Discoserra*. The neopterygians would later give rise to the dominant teleosts in the Mesozoic era.

Perhaps one of the earliest descriptions of ray-finned fishes was by Greek philosopher Aristotle (384–322 BC), who wrote in his *Historia animalium* that "the special characteristics of the true fishes consist in the branchiae [gills] and the fins, the majority having four fins, but the elongated ones, such as the eels, having only two. Some, such as the *Muraena* [moray eel], lack fins altogether." Thus one of the oldest scientific names of an actinopterygian fish came into existence. Later, in sixteenth-century Europe, the church sent scholars such as Frenchman Pierre Belon (1517–64) to university, after which he published his tome *La nature et diversité des poissons (The Nature and Diversity of Fishes)*. At about the same time another Frenchman, Guillame Rondelet, became a professor at Montpellier University and a specialist in studying fish (taken to mean ray-finned fishes in general). Rondelet's work *Universe Aquatalium Historiae* contained many fine descriptions of ray-finned fishes, but he also included anything else that lived in the sea including whales, crocodiles, and shellfish.

Famous Swedish naturalist Carl von Linné (1707–78), also known as Linnaeus, is considered the father of modern taxonomy. He was the first person to apply a binomial system to the naming of plants and animals. His classic work, *Systema Naturae* published in 1758, was to set the stage for the great works describing and naming the diverse numbers of ray-finned fishes that were soon to be discovered. However, it was French scientist Louis Agassiz who, in his classic work of 1833–44 *Recherches sur les poissons fossiles*, first recognized, through comparison of fossil and living ray-finned fishes, the division of three main levels of organization within the group—the Chondrostei, Holostei, and Teleostei. Although Agassiz's classification of the Ganoidei gained much acceptance, his grouping contained many ray-fins as well as lungfishes and some crossopterygians and the odd acanthodian or two.

The question of whether the ray-finned fishes formed a natural grouping was not really formalized until 1871, when American paleontologist Edward Drinker Cope defined his grouping as the Actinopteri to distinguish the ray-fins from the lobe-finned fishes, and this was further set in concrete by Arthur Smith-Woodward, of the British Museum of Natural History, who coined the term Actinopterygii in 1891 (being the equivalent of Cope's term). British biologist Goodrich followed Woodward's usage of Actinopterygii, and it became adopted as the common name for this diverse group of living and fossil fishes.

Today we know of at least 28,000 species of living actinopterygians, and new species are being discovered every year as nets bring up yet more unrecognizable fishes. The ray-fins technically belong in the subclass Actinopterygii (literally meaning "ray wing"), as the fins of these fishes are supported by stiff bony spines or fin rays. The actinopterygians not only are the most diverse of all living vertebrates but also have penetrated the most hostile of all environments on earth, from the deepest ocean depths (–11,000 m) to high mountain streams (+4,500 m), steaming hot volcanic springs (43°C) to freezing Antarctic waters (–1.8°C). They contain the smallest of all adult vertebrates, with one species reaching a mature size of only 7.5 mm, whereas the extinct ray-fin *Leedsichthys* may have grown to sizes in excess of 12 m.

The fossil record of actinopterygians is spectacular—from the oldest remains, scales of fish alive 410 million years ago, to complete Devonian species preserved in three dimensions from the Gogo Formation of Western Australia, and, generally speaking, an impressive variety of fossil species known throughout the rest of geological time right to the present day. As there are so many different families and groups within the Actinopterygii, this chapter focuses largely on the early origins and radiations of the group, with reference to major advances in their evolution that begat their most successful lineages. Many primitive forms of ray-finned fishes live today, as survivors from different stages in the progressive story of the group's evolution. It is largely from the anatomical studies of these primitive forms that the story of the actinopterygian evolution can be accurately reconstructed.

Devonian Ray-Finned Fishes

Fossil scales of the ray-fin *Ligulalepis yunnanensis* date back to the Late Silurian of southern China, and similar scales have been found in Early Devonian Limestones

Basic Structure of a Primitive Ray-Finned Fish

The primitive Devonian actinopterygians are characterized by a long body with a single dorsal fin; all other early osteichthyans had two dorsal fins. Specialized features unique to actinopterygians are that the lower jawbone has a large dentary bone (the one bearing teeth), which has a mandibular sensory-line canal enclosed within it, and that the jugal bone of the cheek has deep pit-line sensory organs.

The head of primitive ray-fins features a long gape with many sharp teeth set on the maxillary, premaxillary (upper jaws), and the dentary (lower jaw). The braincase is well ossified from several centers that are divided in maturity by ventral and occipital fissures. The midline of the palate has a long-toothed parasphenoid bone, and the inner surfaces of the jaws (both upper and lower) are covered by many smaller toothed bones. The skull roof pattern has large frontal bones containing an open pineal foramen, paired parietals and nasals, a single large median rostral (or postrostral) bone, and the spiracular slit is still open in all Devonian and some subsequent species. The cheek is long, with a series of circumorbital bones carrying the infraorbital sensory-line canal, and there is a large preopercular bone above the maxillary. The operculogular series has a long opercular, large subopercular, and a series of 12 or more branchiostegal rays. The shoulder girdle has a postcleithrum bone present, and the skeleton of the pectoral fin has a single row of internally ossified fin radials, as opposed to the longer armlike skeletons in lobe-finned fishes. The fins are supported by many rows of segmented lepidotrichia and may have specialized "cut-water" scales on the fin's leading edges (fringing fulcra). The axial skeleton is poorly ossified in early ray-fins, consisting of perichondral shells of bone above the notochord, thus only neural and ventral arch elements are preserved with no tail skeleton ossifications. Later actinopterygians, like the successful teleosteans, evolved well-ossified bones in the tail.

The body scales had a well-developed peg-and-socket articulation to lock each scale into its rows and had layers of a shiny surface tissue, ganoine, on dermal bones and scales. This type of scale is called the ganoid type and has layers of thin, laminar ganoine over a dentinous layer with vascular canals and a basal spongy bone layer. In later actinopterygians, the ganoine is lost and the scales may take on rounded shapes. The teeth of all actinopterygians, apart from *Cheirolepis,* a very primitive form, have dense caps made of acrodin, a tissue formed of compact dentin. The external ornament of ray-finned fish bones is generally of fine parallel ridges and tubercles, each capped by shiny ganoine.

***Cheirolepis*, the earliest actinopterygian—or ray-finned fish—known from relatively complete remains. This specimen comes from the Middle Old Red Sandstone of Scotland (Middle Devonian age).**

A reconstruction of *Cheirolepis*, the most primitive ray-finned fish known from complete remains. Note the long body, small scales, and long gape. The rigid cheekbones form an immovable plate over the jaw muscles, which in later organisms became free, allowing a wide range of feeding mechanisms to evolve. (Courtesy Brian Choo)

in Australia. There is some debate about whether this material is truly actinopterygian or stem osteichthyan (Friedman and Brazeau 2010), so these fishes are dealt with elsewhere in this book (see Chapter 10). The oldest skull roof of a ray-fin of similar age is from Arctic Russia, belonging to *Orvikuina*. This skull roof does not differ much from that of other Devonian ray-fins, such as the Middle Devonian *Cheirolepis* and the Late Devonian *Mimipiscis, Moythomasia*, and *Gogosardina* from Gogo in Western Australia.

Cheirolepis had a micromeric cover of very small scales, and its skull roof still bore a number of small bones around the snout, a condition lost in later actinopterygians. Its pectoral fins were supported by a short robust lobe. Pearson and Westoll's (1979) description of *Cheirolepis trailli* alluded to its having a strong yaw, or sideways movement, of the head as it swam.

Howqualepis, from the Middle Devonian of southern Australia, was unusual in that it had teeth set into the median snout bone, the postrostral, separating the front of the upper jawbones. Like *Cheirolepis,* it had a long body and relatively acute snout. It inhabited ancient lakes and was a voracious predator, armed with enlarged teeth at the front of its lower jaws.

The best preserved Devonian ray-fins have been found in the limestone concretions of the Gogo region, Western Australia. These show exquisitely preserved braincases and skull features. Two genera, *Moythomasia* (also known from Germany and elsewhere) and *Mimia* were originally described in meticulous detail by Brian Gardiner (1984a) and in recent years have been further described with new species recognized by Brian Choo. The genus *Mimia* will soon be referred to a new genus (work by Brian Choo) because the original genus name is occupied. The recently discovered *Gogosardina* has many rows of tiny scales approaching the micromeric condition seen in *Cheirolepis* and like that form also had

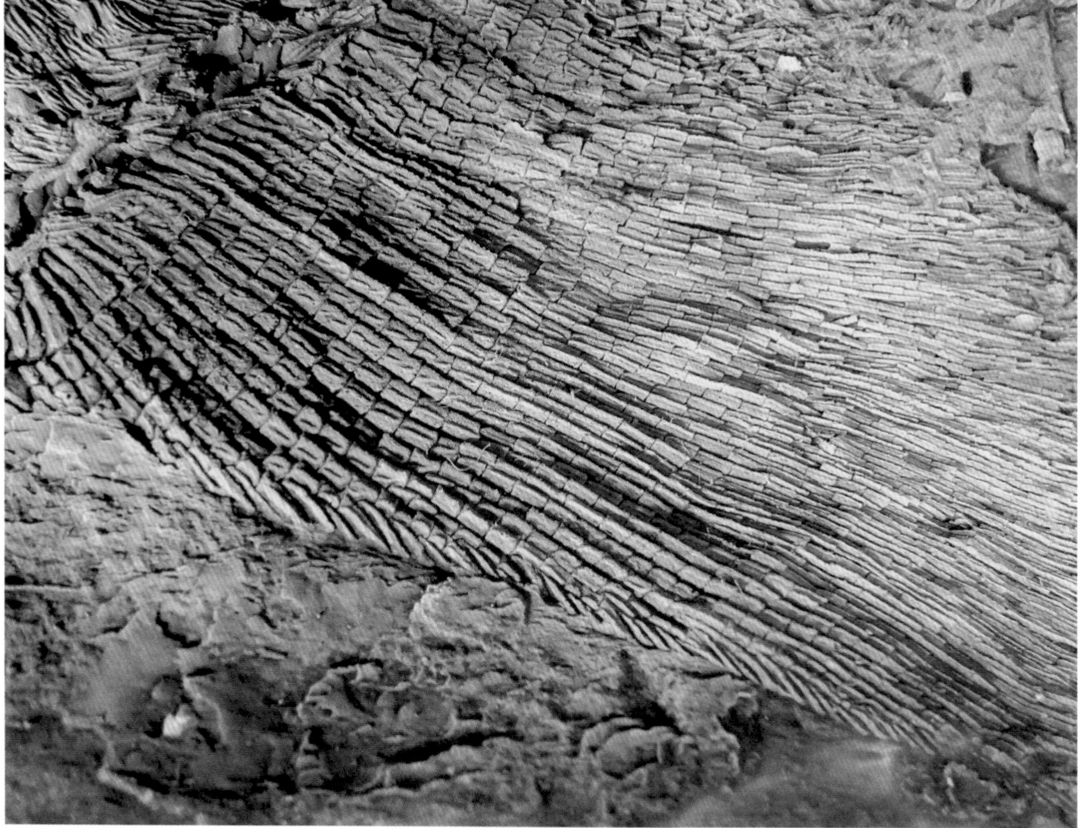

a mosaic of small snout bones. Its other skull features show it is more closely related to *Mimia* than to *Cheirolepis*.

In general, the Devonian ray-fins are quite a homogeneous group, with little variation apart from those having a micromeric scale cover in which there are several

(*Top*) A complete specimen of *Howqualepis*, from Mount Howitt, Victoria, showing the outline of the body and fins. These were freshwater fishes similar to modern-day trout that probably hunted invertebrates and smaller fishes.

(*Bottom*) Close-up detail of *Howqualepis*, showing the tightly articulated ray fins that support the anal fin.

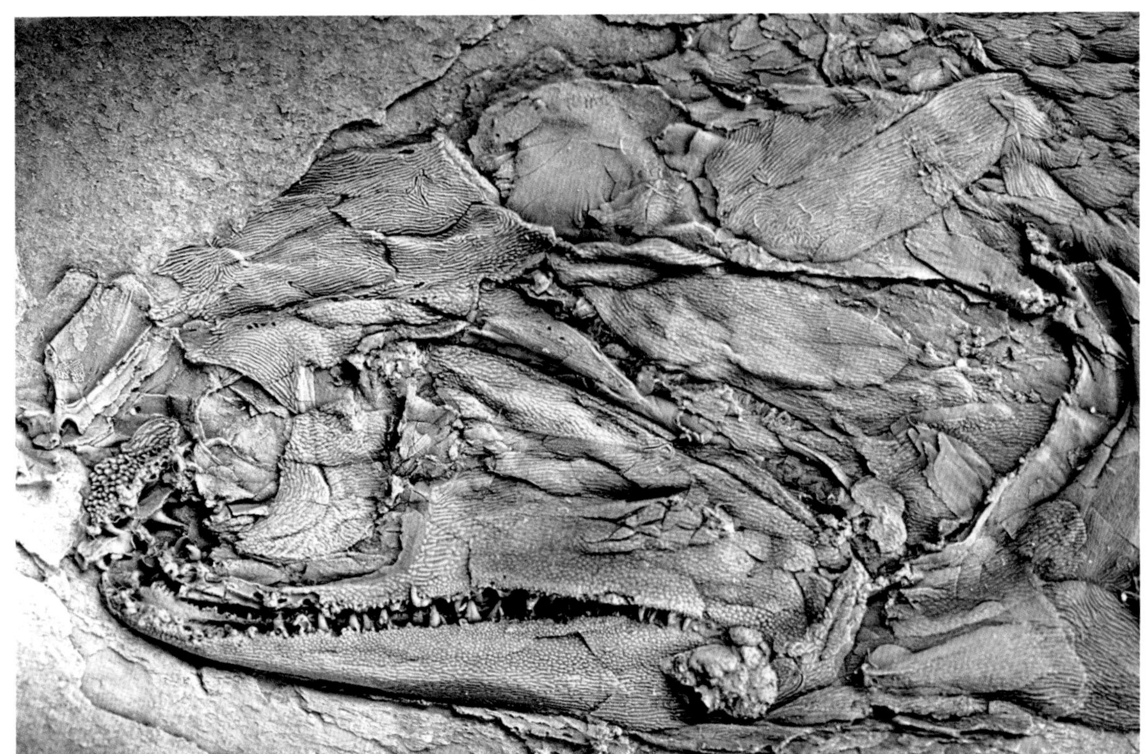

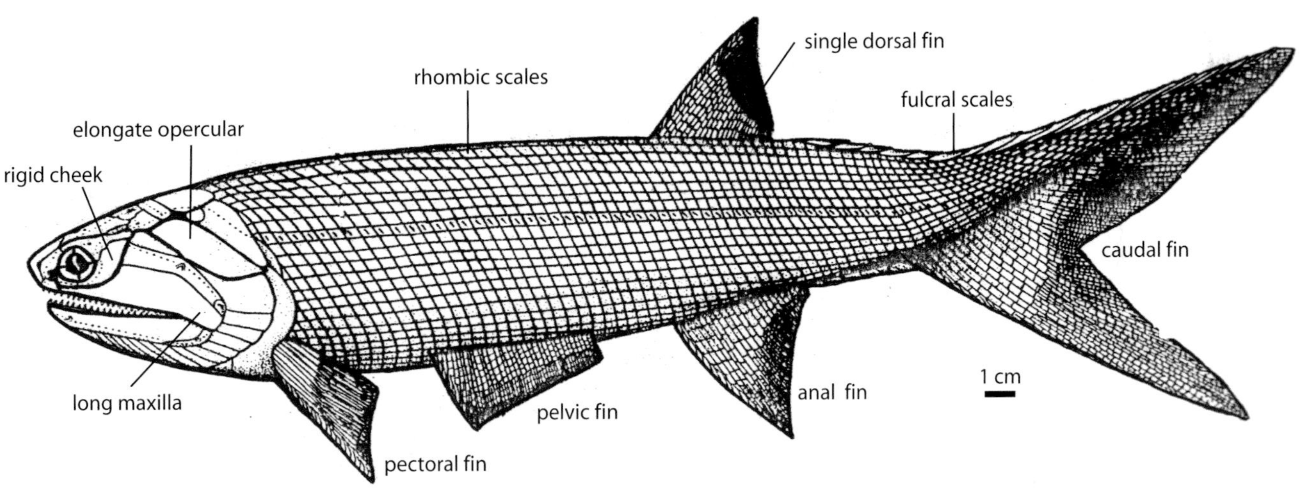

(*Top*) Detail of the head of a primitive Devonian ray-finned fish, *Howqualepis rostridens*, from the Late Devonian Mount Howitt site, Victoria, Australia.

(*Bottom*) Basic anatomical features of a basal ray-finned fish, *Howqualepis*.

scales rows to each body segment. Although most of the Devonian actinopterygians were fishes smaller than 15 cm or so, the largest, *Tegeolepis* from the Cleveland Shale of North America, was another micromerically scaled fish that reached nearly 1 m in length. *Tegeolepis* had very large teeth and a long rostrum.

The main evolutionary trends in the Devonian ray-fins are the shortening of the gape and enlargement of the opercular bones, the stabilization of the skull roof as many small bones near the snout are replaced by a set pattern of large elements (like median rostral and postrostral bones), and the change from the micromeric squamation to one of larger, rhombic scales (although the nature of the primitive state here is still disputed

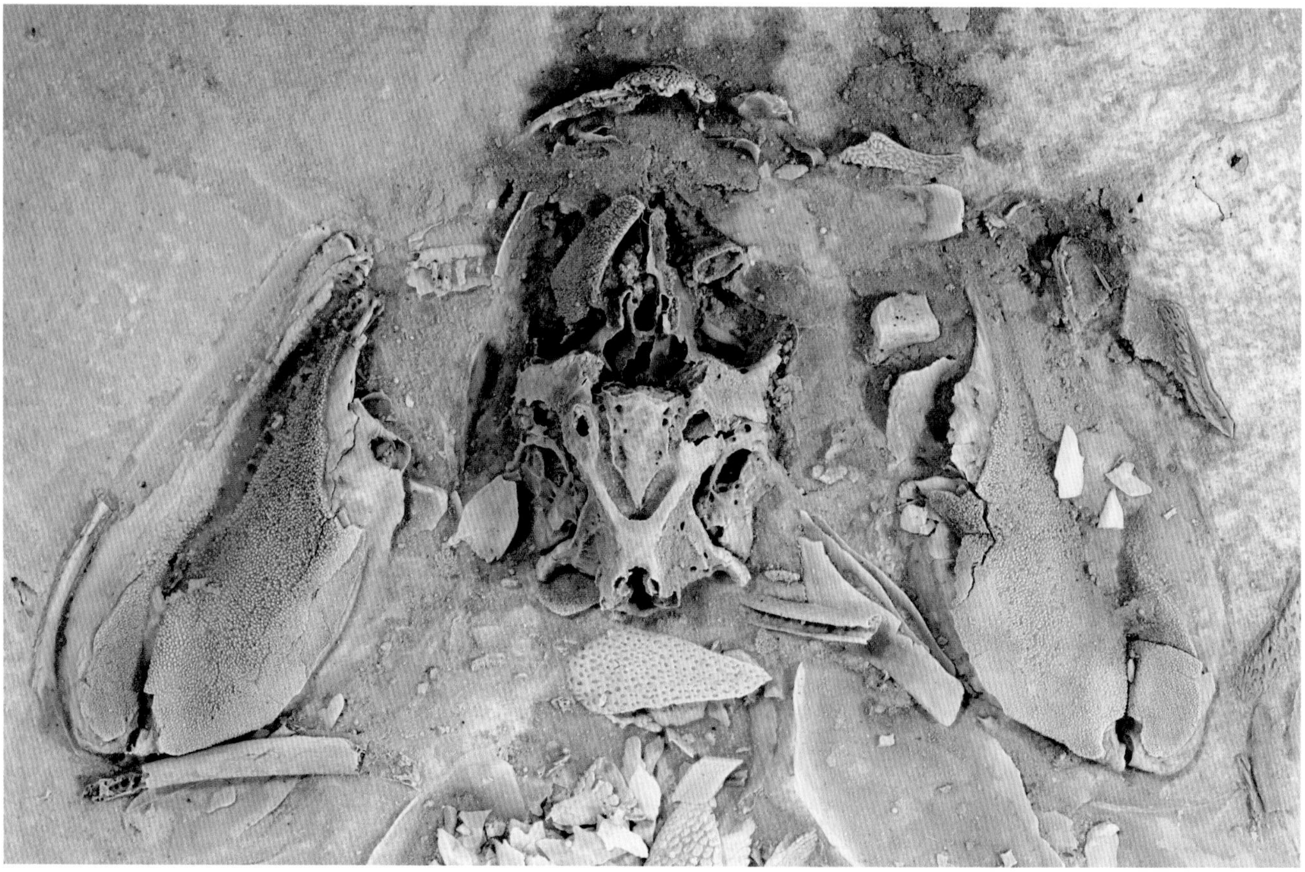

(*Top*) Reconstruction of *Howqualepis*, a fast swimming lake-dwelling predator that inhabited southern Australia 380 million years ago. (Courtesy Brian Choo)

(*Bottom*) The acid-prepared skull of *Moythomasia durgaringa* from the Late Devonian Gogo Formation, Western Australia. Note the delicate preservation of the braincase in the center of the photo. (Courtesy Brian Choo)

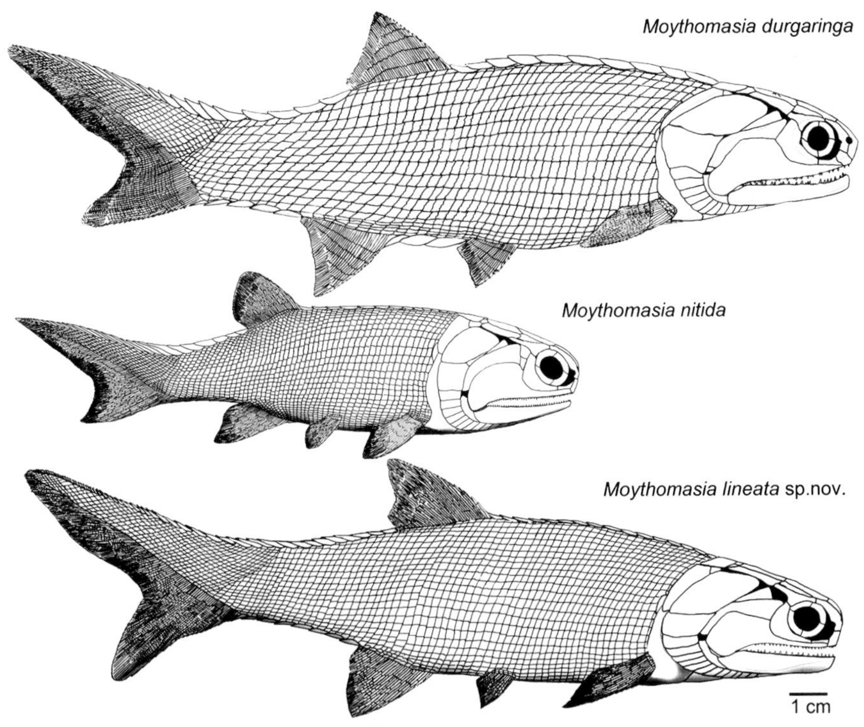

(*Left*) Comparative size and anatomical features of three species of the Late Devonian ray-finned *Moythomasia*. (Courtesy Brian Choo)

(*Bottom*) Reconstruction of *Moythomasia durgaringa* from the Late Devonian Gogo Formation, Western Australia. (Courtesy Brian Choo)

as some early forms like *Ligulalepis* had larger rhombic scales). The braincase was apparently not well ossified in the most primitive fishes such as *Cheirolepis*, but it was both perichondrally and endochondrally ossified in Late Devonian forms found at Gogo. By the Late Devonian, the first genera appeared that became widespread across the globe, such as *Moythomasia*, known from Western Australia and Europe.

There are two living representatives of this early actinopterygian radiation, the freshwater African reedfishes, *Polypterus* and *Calamoichthys*. These are elongated fishes with numerous spines forming the dorsal fin. In

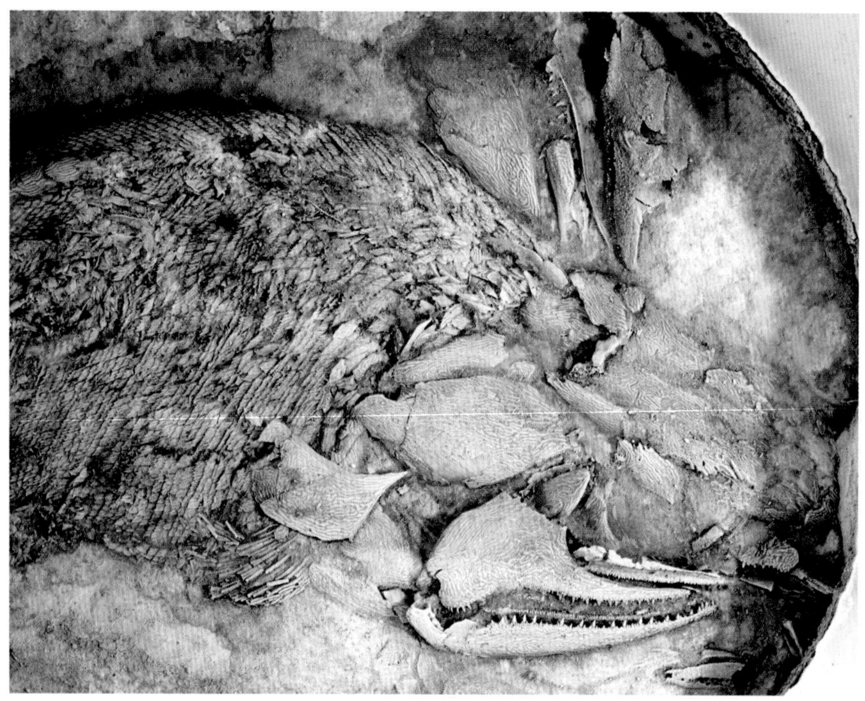

The skull of an as-yet-undescribed form of paleoniscoid fish from the Gogo Formation, Western Australia. (Courtesy Brian Choo)

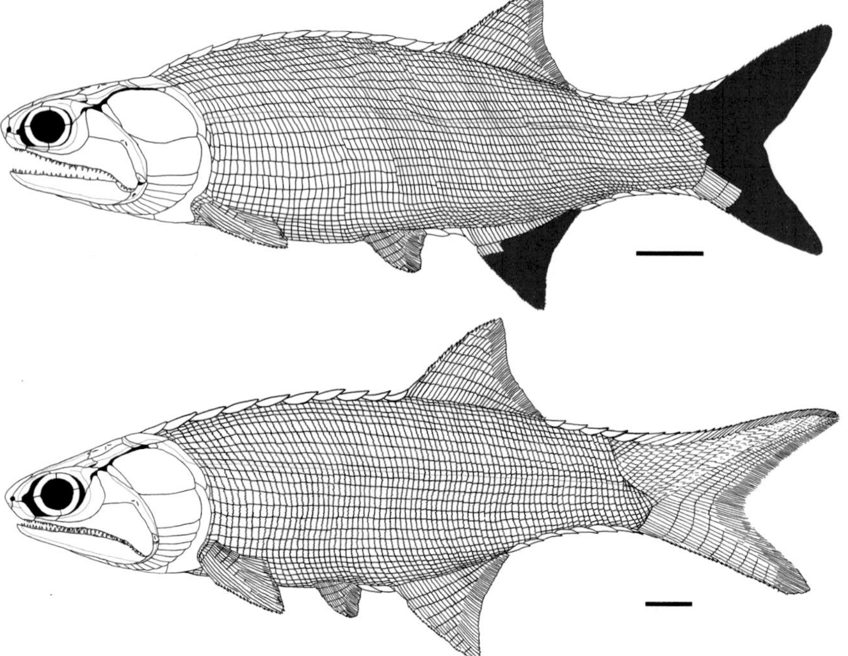

Comparison between two species of Late Devonian ray-finned fish, *Mimipiscis*. Bar = 1 cm. (Courtesy Brian Choo)

their early growth stages, the cheekbones around the eye fuse with the upper jaw. They are very primitive in that unlike all other living and fossil ray-fins they lack several anatomical features, such as a surangular bone in the lower jaw, a pectoral fin internal skeleton with the leading bone perforated, and a hemipoetic organ above the medulla oblongata in the brain.

The Rapid Rise of the Primitive Ray-Fins

The Early Carboniferous fish faunas of the world are dominated by the abundance of actinopterygians and sharks, although in most cases the latter are represented in the fossil record only by teeth and scales. The ray-fins are well preserved as whole-body fossils in

(*Left*) Part of the body and pectoral fin of *Tegeolepis*, a very large (1 m) predatory ray-fin found in the Cleveland Shale of Ohio. (Courtesy Cleveland Museum, Ohio)

(*Bottom*) The Late Devonian ray-fin *Tegeolepis* lived in the shadow of the gargantuan 6-m-long dinichthyid placoderms and 2-m-long cladoselachian sharks. (Courtesy Brian Choo)

***Polypterus*, the reedfish, inhabits lakes and rivers of Central Africa and is considered to be a living representative of the most primitive known group of ray-finned fishes, the paleoniscoids.** (Courtesy Rudy Keiter)

many deposits of the Carboniferous and Permian Periods around the world, most notably from the sites near Edinburgh (Scotland), in the Bear Gulch Limestone of Montana (United States), in Germany and Russia, and in Victoria (Australia). More than 50 families of ray-fins, comprising hundreds of recorded species, are known from the Late Paleozoic, all of which are generally lumped as "palaeoniscoids" or ray-fins of primitive organization (formerly as in "Chondrostei"). Typical troutlike fishes such as *Mansfieldiscus* from Australia indicate some forms underwent little change since the Devonian Period.

Some particularly interesting families with deep-bodied shapes and highly specialized feeding mechanisms began to emerge by the Late Paleozoic, such as the platysomid and bobasitraniid groups. One of these, *Ebenaqua ritchiei*, from the Permian of Queensland (Australia), is exquisitely preserved and known from hundreds of well-preserved specimens. It had a very small mouth, large dorsal fin, and greatly reduced anal and pelvic fins. It most likely fed on algae or weeds in the coal swamps it inhabited.

These and many other palaeoniscoids were until recently grouped in the Chondrostei, along with the two families of living primitive ray-fins, the sturgeons (Acipenseridae) and paddlefishes (Polyodontidae). Today the term Chondrostei is used only to include these two families, which both have good fossil records spanning back to the Late Cretaceous. These chondrosteans are characterized by having braincases that lack eye muscle insertion pits (myodomes); the upper jawbones (premaxillae, maxillae, and dermopalatines) are fused together and the upper jaw cartilage meets in the midline at the front of the mouth. If you looked at picture of a paddlefish, *Protosephurus* from the Early Cretaceous of

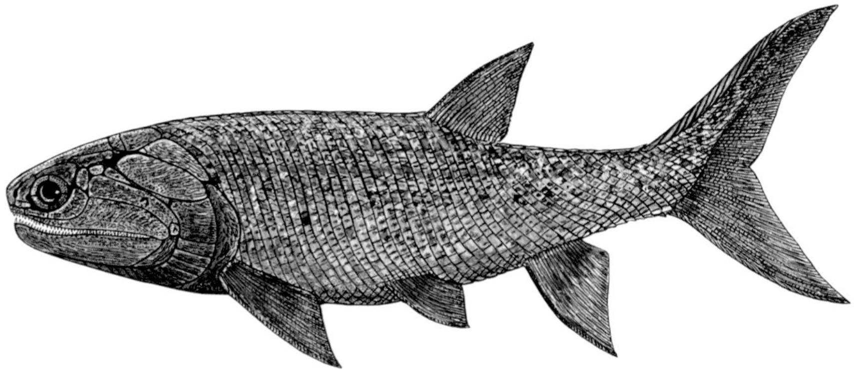

(*Left*) Reconstruction of *Mansfieldiscus* a trout-sized river-dwelling paleonisicoid fish from the Carboniferous deposits of Mansfield, Victoria, Australia.

(*Bottom*) A specimen of *Ebenaqua ritchiei* from the Permian Blackwater Coal deposits of Queensland, Australia.

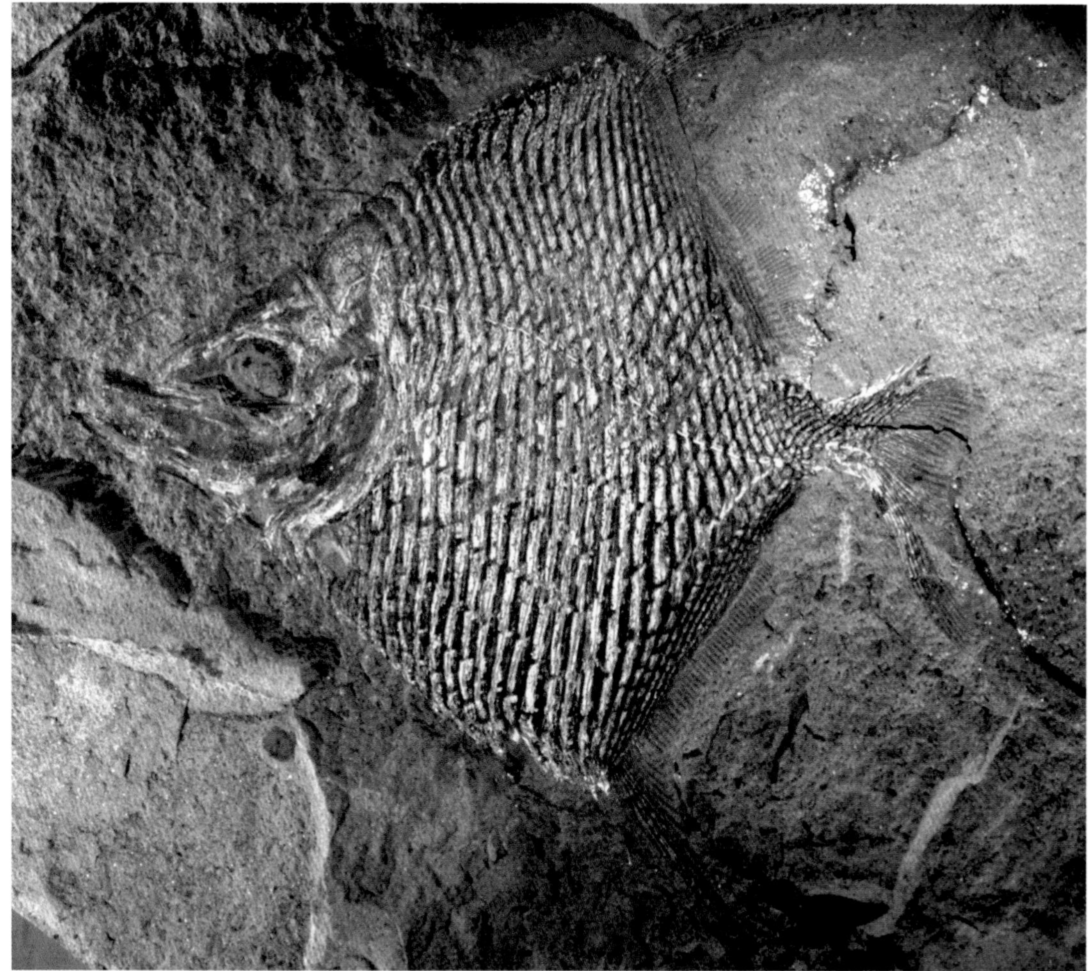

China, you could easily see that the group has changed little in appearance over the past 120 million years. Paddlefishes are today among the largest filter-feeding ray-finned fishes.

There are many other groups of primitive ray-finned fishes that enjoyed a great radiation in the Late Paleozoic and Early Mesozoic, such as the perleidiform and redfieldiform fishes. These include a variety of fossil species that have been found in Permian, Triassic, and Jurassic deposits around the world. Most of the well-preserved fishes from the Triassic of the Sydney Basin, Australia, fall into one of these groupings, such as *Dictopyge* (a redfieldiform), *Cleithrolepis,* and *Manlietta* (perleidiforms).

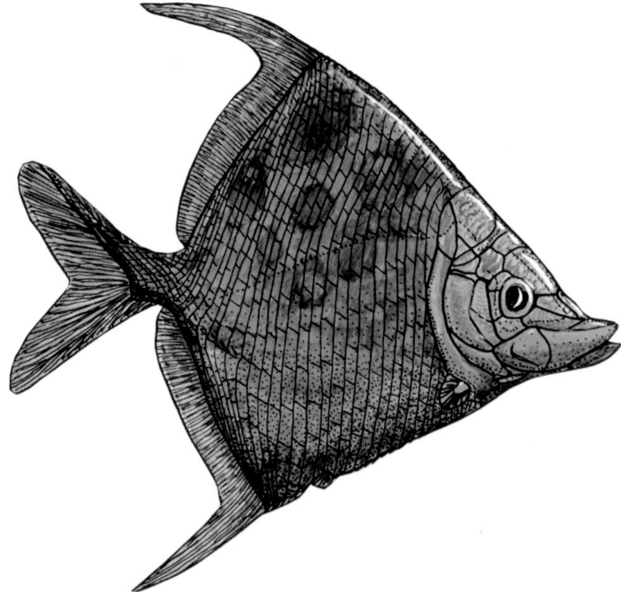

(*Left*) Reconstruction of *Ebenaqua ritchiei*, a deep-bodied bobasatranid fish from the Permian coal measures of Southern Queensland, Australia.

(*Bottom*) A living sturgeon (*A*), *Acipenser*, one of the most basal of all known ray-finned fishes. Its largely cartilaginous internal skeleton caused early anatomists to name their group the Chondrostei, or cartilage fishes. An Eocene garfish (*B*) from the Green River Shales of Wyoming. Garfish like this have not changed significantly since the start of the Cenozoic Era. (Panel A, courtesy Rudy Kuiter, Museum Victoria)

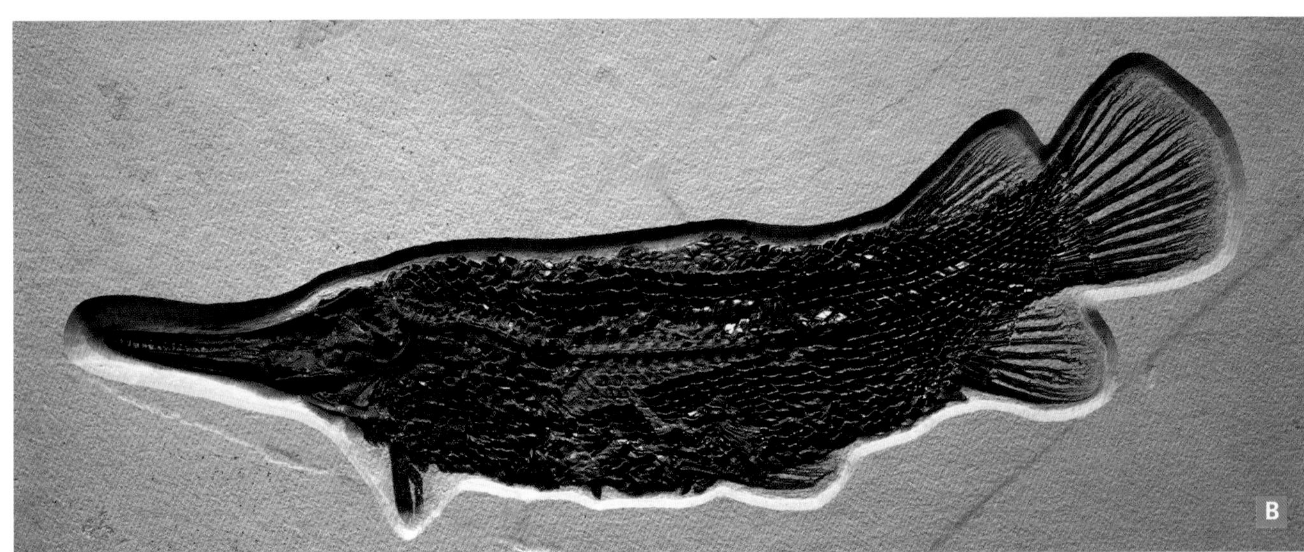

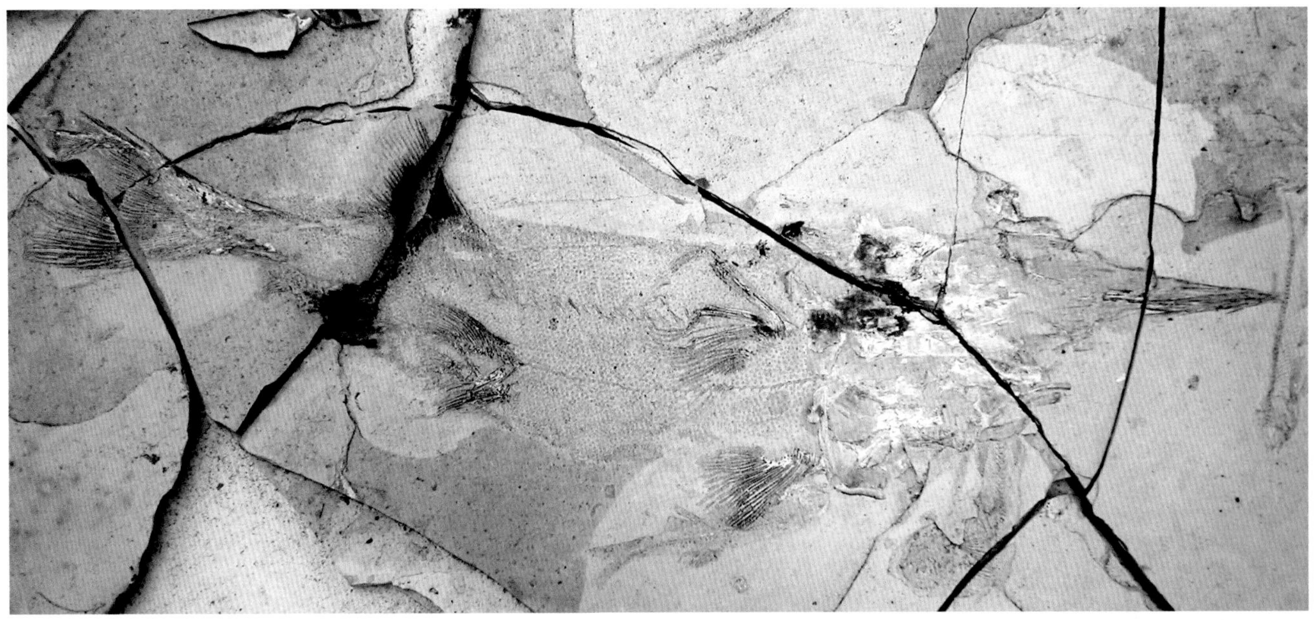

(*Top*) *Protosephurus*, an Early Cretaceous paddlefish from northern China.

(*Bottom*) *Cleithrolepis*, a deep-bodied ray-finned fish as found in the Triassic sandstones near Gosford, New South Wales, Australia.

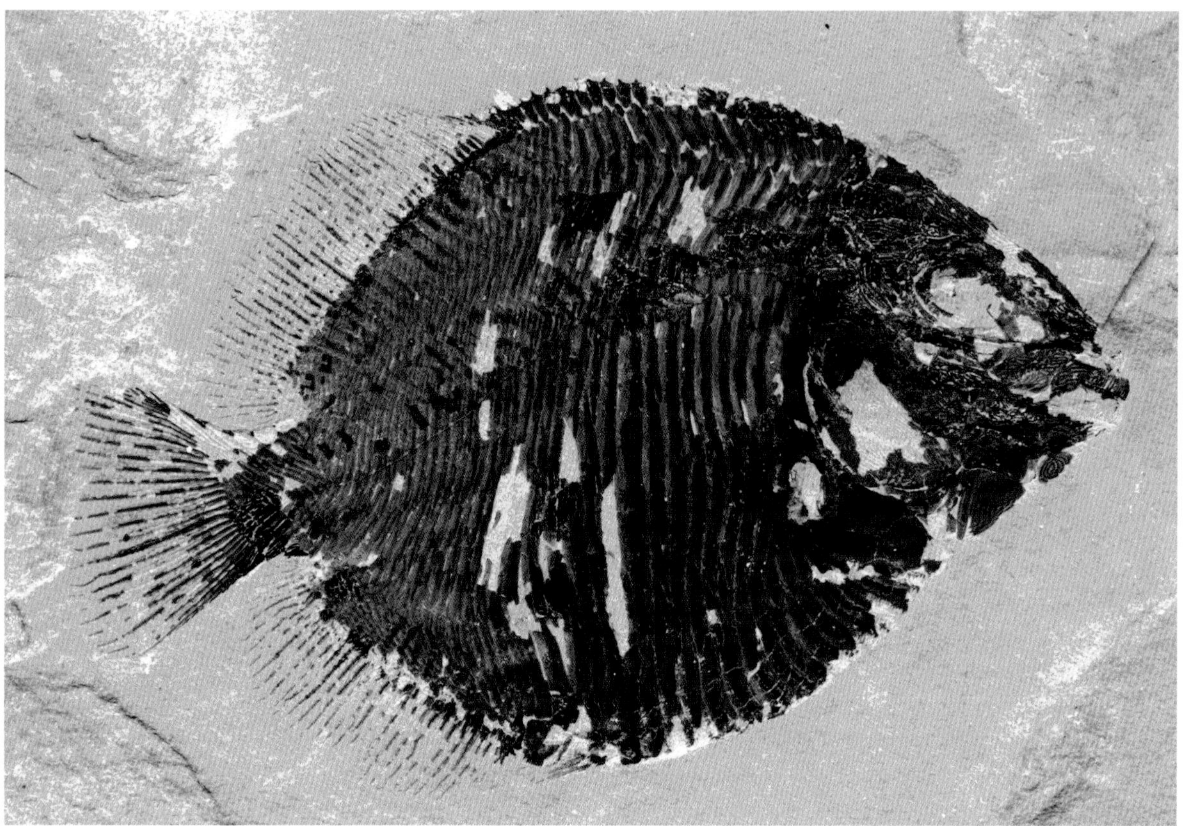

***Discoserra*, from the Early Carboniferous Bear Gulch Limestones of Montana, is now regarded as the earliest member of the modern radiation of ray-finned fishes known as neopterygians.** (Courtesy of Matt Friedman, Oxford University)

Neopterygii: The Dawning of Success

By the Late Paleozoic, there were many successive higher levels of actinopterygian evolution. It was then that actinopterygians acquired a vertical suspensorium, where the lower jaws articulate with the upper jaws by a vertically oriented quadrate bone. Also the cheekbones that were firmly united with the upper jaw in primitive forms became freed, allowing a great variety of feeding mechanisms to evolve. Many accessory bones evolved in the cheek region, giving endless patterns of bones in the heads of the different groups. However, the biggest innovations were in the fin and tail skeletons.

The advanced group of actinopterygians called Neopterygii have fin rays of the dorsal and anal fins equal in number to their support bones, and the upper jawbones are fused in the midline. Inside the mouth, they have well-developed pharyngeal tooth plates that have been consolidated into stout bones for grinding up food as it passes into the back of the gullet. Within the Neopterygii are two main groups, the Ginglymodi, containing the garfishes, and the Halecostomi, containing bowfins (amiids) and teleosteans. Grande and Bemis (1998) wrote a seminal work on the amiid fishes combining fossil and extant data sets to create a robust phylogeny of the group.

Recent research has shown that the Neopterygii might have had an early origin as far back as the Carboniferous Period with forms like *Discoserra* from the Bear Gulch Limestone of Montana (Hurley et al. 2007). From within the early Neopterygii emerged a small group of tiny fishes around the middle of the Triassic Period, the first of the teleosteans. The subsequent diversification of the teleosteans (see Chapter 9) was the greatest radiation event in vertebrate evolution and one that survived several global mass extinction events almost unscathed.

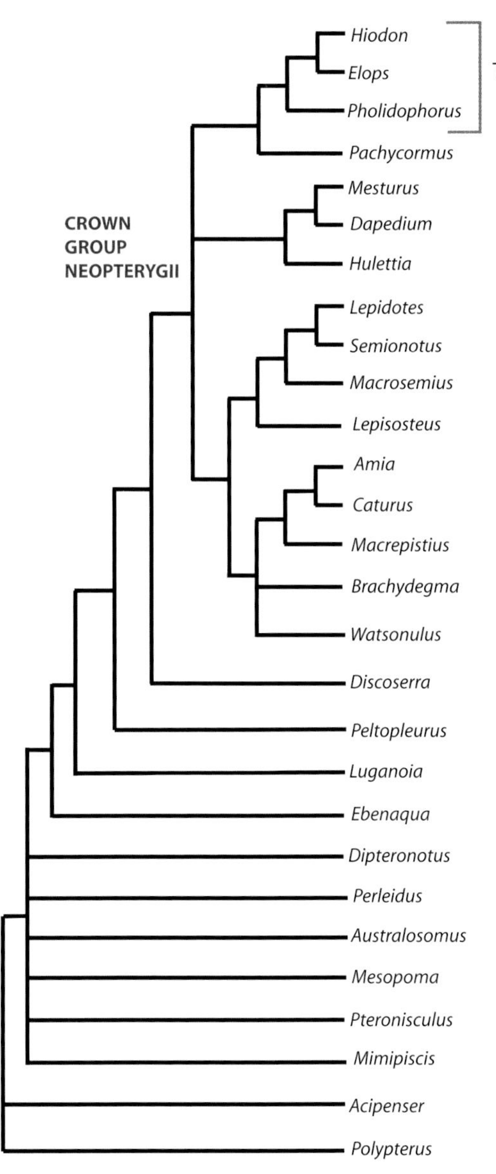

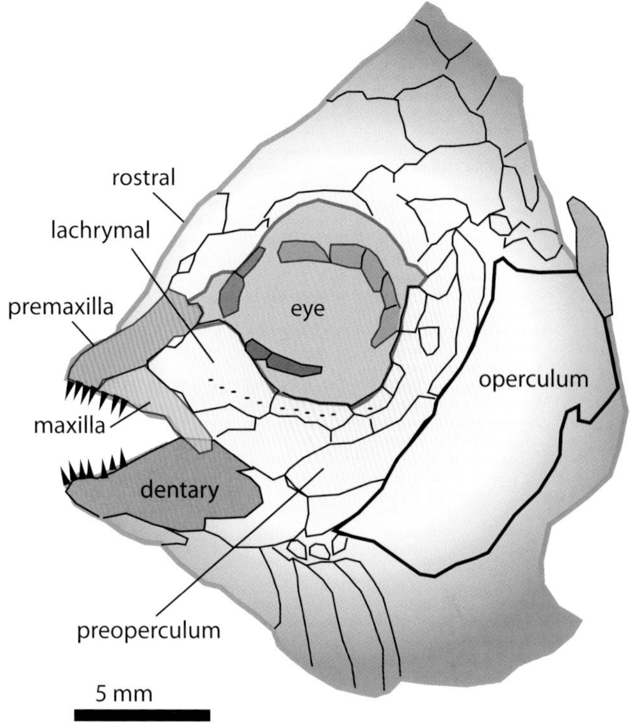

(*Left*) A recent scheme of hypotheses showing the major evolutionary lineages of ray-finned fishes leading to neopterygians and teleosteans. (From Hurely et al. 2007)

(*Above*) Fossilized body parts, including skulls, a jaw, and a tail, of *Discoserra*, the oldest neopterygian known.

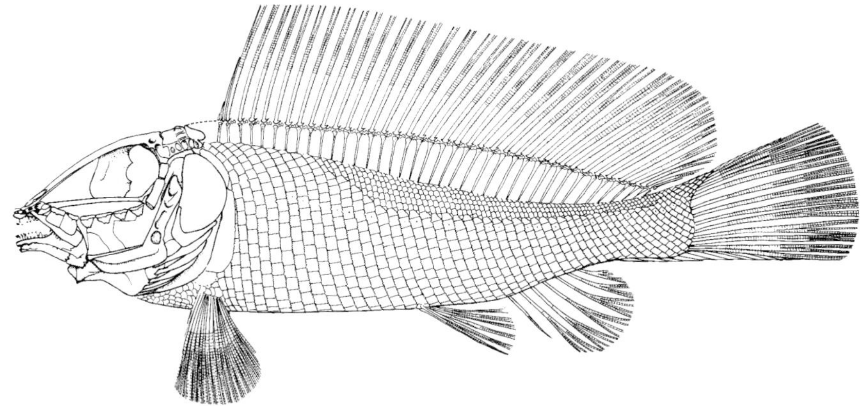

***Macrosemius*, a halecostome fish from the Late Jurassic / Early Cretaceous. Halecostomes were a major group of ray-finned fishes in the Mesozoic Era that were closely related to the gars.**

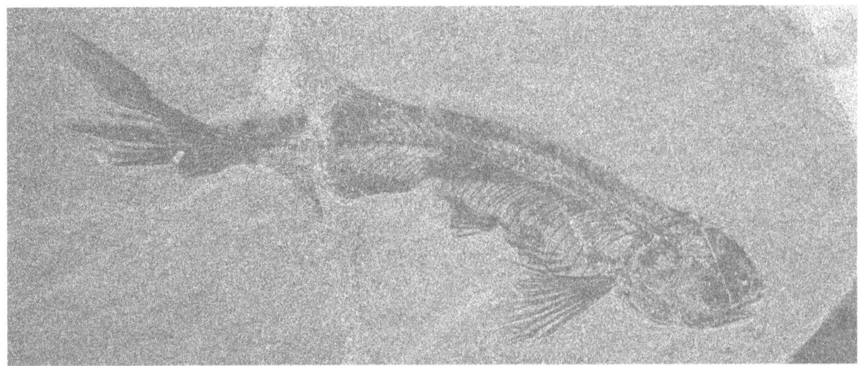

CHAPTER 9

Teleosteans, the Champions

The origin and diversification of the world's most successful vertebrates

During the early part of the Mesozoic Era, the ray-fins evolved as the most efficient swimmers and feeders, their cheekbones having been decoupled from their toothed jawbones and their tail skeletons having been greatly improved. The latter feature characterizes the teleosteans, fishes that became the most successful group of vertebrates to modern times, with some 28,000 species, accounting for about 96% of all living fishes. Teleosts had their humble origins back just after the age of the dinosaurs, with fishes like *Pholidophorus* possibly appearing in the Late Triassic some 215 million years ago. By the Jurassic they were well established in both marine and freshwater environments. The percomorphs, the largest of the living groups of teleosts, evolved by the Late Jurassic Period. By the dawn of the Cenozoic, many of the living families of fishes had appeared. The first teleost fishes specially adapted for life on coral reefs were evident by the Eocene. The hectic roller-coaster ride of actinopterygian evolution had finally begun slowing down, although great diversification keep going on particularly at species and generic levels to result in today's tally.

As we discussed in the last chapter, teleosts inhabit the widest range of habitats for any vertebrate group on Earth, from high mountain hot springs down to the deepest abyssal depths of the oceans and from hot soda lakes to Antarctic chill. Certain species live in caves and have lost eyes, whereas other forms have lost their scales or sometimes even their paired fins. The group shows an extraordinary range of plasticity in body form, shape, and feeding mechanisms, and it is this key ability to adapt and change that has made them the evolutionary success they are today.

Teleosts have been defined as a natural (monophyletic) group based on some 27 specializations of their anatomy by de Pinna (1996), although each person who attempts to define teleosts needs to be clear on where they are drawing the boundary line. Including fossil forms complicates the demarcation and becomes an arbitrary exercise in where one wishes to make the boundary. There has been some controversy over which fossil groups should be included in the teleosts, but most scientists today are happy to define the clade as including all groups above the level of *Pholidophorus* from the end of the Triassic. Key features defining the teleosts include the presence of certain bones in the tail skeleton, called uroneurals, which function to stiffen the dorsal lobe of the tail and support a series of dorsal fin rays. This simple innovation gave the group greater swimming power

The swordfish *Xiphias* (*A*) is representative of the fishes that are at the peak of ray-finned fish evolution, the teleosteans. Through modified tail fin skeleton and highly flexible feeding mechanisms, they have radiated into becoming the dominant group of backboned animals on the planet today. The Late Jurassic giant, *Leedsichthys* (*B*), was a filter-feeding pachycormid teleost whose remains have been found in the Oxford Clay of Great Britain. Despite some claims of its reaching 20 m or more, specialists who study these fishes now put its maximum size at around 10–12 m. (Courtesy SeaPics; Brian Choo)

and allowed a great variety of body shapes to evolve. But teleosts also have extra features that other fishes lack. They underwent modifications to the jaws, making the front toothed bones of the upper jaw, the premaxillae, free to move independently of the main upper jawbone, the maxilla. Their large ventral gill arch bones, the basibranchials, carried unpaired tooth plates, and teleost head muscles also had many unique characteristics.

Early Beginnings of Teleosts

In a review of basal teleosts, Gloria Arratia (1997) stated that "the question of when the teleosts appeared is still unanswered because there is no agreement about which are the most plesiomorphic (basal) teleostean taxa." The elopomorph and osteoglossomorph fishes are widely regarded as the most primitive of all living teleosts, but there is recent debate over which of these groups is more basal. Arratia believes that elopomorphs are the living sister group to all other teleosts, and many follow this view. Fossil evidence points to the herring-like *Pholidophorus*, from the Late Triassic / Early Jurassic of Europe, as being the most primitive of all teleosts, but this is open to further discussion. The oldest unambiguous members of the group are of Early Jurassic age and include such basal forms as *Proleptolepis* and *Cavenderichthys*.

Many primitive fossil teleosts are known from the Mesozoic Era. *Leptolepis*, for example, a common genus known from many species around the world in Jurassic and Cretaceous freshwater deposits, has a troutlike appearance but when the head is examined in detail it shows many primitive features. The premaxilla is tiny and still tightly bound to the maxilla, and there are additional supramaxilla bones above the maxilla. Fishes like *Leptolepis* were highly successful, being able to migrate around the world and no doubt capable of living in both marine and freshwater environments, perhaps favoring

Pholidophorus latisculus, perhaps the most basal of all teleosteans, from the Late Triassic of Italy. (Courtesy Anna Paganoni, Museo Civico di Scienze Naturali, Bergamo, Italy)

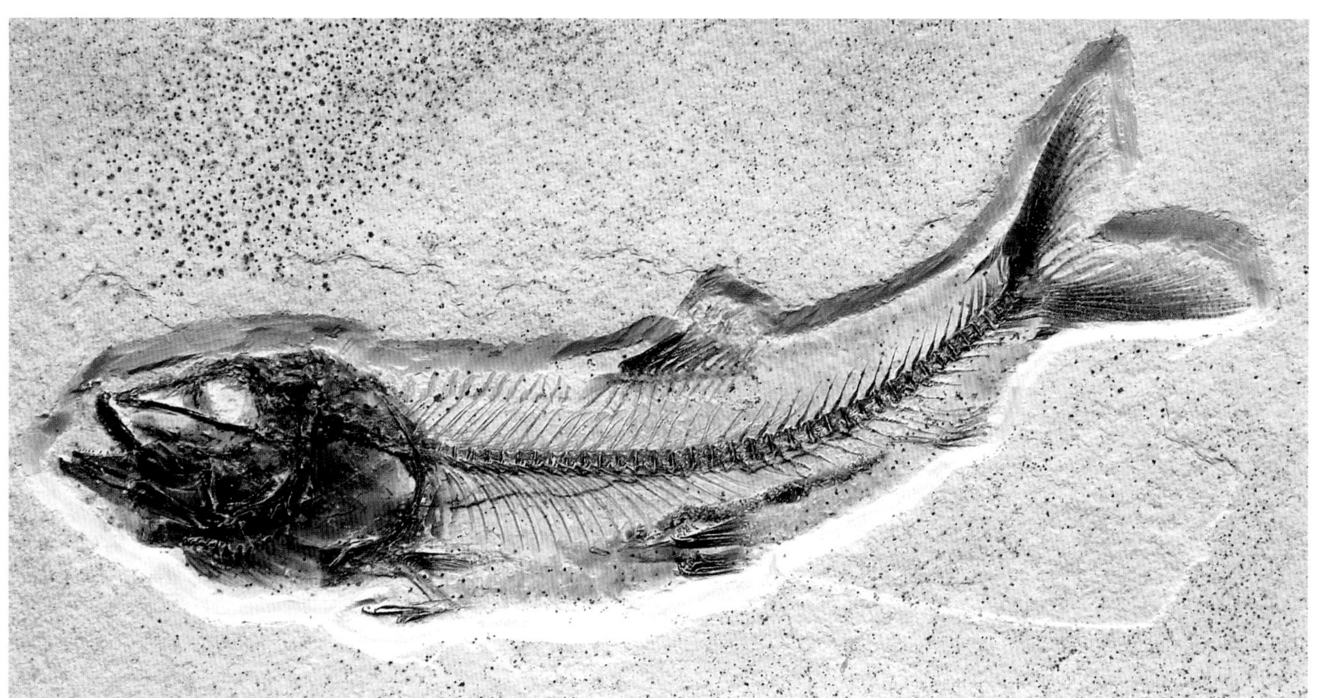

(*Top*) An unnamed basal teleost form the Early Jurassic of Germany. These herring-like fishes were more advanced than their ancestors in having modified tail-fin skeletons (Courtesy Museo Civico di Scienze Naturali, Bergamo, Italy)

(*Bottom*) *Cavenderichthys talbragarensis*, from the Late Jurassic of New South Wales, Australia, one of the first of the advanced ray-fins (teleosteans).

The genus *Leptolepis* was once though to be the most widespread basal teleost, found in sites around the world from Late Jurassic / Early Cretaceous age. Major revisions have now reclassified many of the species into other genera. Others like this *Leptolepis koonwarriensis*, from the Early Cretaceous of Koonwarra, Victoria, Australia, remain to be restudied.

the latter for breeding, like many living fishes such as the salmon.

Osteoglossomorphs: Bony Tongues

The osteoglossiform (from Greek words meaning "bony tongues") branch of osteoglossomorphs are omnivorous-to-predatory freshwater fishes and among their members are some of the largest and most spectacular piscine predators in our great rivers today, like the 2.5-m *Arapaima gigas* of the Amazon River in South America. Fossil forms like *Phaerodus* from the Eocene Green River Shales of Wyoming are common examples of the types of fishes that once inhabited Tertiary lake and river deposits.

The early diversification of fossil osteoglossomorphs may include the giant predatory marine forms like the Ichthyodectiformes. They were active, free-swimming hunters that may have preyed on other fishes as well as unwary smaller marine reptiles, including young ichthyosaurs and plesiosaurs. Large, well-preserved skulls of these have been found in central Queensland (for example *Cooyoo australis* and *Pachyrhizodus marathonensis*) and, when prepared in three dimensions, they reveal magnificent details.

Xiphactinus was an open ocean hunter that reached sizes of almost 6 m. The most famous example of this fish is *Xiphactinus audax*. One specimen, found in Kansas, in the United States is worthy of note, as it is some 4 m long and contains another 2 m fish (*Gallicus*) inside it, its last meal. It is now on display at the Sternberg Mu-

The Mystery of the Fossil Flatfishes

An age-old puzzle of evolution, one first pondered over by Charles Darwin himself, was how the asymmetrical flatfishes, such as the halibut, sole, and flounder (order Pleuronectiformes) might have evolved from symmetrical free-swimming fishes. Research by Matt Friedman published in the journal *Nature* in 2008 has solved this mystery. Friedman found that two fishes, *Amphistium* and *Heteronectes*, from the Eocene of Europe, had incomplete orbital migration, where one eye had moved farther around to the one side of the face than the other. Thus it was that the evolution of profound cranial asymmetry was a gradual process with intermediate stages leading to modern flatfishes captured by fossils such as *Heteronectes*.

Heteronectes, an Eocene fish that shows an intermediate phase of evolution between regular fish with symmetrical heads and the modern-day flatfishes with marked asymmetry. (Courtesy Matt Friedman, Oxford University)

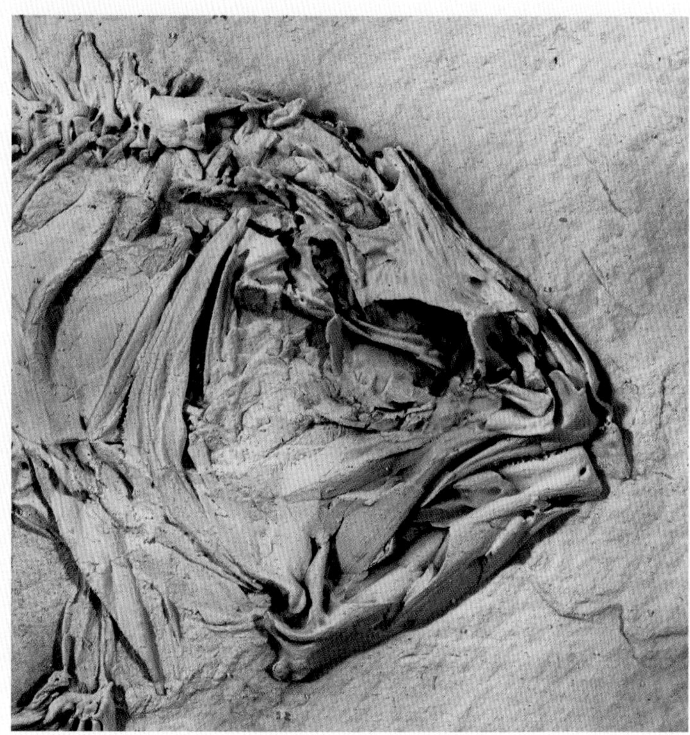

seum in Hays, Kansas. Famous American paleontologist and dinosaur hunter, Edward Drinker Cope, wrote about this fish in 1872 as follows: "The head was as long or longer than that of a fully grown grizzly bear, and the jaws were deeper in proportion to their length. The muzzle was shorter and deeper than that of a bull-dog. The teeth were all sharp cylindric fangs, smooth and glistening, and of irregular size. . . . Besides the smaller fishes, the reptiles no doubt supplied the demands of his appetite."

One of the most abundant small fish fossils of this age comes from the shaley lake deposits of the Jehol Bi-

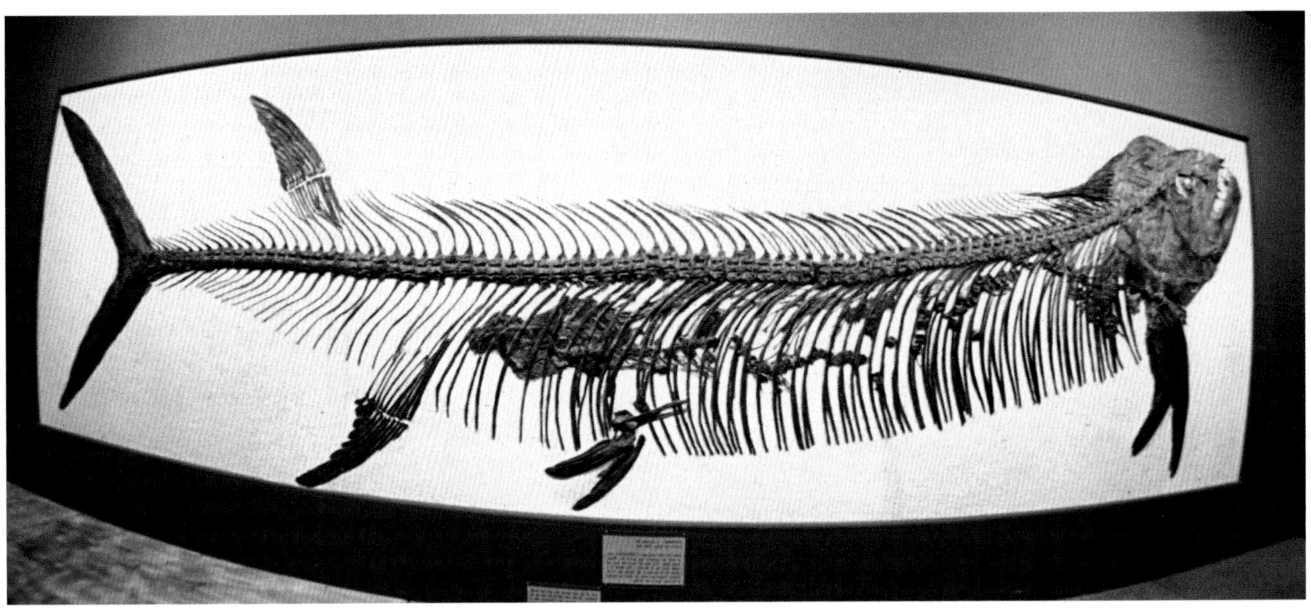

ota in northern China (Liaoning Province). These little sardine-like fishes, named *Lycoptera davidi* are regarded by some researchers as the sister taxon to all living osteoglossomorphs.

The hiodontids, or "mooneyes," are another primitive group of living osteosglossomorphs. These feisty freshwater fishes are well known by North American

(*Top*) This gigantic specimen of the predator Cretaceous fish *Xiphactinus audax* had eaten another large fish called *Gallicus*. It was found by Charles Sternberg in 1870 and measures about 4.2 m in length. This famous display of "the fish within a fish" can be seen at the Sternberg Museum in Hays, Kansas.

(*Bottom*) A reconstruction from the Cretaceous Period of *Xiphactinus* eating a *Gallicus* half its own size (Courtesy Brian Choo)

The head of *Calamopleurus*, from the Early Cretaceous Santana Formation of Brazil. *Calamopleurus* was a fast-swimming predator. (Courtesy John Maisey, American Museum of Natural History).

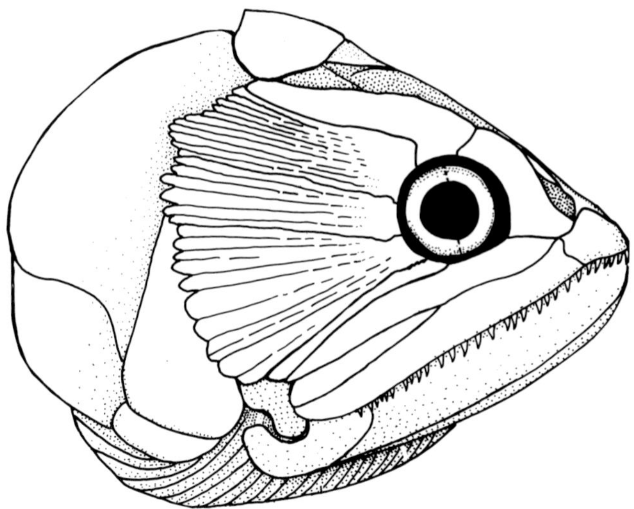

The head of *Pachyrhizodus*, a 2-m-long fish that inhabited the inland seas of Australia around 100 million years ago.

anglers for their golden eye colors. They are represented by various fossil forms, such as *Plesiolycoptera* and *Yanbiania* from the Late Cretaceous of China and *Hiodon* (formerly *Eohiodon*) from the Eocene Green River Shales of Wyoming.

Among the other extant families of osteoglossomorphs are the bizarre elephant fishes or mormyrids of tropical Africa and the Nile, so named for their downward-projecting snout that resembles an elephant's trunk. Some of these fishes can transmit and detect weak electric currents, a useful ability as they are active primarily at night.

Elopmorphans: Elongating Evolution

The elopomorphan fishes encompass some 856 species. Many are well known, such as the tarpons (*Megalops*), bonefishes, and many kinds of eels (anguilliforms), including a variety of specialized deep-sea monsters like the gulper eels. Modern molecular work on this group has shown them to be a natural (monophyletic) grouping in recent years (Inoua et al. 2004). They have fossil record spanning back to the Jurassic, with forms like *Anaethelion* and various *Elops*-like species described from deposits in Germany (Arratia 1997). The main features that define these forms and link them to modern representatives of the subdivision are the presence of elongated jaws with numerous needle-like teeth and having a special ribbon-like larval phase and a certain compound neural arch bone in the tail.

Albuloids, represented by the living genera *Albula* and *Pterothrissus*, make up the primitive sister taxon to the group containing eels. *Albula* has a fossil relative *Lebonichthys*, which occurs in the Late Cretaceous of Lebanon, and *Pterothrissus* has a relative from around that era that is very similar to it named *Isteius*. The more primitive fossil relatives of this clade could include *Brannerion* from the Early Cretaceous Santana Formation of Brazil and *Osmeroides* from the Middle-Late Cretaceous (Forey et al. 1996).

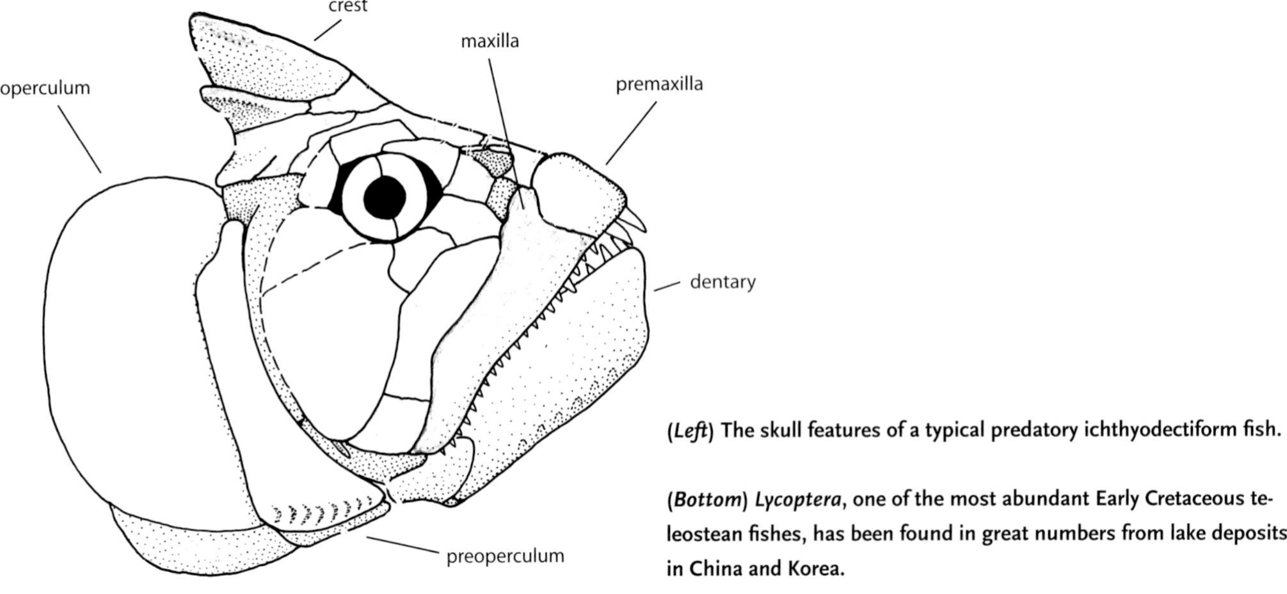

(*Left*) The skull features of a typical predatory ichthyodectiform fish.

(*Bottom*) *Lycoptera*, one of the most abundant Early Cretaceous teleostean fishes, has been found in great numbers from lake deposits in China and Korea.

Anaethelion was a small herring-like fish that lived in the shallow seas of Germany around 150 million years ago; its remains have been found in the Solnhofen deposit of Bavaria. Similar closely related fishes occur in the same deposit that resemble the modern *Elops* but are slightly different in skeletal features. Ribbon-like larval fishes in the same deposits testify that these fishes did develop in a similar way as modern elopomorphans. The leptocephalus larva is unique to these fishes today.

The eels (Anguilliformes) have undergone a great ra-

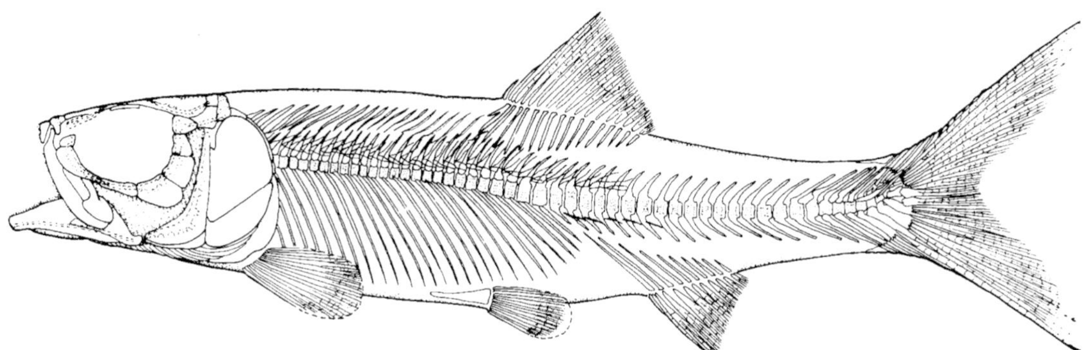

(*Left*) ***Heterotis niloticus***, **the African bonytongue, is a living osteoglossomorph fish. Often kept in aquariums, in the wild it is a filter-feeder.** (Courtesy Rudy Kuiter, Museum Victoria)

(*Bottom*) ***Anaethlion***, **a small Jurassic teleost from Germany. This restoration of its skeleton by Gloria Arratia shows the freeing up of the bones in the cheek to allow for protrusible feeding action.**

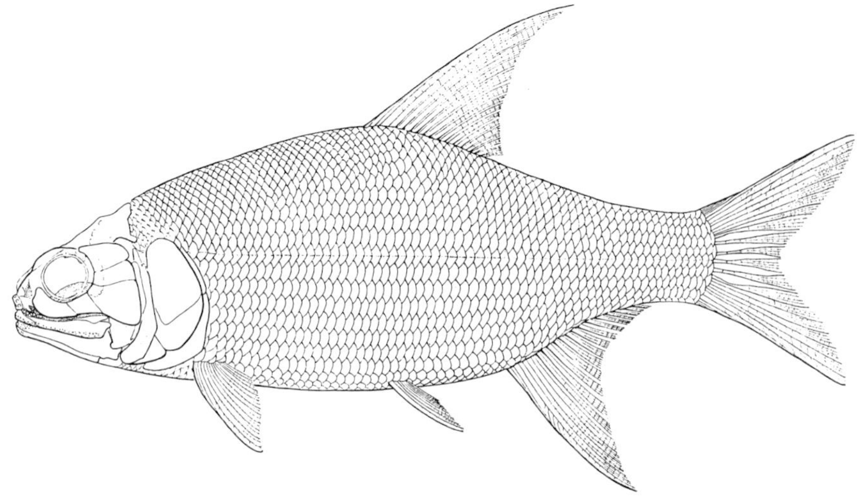

A restoration of the skeleton of *Brannerion* from the Early Cretaceous of Brazil. (Courtesy John Maisey, American Museum of Natural History)

The tail skeleton of *Brannerion*, a Cretaceous teleost, shows the one of the important specializations that enabled the group to become so successful—enlarged tailbones called hypurals. (Courtesy John Maisey, American Museum of Natural History)

diation into many families since their first appearance in the fossil record about 90 million years ago in the Cretaceous, represented by forms such as *Anguillavus, Urenchelys,* and *Enchelion,* from the Late Cretaceous of Lebanon, although none of these can be confidently placed in any living families of eel. Primitive features can be seen in some of these early fossil eels, such as the retention of paired pelvic fins and separate tail (caudal fin) on *Anguillavus*. Beautiful examples of eels of the modern type occur in the shallow marine Monte Bolca deposits of Italy, dated at around 50 million years ago. Today eels are widespread on the tropical reefs of the world, where morays of various shapes and colors dart in and out of the crevices ambushing prey. Garden eels (heterocongrinids) bury themselves in tropical reef sands and pop out to filter-feed on plankton. Eels

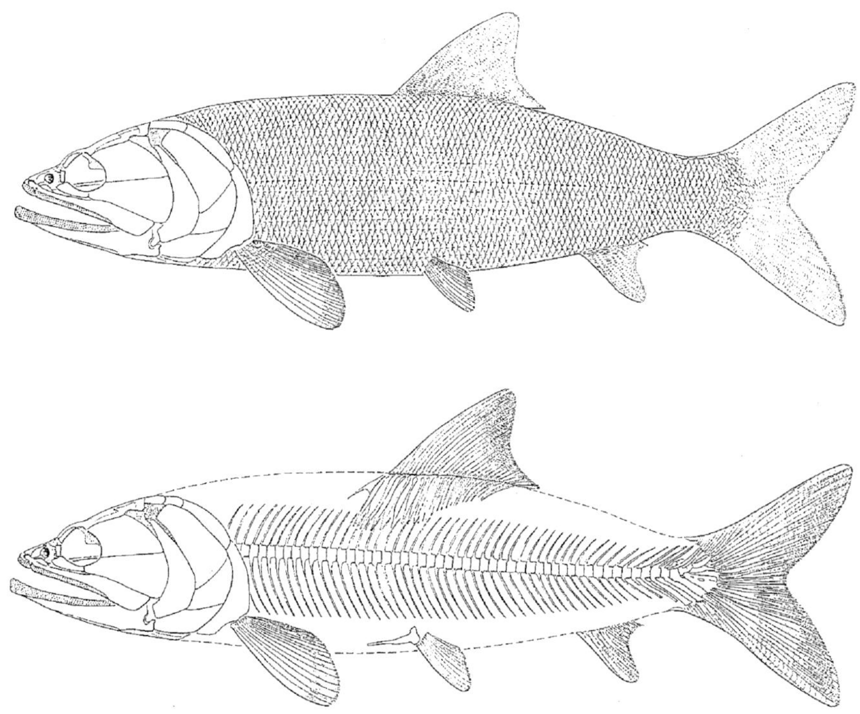

***Paraelops*, a Cretaceous member of the Elopiform teleosteans.** (Courtesy John Maisey, American Museum of Natural History)

The moray eel, *Muraena*, was one of the first fishes ever described and its method of reproduction was researched by Aristotle around 325 BC. Eels have a fossil record originating in the Mid-Late Cretaceous Period. (Courtesy iStock)

are highly specialized in their reduction or loss of body scales and loss or reduction of paired fins. Moray eels are highly specialized in other ways, like having large teeth on their internal gill arch bones (ceratohyals and epihyals) that can project out when they bite to make up another set of "biting jaws" for gripping prey and pulling it back into their mouths, making them unique and formidable predators.

Ostarioclupeomorphans: An Empire of Catfishes and Their Kin

This unwieldy name merges two major groups of teleosts together, the Clupeomorpha and the Ostariophysi, based on strong supporting evidence from molecular data, anatomy, and paleontological finds (Arratia 1997). The main anatomical feature of the group is the presence of a Weberian apparatus, whereby the swim-bladder has a direct connection to the braincase to transfer and magnify sounds. A range of fossil forms, including the well-known *Diplomystus* from the Eocene Green River Shales of Wyoming, are placed in an extinct family the Paraclupeidae. This grouping of extinct genera such as *Armigatus* and *Sorbinichthys* ranged from the Lower Cretaceous to the Eocene.

Clupeomorphans are a very important food group of fishes that contain the modern herrings, sardines, pilchards (Clupeidae), and anchovies (Engraulidae). Fossil representatives of these tasty little teleosts include Miocene anchovies from Cyprus (Grande and Nelson 1985) and a range of fossil herrings dating back to the Paleocene. One of the commonest fossil fishes in the

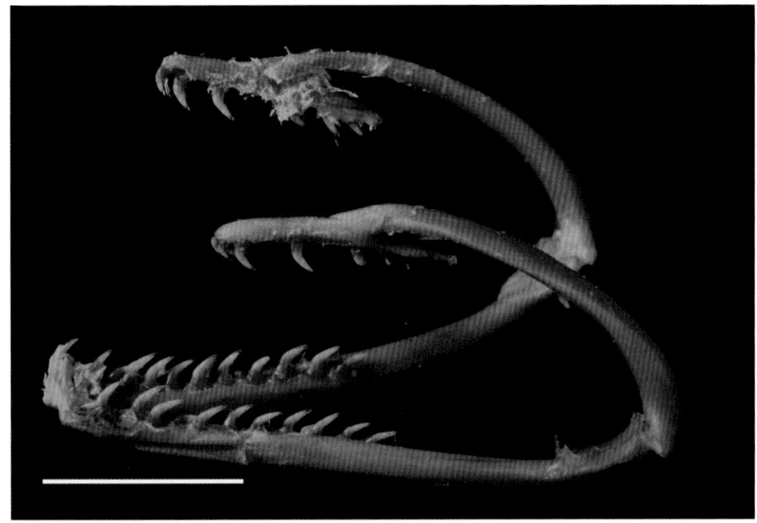

(*Left*) Moray eels have a second set of jaws in their throats that aid in securing their prey and dragging it into the throat after it is seized.

(*Bottom*) *Diplomystus*, a common clupeomorph fish found in the Eocene Green River Shales of Wyoming.

world, gracing museums and fossil shops worldwide, is the herring *Knightia*, from Paleocene-Eocene of western North America and China. It is provisionally placed as a member of the Pellonulines, or freshwater herrings, and as the sister taxon to the living herring *Clupea* by Chang and Maisey (2003). Whole schools of these fishes occur covering bedding planes of Green River Shales were unearthed in commercial fossil quarries in Wyoming. In this respect, so many individuals of the species *Knightia eocenica* have been excavated and sold that almost all fossil collectors in the world probably have one or more in their collection.

Ostariophysians include the largest and most diverse group of freshwater fishes alive today, the varied named catfishes. The group contains nearly 8,000 species, of which the cyprinids, or carps, have some 2,400 species, and the whiskered catfishes (Siluriformes) include some 2,800 species, with nearly 900 species in the armored loricariid catfishes alone. Although many species of the group are edible, by far the major use of the group is in the aquarium trade, with many colorful varieties and breeds available for sale around the world. Many of these pet species originated in the Amazon basin.

Fossil ostariophysians are common in Tertiary de-

A shoal of the little prehistoric relative of the herring, *Knightia eocenica*, from the Eocene Green River Shales of Wyoming.

posits of the world. The oldest member of the group is probably a small fish, just under 16 cm long, called *Tischlingerichthys* from the Late Jurassic of Germany (Arratia 1997).

Euteleosts: An Unparalleled Radiation of Forms

This group contains all the remaining ray-finned fishes, even though there is not strong evidence that this is a natural or monophyletic grouping (Nelson 2006), there is a case to be argued that the caudal fin has a unique structure, as argued by Johnson and Patterson (1996). The major groups included here are the Proacanthopterygii (about 366 living species) and the Neoteleostei, comprising some 17,000 species, most belonging in a single order, the Perciformes (10,000 species). Evolution within this latter group is seen by the acquisition of sharp spines in all the fins, a successful defensive mechanism that ensured their survival throughout the past 100 million years of their evolution.

The group is also characterized by having lost certain bones in the skull (orbitosphenoid, mesosphenoid, epipleural, epicentral bones). The adult fishes appear to have lost the presence of bone cells in their skeletons. The group Euteleostei can be defined by Arratia (1997) as a monophyletic group based on features of the caudal skeleton. In terms of evolution, we could interpret that the changes in the tail made them capable of advanced locomotory ability or gave them faster and more efficient movement through the water and an edge on all other groups of fishes. In this respect, the great radiation of euteleosteans is in parallel with the great radiation of dinosaurs that dominated the Mesozoic land faunas, as their prime adaptation was in the modifications of the foot and ankle for faster movement.

The fossil record of this great radiation of euteleosts starts out with small-herring-like forms such as

Orthogonikleithrus and *Leptolepides*, both known from the shallow seas of Germany during the Late Jurassic (c. 150 million years ago). *Leptolepidoides* was another small salmon-like form from the Late Jurassic of Germany. Throughout the world, small fishes named as *Leptolepis* occur in freshwater and shallow marine deposits. Further study of the many species has now recognized them as being separate genera from different regions (for example, the Late Jurassic species *Leptolepis talbragarensis* from New South Wales, Australia, is now referred to as the new genus *Cavenderichthys*; Arratia 1997). By the Early Cretaceous many representatives of modern teleost families appeared. The first members of the order Salmoniformes, which today contain the delectable salmon and trout species, can be traced back to forms of this age, like *Kermichthys*.

Fish fossils showing well-preserved examples of the diversification of primitive ray-finned fishes and early teleosts are found from a famous locality in South America where limestone nodules wash out of eroding hills. The area is the Araripe Plateau near Jardim in northeastern Brazil, and, like fossils from Gogo and central Queensland in Australia, the fish can be prepared by weak acetic acid to reveal extraordinary details of their skeletons. Indeed some specimens are so well preserved that they show mineralized muscle and even phosphatized stomach tissues. Many of the commonest fossils here, for example *Vinctifer* and *Rhacolepis*, are frequently sold by fossil dealers around the world.

The Neoteleostei are characterized by features of the head musculature—having a retractor dorsalis muscle present that could operate the upper pharyngeal tooth plates—and they all have a median rostral cartilage between the premaxillae and the braincase, as well as

Outcrops of the Santana Formation, Brazil, where many species of Early Cretaceous fishes, invertebrates, plants, and land animals have been found. (Courtesy John Maisey, American Museum of Natural History)

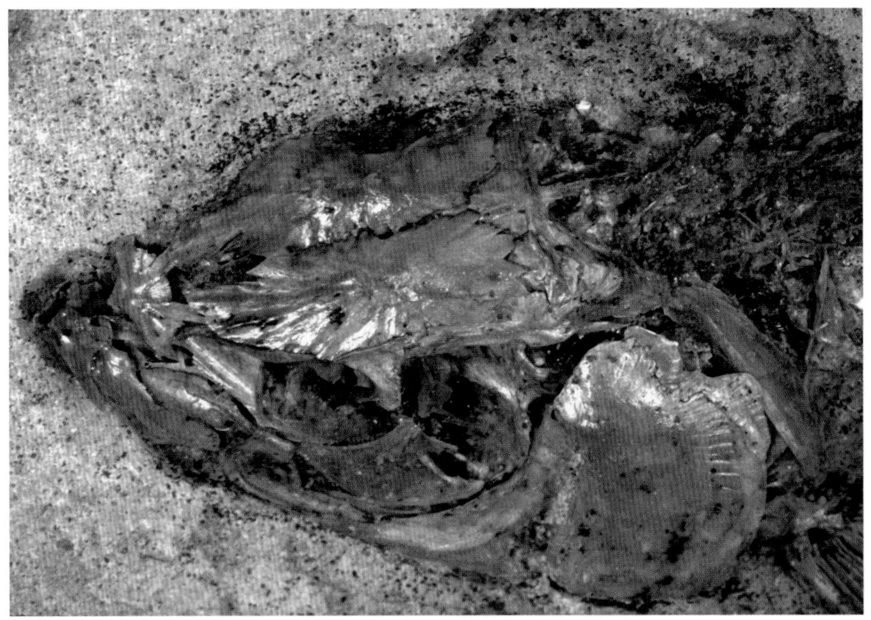

(*Left*) The head of *Tharrias*, an Early Cretaceous teleost from Brazil. (Courtesy John Maisey, American Museum of Natural History)

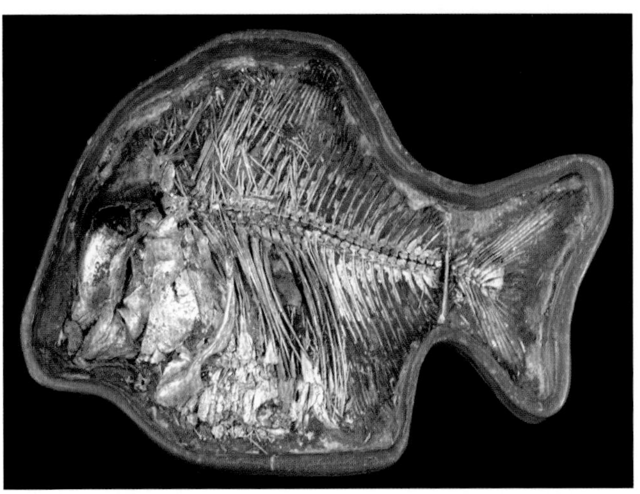

(Bottom *Left*) *Neoproscinetes*, a deep-bodied pycnodont teleost from the Santana Formation, Brazil. The specimen has been acid-prepared. (Courtesy John Maisey, American Museum of Natural History)

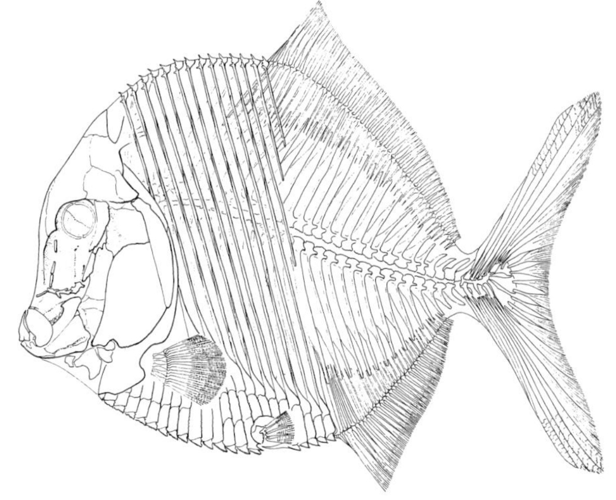

(*Bottom Right*) Drawing of the skeleton of *Neoproscinetes* from the Cretaceous of Brazil. Note the robust jaws that evolved for crushing hard prey items and the elongated rodlike scales that cover the front half of the body. (Courtesy John Maisey, American Museum of Natural History)

specialized features of their pectoral fins and upper jaw musculature. These few innovations gave the group an ability to better manipulate and break down food inside the mouth by being able to retract and move forward the pharyngeal tooth plates situated at the front of the gullet. Food was captured or bitten off by the mouth using the toothed bones of the upper and lower jaws, then crushed inside the gullet by the powerful grinding mills, the pharyngeal tooth plates.

The most successful group of fishes within the neoteleosts is the "spiny fins," or Acanthopterygii. These are characterized by one further refinement of the pharyngeal tooth plates: having enlarged second and third epibranchial bones (in the gill arches) and insertion of the retractor dorsalis muscle onto just the third pharyngobranchial bone. In simple terms, the acanthopterygians fishes can protrude their jaws in a number of efficient ways, turning their mouths into versatile prehensile devices. There are 13 orders of fishes within this group, testifying to the success of this feeding mechanism. The most successful of the acanthopterygians are the Percomorpha, a poorly defined group including many of the living families of fishes (some 13,000 species). Within this are the perchlike fishes (order Perciformes), comprising approximately 20 suborders, 150 families, and several thousand species of fish, making it the largest

(*Left*) Lorenzo Sorbini of the Verona Natural History Museum at the famous Monte Bolca site in Italy, where many hundreds of complete Eocene fish fossils have been found. Sorbini, who passed away in 1998, was a specialist in Cenozoic fishes.

(*Bottom*) *Eobothus*, an Eocene flatfish related to modern-day soles and halibuts, from the Monte Bolca site, Italy. (Courtesy Lorenzo Sorbini)

order of vertebrates. However the Percomorpha group cannot be well defined and might represent a taxonomic "duffel bag" in which many similar looking species are placed, without our really knowing if they are closely related.

The famous Green River Shales of Wyoming in the United States have yielded hundreds of different species of ray-finned fishes that lived in the Eocene Epoch about 45–50 million years ago, in a series of rivers and lakes. This window into the past shows us the whole evolutionary succession of ray-fins, from primitive chondrosteans and gars through to advanced neoteleosteans.

A deep-bodied perciform teleost from the Monte Bolca site, Italy. Perciforms include the greatest diversity of living fish species. (Courtesy Lorenzo Sorbini)

Another well-known fossil site of Eocene age, but representing marine habitats, is located in the mountains of northern Italy near the village of Bolca. These superb fish fossils were collected as far back as the sixteenth century, as noted by Andrea Mattioli in 1555. A museum that included many fish fossils was set up in Verona by Francesco Calceolari in 1571. Today the Natural History Museum in Verona houses the finest collection of Eocene marine fish fossils ever assembled. Again, many different fossil species have been found, most representing families of living teleosts, as well as remarkably preserved fossil sharks and rays. So far more than 150 fish species have been recorded from the Bolca sites, and much of the original work describing these fishes was been done by Lorenzo Sorbini, who before he passed away was curator at the Verona Natural History Museum. Examples of Bolca fishes from extant families are the anglerfish *Lophius* and the flattened flounder *Eobothus*. Others, like *Exellia*, represent their own unique families. David Bellwood (1996) of James Cook University has shown that several families of teleosts endemic to coral reefs first appear in the Monte Bolca assemblage, indicating this is the oldest record of reef-adapted fishes.

It is clear from the Eocene fossils found in Wyoming and Italy that many of the modern groups of fishes, represented at family level, had appeared by the early part of the Cenozoic. Most of the evidence for their early appearances comes from the ear-stones (otoliths) found in marine sediments. Many hundreds of these can be extracted from a bulk sample of marine sediment using the correct sieving procedures, and they have been used around the world to identify the nature of Tertiary fish faunas from varying latitudes. By the latter part of the

Today many anglerfish are denizens of the deep-sea abysses, but this species, *Lophius* (*A*), lived in tropical warm waters around Italy some 40 million years ago (Monte Bolca site). The teleosts radiated into a broad variety of species, adopting wide range of body shapes and defensive strategies. This zebrafish (*B*), a perciform, uses its sharp poison-tipped spines to ward off attackers.

Cenozoic, we have records of mainly extant families of teleosts wherever fish fossil are found, the vast majority of recently extinct species being described from earstones in marine deposits (Nolf 1985).

Teleosts evolved and diversified rapidly and today this legacy holds them up as the champions of vertebrate biodiversity. But they are more than just a highly speciose lot. They also make up a vast amount of the marine biomass and contribute toward the food chains of probably most marine organisms, one way or another. The recently discovered vulnerability of many commercially fishes species like for example the Atlantic cod, orange roughy, and Patagonian tooth fish (Chilean sea bass) should make us all concerned about how much food we can rely on in future from teleosts and what measures we need to rapidly put in place to ensure their sustainable future. Recent research monitoring marine reserves on the Great Barrier Reef gives us hope as fish biomass was observed to double in only 5 years if nonfishing zones are enforced (McCook et al. 2010).

(*Top*) ***Phaerodus testis***, an osteoglossid from the Eocene Green River Shales, Wyoming.

(*Bottom*) *Hsianwenia*, from the Pliocene Qiadam basin of China, is a most unusual cypriniform fish whose skeleton has become heavily ossified, because it lives in arid climatic conditions where calcium levels in the lake water were once very high. (Courtesy Meeman Chang)

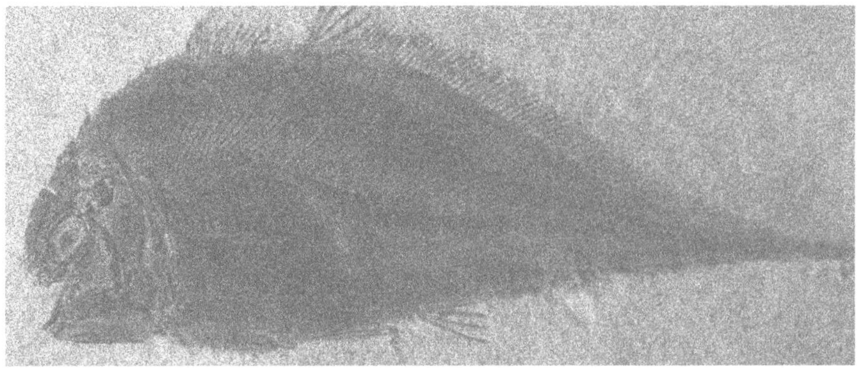

CHAPTER 10

The Ghost Fish and Other Primeval Predators

Lobe-finned sarcopterygians and where they fit

The lobe-finned fishes, once united in a group called the Crossopterygi, were a major group of large predators in the Devonian Period but are today represented only by the coelacanth. Recent discoveries have forced discarding the concept of crossopterygians as a natural group, and we now refer to the early enigmatic forms simply as stem group, or cousins to, sarcopterygians. These include the mysterious Silurian ghost fish, *Guiyu*, from China, together with strange predatory forms with large lower jaw tooth whorls like *Psarolepis* and *Achoania* and the ferocious dagger-toothed fishes (onychodontids). The first coelacanths appeared in the latter part of the Early Devonian Period, possibly represented by a form called *Styloichthys* from China. Coelacanths developed specialized suctorial feeding mechanism that would set them aside from all other lobe-fins and ensure their survival long after the others had been extirpated and their remains preserved in stone.

One of the most remarkable biological discoveries of the twentieth century took place in South Africa in the late 1930s. A strange-looking fish about 1.5 m long was caught off the mouth of the Chalumna River, near East London, South Africa, on 22 December 1938. The following day Marjorie Courtenay-Latimer, a young ichthyology curator at the East London Museum, saw the fish while searching for unusual specimens at the dockside. The fish was bizarre—it had muscular fleshy lobes for the pectoral, pelvic, second dorsal, and anal fins. The tail was symmetrical with a long central lobe and a tuft at the end.

Once Courtenay-Latimer saw that this was an unusual fish, she then realized the problem she would have in getting it back to her laboratory. The fish was already beginning to smell as decay had set in, but wrapping it in sackcloth she managed to persuade a taxi driver to take her and the prize fish back to the museum. Some days later when J. L. B. Smith, Senior Lecturer in Chemistry, inspected the fish, he was able to identify it as a member of a group called coelacanths, thought to be extinct for more than 50 million years! Smith named the fish *Latimeria chalumnae* in honor of Miss Latimer and the location where the fish came from.

By this time the specimen had been gutted, little scientific information could be obtained from it, so the search began for another coelacanth. Not until 1952 did the next specimen come to light, this time from the

(*Top*) The coelacanth *Latimeria* is the only living genus of a once-flourishing group of primitive lobe-fined fishes that include coelacanths, onychodontids, and other strange basal sarcopterygians. Coelacanths were thought to be extinct at the end of the Cretaceous Period until this living species was found in 1938 off the coast of South Africa, and now there are two extant species known.

(*Left*) A late Silurian fish site in Yunnan, China, where the oldest near-complete body fossil of *Guiyu*, a basal osteichthyan fish, was found.

Comoro Islands northwest of Madagascar, about 2,000 km north of the first specimen's location. Since then a good number of coelacanths have been caught, mostly around the Comoro Islands. The anatomy of the coelacanth was first described in detail by French professors Millot and Anthony, and in recent years many other scientists have studied detailed aspects of its biology. Perhaps the greatest advances in coelacanth studies have been made by Hans Fricke of the Max Planck Institute in Germany, who filmed several living coelacanths in their natural habitat using a small submersible submarine, at depths of between 117 and 198 m. It had taken Fricke nearly 40 dives at 30 different sites to find the elusive animal in its natural home. *Latimeria* is quite a slow-moving fish, which often drifts in the currents and adjusts its position using highly mobile pectoral and pelvic fins. The fins can move like the gait of a land animal—pectorals and pelvic have the ability to move in opposite directions. Coelacanths live at depths between 100 and 400 m and feed on other fishes and the occasional cuttlefish. In the mid–1990s a truly remarkable discovery was made in Borneo that another species of coelacanth was alive off the coast of Sulawesi. This new species, *Latimeria mindenaoensis* gave credence to the concept that the two populations had split off several millions of years ago.

The significance of the original discovery made by Latimer and Smith in the late 1930s can really be appreciated when we consider what was known about fossil coelacanths. They were first recognized from fossils by French scientist Louis Agassiz, and in his famous books of 1843–44, *Recherches sur les poissons fossiles,* he described *Coelacanthus granulatus* from the Late Permian of Germany. A large number of fossil coelacanths were then identified in rocks ranging from Middle Devonian Age to the end of the Mesozoic Era, but the fact that no Cenozoic fossil coelacanths had been found suggested that the group became extinct along with the dinosaurs at the end of the Cretaceous. The coelacanth was first promoted as a "missing link" between fishes and land animals. In fact the link it does represent is as the only surviving member of a once large and diverse group of fishes once known as the crossopterygians (or "tassel-finned fishes") and now known as sacropterygians. Within the Devonian, members of this group are to be found the closest fish groups to the earliest land animals, the amphibians. In this chapter we deal with the stem sarcopterygians, those at the base of the evolutionary tree before it divided into the two major groups, Dipnomorpha (Chapter 11) and Tetrapodomorpha (Chapters 12 and 13).

The Chinese Ghost Fish and Its Kin

One of the most amazing fossil discoveries of the new millennium has to be the finding of the first relatively complete osteichthyan fish from the late Middle Silurian Period. The team led by Zhu Min had earlier noticed scales of osteichthyan fishes from the site, and so they kept searching until more complete remains were found in 2008 and described by Zhu and his team (2009). The fish named *Guiyu,* meaning "ghost fish," fills in a blank or ghost lineage in the record of fossil osteichthyans. It was a small fish, up to 28 cm long, and had a hinged braincase, rhombic scales with pegs attaching them to each other, and peculiar spines along the back and preceding the paired pectoral fins. The cheek was solid, made of one large preopercular over a maxilla, and the front of the lower jaw bore a large tooth whorl. *Guiyu* resembles placoderms in having a median dorsal plate on its back and spinal plates on its front fins. In having the divided braincase, it is clearly placed the base of the Sarcopterygii.

A poorly known form from the Lower Devonian of Yunnan, named *Meemania* in honor of Chang Meeman, has been studied from two skulls. It shows the beginnings of the pattern seen in other sarcopterygian skulls, but retains a placoderm-like large rostral-pineal bone. The dermal bones have the peculiar feature of layers of ganoine with cosmine on top of them, a sort of intermediate stage between actinopterygians and sarcopterygians.

Other early sarcopterygians have been found from

The scales (*A*) of *Guiyu*, from the Silurian of China, resemble those of ray-finned fishes in having peg-and-socket articulation but are more like other sarcopterygians in their tissue structure. The near-complete body and head of *Guiyu* (*B*) from the Silurian of China, which was discovered by Zhu Min and his colleagues in mid-2008. (Courtesy Zhu Min, Institute of Vertebrate Paleontology and Paleoanthropology, China)

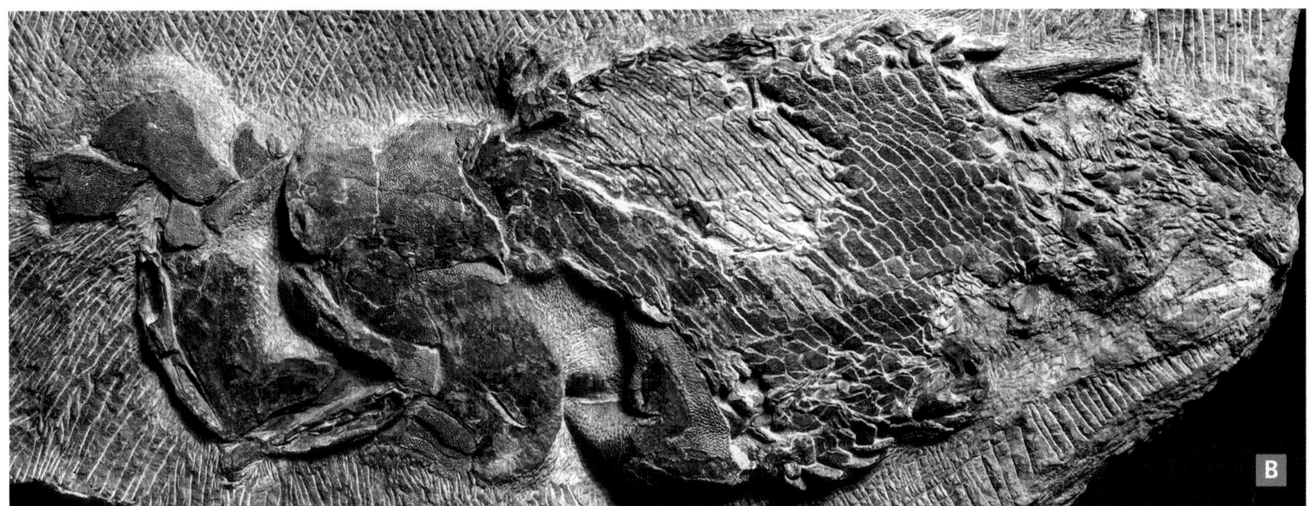

the Late Silurian / Early Devonian of Yunnan, China. These include forms similar to *Guiyu*, such as *Psarolepis* and *Achoania*. Both have bone covered by a shiny surface layer of cosmine with large pores and have grooves for tooth whorls set in the lower jaws. *Psarolepis* was first described from an isolated lower jaw; several field seasons of digging at the locality have now produced much of the skeleton. It too was small, under 30 cm long, but had very formidable jaws with large fangs at the front of the mouth. Its nostrils were located high on the snout, and teeth lined the median rostral bone at the front of the head, separating the toothed upper jawbones. *Achoania* differs from *Psarolepis* in its larger palatal bone, the parasphenoid, and shape of the braincase. A third closely related form called *Styloichthys* sits somewhere near the base of the split between the coelacanths and higher sarcopterygians, according to Zhu Min and colleagues. Matt Friedman (2007) has put forward an alternative view that it could be the sister taxon to coelacanths. *Styloichthys* has deep lower jaws with very large muscle fossae and short dentary bones, and its large radiating pattern of denticles seen on the prearticular is also observed on some later coelacanths.

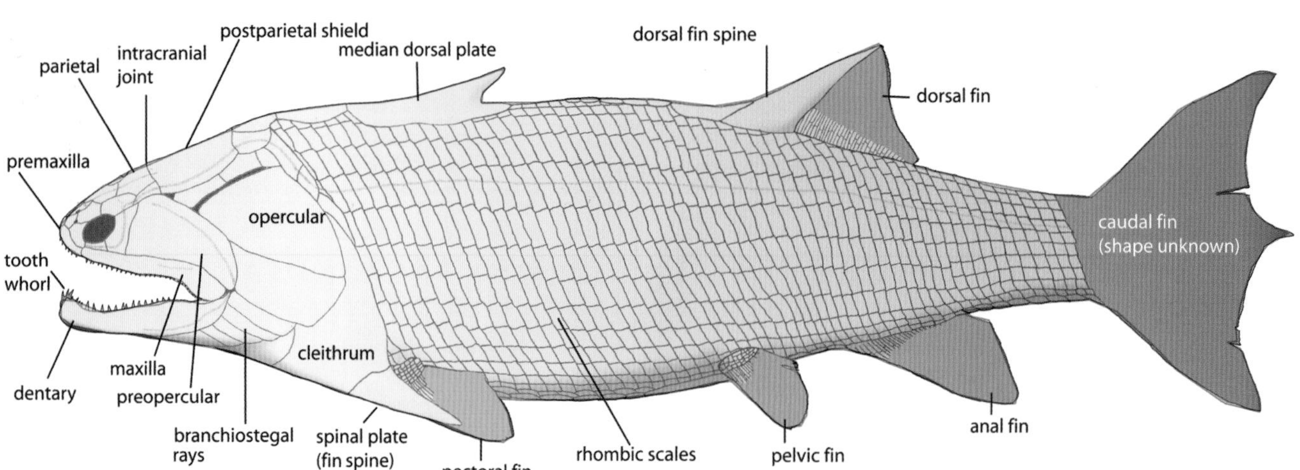

(*Top*) A reconstructed scene showing *Guiyu* hunting in the warm marine seas of southern China in the late Silurian. Yunnanolepid antiarchs are scattering in the background. (Courtesy Brian Choo)

(*Bottom*) A restoration showing the main anatomical features of *Guiyu*, one of the earliest osteichthyan fishes.

(*Top Left*) The skull roof of *Meemania* from the Early Devonian of Yunnan China. This fish has shiny dermal bone tissue with pores (cosmine) that also has layered structure (ganoine) similar to ray-finned fishes. (Courtesy Zhu Min, Institute of Vertebrate Paleontology and Paleoanthropology, China)

(*Top Right*) This is the skull roof of *Psarolepis*, an early predatory sarcopterygian from the Late Silurian / Early Devonian of Yunnan, China. It shows the basal condition for sacropterygians in having a divided braincase. (Courtesy Zhu Min, Institute of Vertebrate Paleontology and Paleoanthropology, China)

(*Left*) The snout of *Psarolepis* shows the high incurrent nostril and solid rostral bones at the front of the head. The large pores in the dermal bone denote an unusual kind of cosmine.

At the beginning of the Devonian, there were many lines of osteichthyan fishes. This is the front part of the head of *Achoania* from China, a sarcopterygian at the base of the lineage that led eventually to lungfishes and tetrapods. (Courtesy Zhu Min, Institute of Vertebrate Paleontology and Paleoanthropology, China)

Younglepis and Powichthys

Outside of China, the oldest stem sarcopterygians date from the lower Early Devonian. By this time porolepiforms and some other strange forms had appeared in the Old Red Sandstones of Spitsbergen and elsewhere in Europe. Two of them were similar, *Youngolepis* and *Powichthys*, and appeared to be closely related but more advanced than *Psarolepis* and its relatives. *Youngolepis* is known from the Early Devonian of China and was named in honor of the famous Chinese paleontologist C. C. Young, whereas *Powichthys* is named after its discovery site, Prince of Wales Island (initials POW), in Arctic Canada. *Youngolepis* is better known than *Powichthys*. Both were predatory fishes, no longer than about 50 cm, with heavily ossified bones in the head and thick rhombic scales over the body. They appear to be more specialized than other early sarcopterygians in having a series of many small bones flanking the main paired bones of the skull roof, a condition also found in lungfishes. To compensate for their small eyes, they had a well-developed lateral-line sensory system to detect prey.

Both of these fishes are known principally from their cosmine-covered skulls, which have a relatively long parietal shield and a short postparietal shield. The braincase in both forms is not fully divided into two components, as in other sarcopterygians, indicating a primitive condition that precedes the division of the braincase into two components in all other sarcopterygians. *Youngolepis* had a cheek with several fused bones but is basically much like that of a tetrapodomorphan in its cheekbone pattern. Both *Youngolepis* and *Powichthys* have many small bones along the flanks of the large central skull roof bones. In addition, the flask-shaped cavities within canals within the cosmine layer have an enameloid tissue dipping into them, a feature seen developed even further in porolepiforms but not at all in other sarcopterygians. Recent work focusing on the histology of the bones and teeth of *Youngolepis* and *Powichthys* suggest that they should be included as sister taxa to the porolepiform group and that they are all at the base of the lineage leading to lungfishes.

The Dagger-Toothed Fishes (Onychondontiformes)

The Onychodontiformes (nail-toothed fishes) are a poorly known group of Devonian sarcopterygians, which feature lower jaws with large dagger-like tooth whorls. Until the description of the well-preserved *Onychodus jandemarrai,* specimens from the Gogo Formation of Western Australia (Andrews et al. 2006), the group was

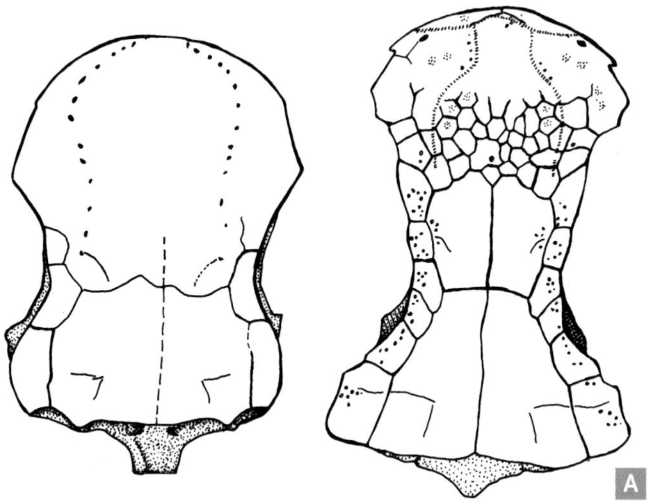

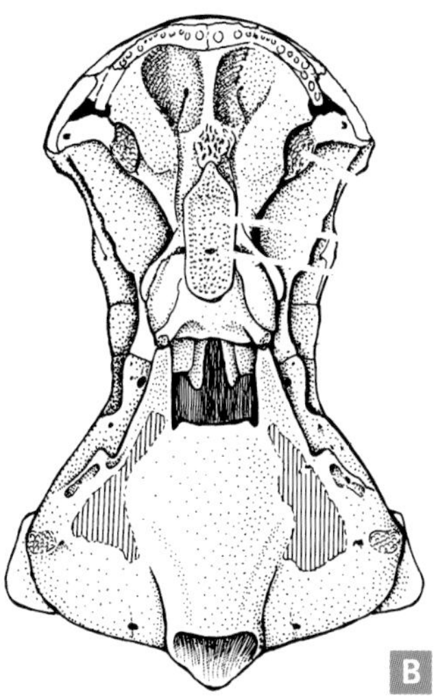

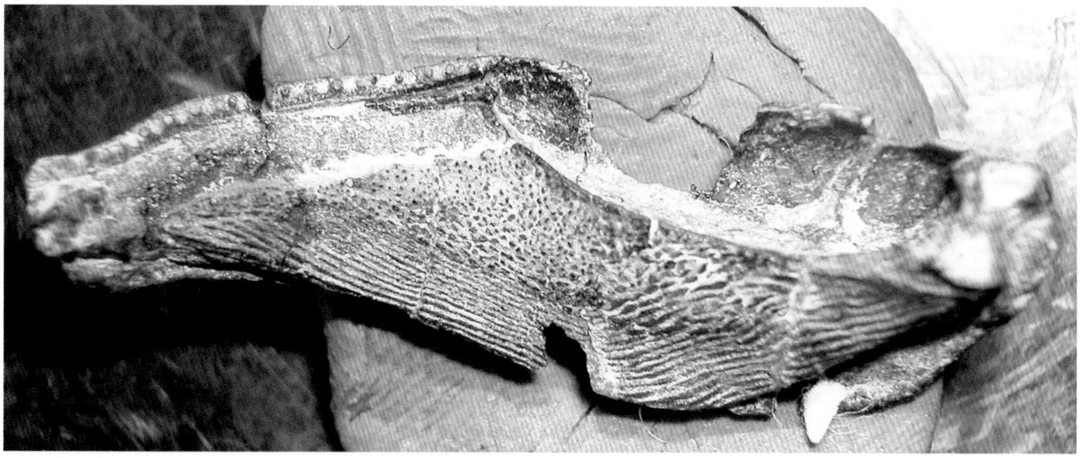

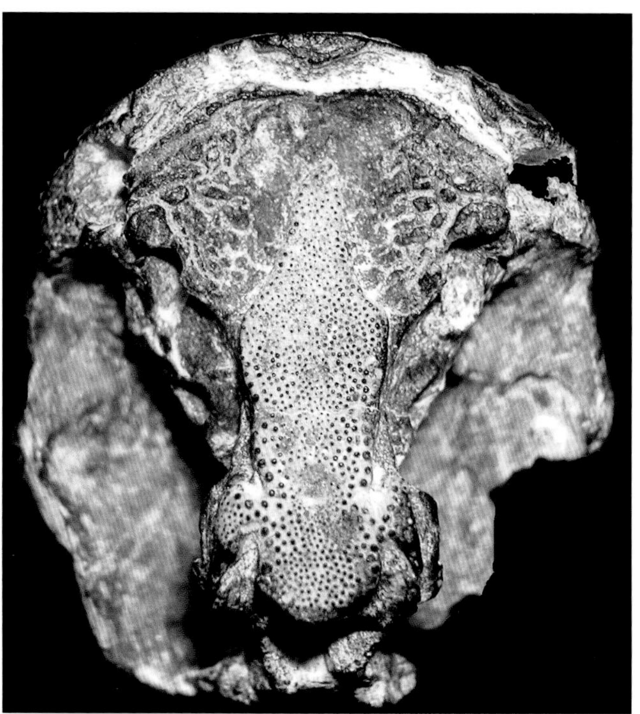

(*Top*) Skulls (*A*) drawn in dorsal view of two Early Devonian sarcopterygians, show the early separation of the head into two units, which divided the braincase. *Diabolepis* (*far left*) from China is the most basal member of the lungfish lineage, whereas *Powichthys* from Arctic Canada was an early member of the porolepiforms. Features of the endocranium in ventral view of *Powichthys* (*B*), a basal member of the dipnomorphan lineage.

(*Middle*) The robust lower jaw of *Styloichthys* from the Early Devonian of China. (Courtesy Institute of Vertebrate Paleontology and Paleoanthropology, China)

(*Left*) The palate of *Youngolepis*, a predatory sarcopterygian from the Early Devonian of China. (Courtesy Institute of Vertebrate Paleontology and Paleoanthropology, China)

Basic Structure of a Sarcopterygian

Sarcopterygian fishes are characterized by being relatively long-bodied with a skull that in primitive forms has a hinged braincase, divided into a front section (ethmosphenoid) and rear section (oticco-occipital). This flexure within the braincase is reflected in the skull roof bones of each group of sarcopterygians; they are said to have a frontal and a parietal shield separated by the intracranial joint. Some advanced members of the groups have the intracranial joint immobilized by fusion of bones between the frontal and parietal shields (for example the panderichthyids discussed in this chapter). The head has large eyes in most species and paired external nostrils in all groups except for osteolepiforms, which have a single external pair of nostrils and a palatal nostril opening called the choana.

The cheek has a regular pattern of bones, with one or more large squamosal bones present, and the jaws have large dentary bones supported by a series of infradentaries. There are well-developed fangs in addition to regular marginal teeth on the toothed bones—except in coelacanths, which have a smaller area of toothed biting jaws and an unusual double-tandem jaw joint (clearly a specialized feeding mechanism). The teeth have enamel present over dentin, and in some groups, like the osteolepiforms, porolepiforms, and rhizodontiforms, there is complex infolding of the enamel and dentin. In cross-section, these teeth appear highly complex and are termed *labyrinthodont teeth*. The gill arches are well ossified and have large ventral gill bones (basibranchials), sometimes with a forward-pointing sublingual bone.

The shoulder girdles, the internal shoulder girdle (scapulocoracoid), and, in tetrapodomorphans, the pectoral and pelvic fin skeletons are all well ossified. This ossification is most highly evolved in the osteolepiforms and rhizodontiforms, which have a powerful humerus that articulates with ulna and radius bones, as in all groups of higher vertebrates (except limbless forms). The bodies are often rather long in tetrapodomorphans and rather conservative in shape; no unusual deep-bodied or flattened forms seemed to have evolved.

The scales and dermal bones in all primitive sarcopterygians have cosmine present, but this is often lost in later lineages. The scales are characteristic for most sarcopterygian groups. Primitive members of the osteolepiforms and porolepiforms have thick, rhombic cosmine-covered scales, but during their separate evolutionary radiations these have changed to rounded non-cosmine-covered scales. The rhizodontiforms, onychodontiforms, and actinistians have rounded thinner scales. The sensory-line systems of crossopterygians are well developed and can be seen as a series of pores and deep pit-line canals on the dermal bones and scales.

known only from whole specimens of a little fish called *Strunius* from Germany and from an assortment of partial skulls, jaws, and bones belonging to other species in the larger genus *Onychodus*, which is now known to have been widespread around the Devonian world.

The oldest onychodontid fishes are from the Early Devonian of China and Australia. *Bukkanodus*, from the 400-million-year-old Fairy Formation of Victoria, Australia, is known by a variety of isolated bones and teeth and clearly shows the typical compressed tooth whorls seen in all onychodontids. The fish was about 15–20 cm maximum length and lived in a freshwater volcanic lake environment.

Qingmenodus, from the same age in southern China, is represented by a rear half of the skull, plus some isolated bones and a tooth whorl. It was a much larger fish, possibly around 70 cm, and shows that early onychodontids had a very long rear division of the skull, with a large basicranial muscle that attached under the braincase. This muscle assisted in movement of the intracranial joint, giving onychodontids and coelacanths (which share a similar arrangement of the muscles) a more flexible skull for biting down hard on prey (Lu et al. 2010).

The largest species of *Onychodus* grew to about 3–4 m in length and may have been lurking predators, much like today's moray eels in reef habitats. The skull was very kinetic, with a large hinge between the two divisions of the braincase that enabled easy movement when the snout was raised. This is clearly an adapta-

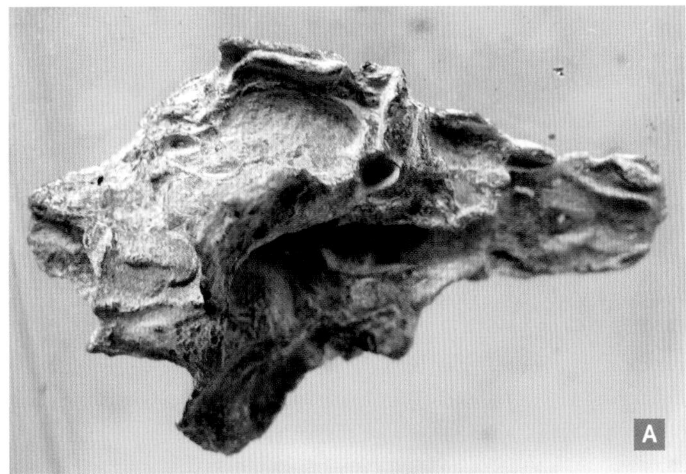

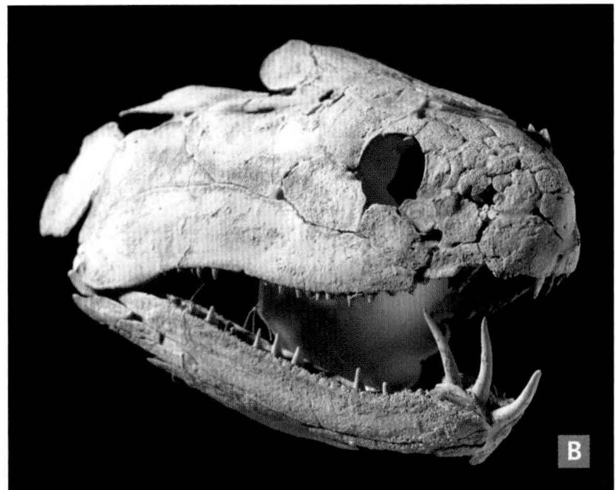

The rear part of the head of *Qingmenodus* (*A*) from the Middle Devonian of China, showing the braincase. (Courtesy Zhu Min, Institute of Vertebrate Paleontology and Paleoanthropology, China). The skull (*B*) of *Onychodus jandamarrai* from the Late Devonian Gogo Formation of Western Australia, showing large tooth whorls at the front of the lower jaw (only one set of whorls fit on this restoration). The fish may have been able to slightly retract these whorls when capturing prey.

tion to allow the large fangs of the lower jaw to the fully utilized. Indeed, when the mouth of *Onychodus* was closed, the large lower jaw fangs almost touch the skull roof.

Other features of onychodontid's skulls that make them different from other sarcopterygians are in the pattern of skull roof and cheekbones. The upper jawbone, or maxillary, is much like that of an actinopterygian in having a large postorbital blade. The cheek features two large almost equal-sized bones, the squamosal and preoperculum, and three infraorbitals flank the eye. The opercular mechanism is typical for sarcopterygians, except that the submarginal bones may be absent. The body of *Onychodus* is long and slender, and from limited fossil evidence it appears that the dorsal fins were placed far at the back near the tail. The pectoral and pelvic fins are not well known from fossil material, although ossification of the humerus indicates that at least the front

The front of the lower jaw of the onychodontid *Luckius* from the Middle Devonian of central Australia. Note the deep groove for the tooth whorl.

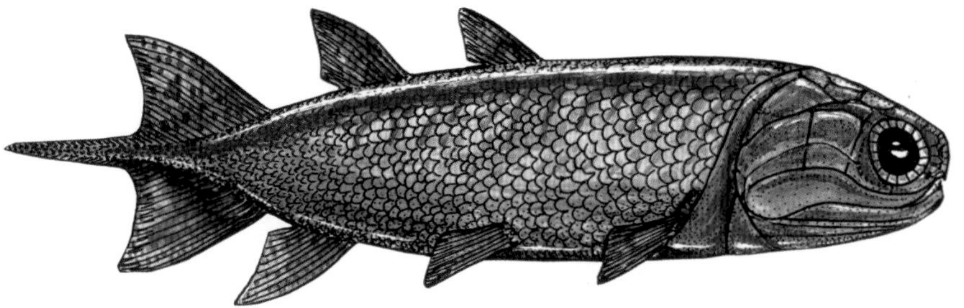

Reconstruction of *Strunius*, a little onychodontid from the Middle-Late Devonian of Germany.

paired fins were quite robust with powerful muscular attachments.

One specimen of *Onychodus* from Gogo actually shows the remains of its prey. The bones of a small placoderm were found lodged in the position of the throat of the partial *Onychodus* skull. As the placoderm bones were facing forward (away from the direction of the *Onychodus*) and these were entirely intact, showing no signs of damage, it has been deduced that the prey was captured by the tail and then swallowed whole. The measurements show that the placoderm would have been about 30 cm long, whereas the *Onychodus* was about 60 to 70 cm in length, thus capable of catching and gulping live fishes half its own size!

Coelacanths (Actinistia): The Tassel-Tails

The Actinistia, or coelacanths, first appeared in the Early Devonian and survive today, represented by the genus *Latimeria*. The peak of diversity for coelacanths was probably achieved during the Carboniferous, when many varied forms inhabited the shallow seas and rivers of the world. The most distinguishing features of actinistians is that they lacked an upper jawbone (maxilla) and had a loose cheekbone arrangement, numerous paired snout bones each with large pores (reflecting the fact that the snout has a special rostral organ), and a lower jaw with a special "double-tandem" articulation and short-toothed dentary bone. The bodies of coelacanths have a long median lobe on the tail with a rear tassel or tuft, and the paired fins and some of the median fins are all strongly lobed. Those fins not strongly lobed in some genera have stout spines supporting the fin web. The shoulder girdle bones are more elongated compared with other sarcopterygians, and most genera possess an additional bone unique to coelacanths, the extracleithrum, attached to the base of the cleithrum in the pectoral girdle.

The earliest fossil coelacanths come from the Early Devonian (400 million years ago) of Australia, known by one isolated but characteristic lower jaw, named *Eoactinistia*. The earliest complete coelacanth skulls are of Late Middle or early Late Devonian age and come from Germany, Canada, and Australia. Forms such as *Diplocercides* from Germany (and recently Australia) show the typical suite of coelacanth characteristics but lack many of the advanced coelacanth features, such as specializations of the braincase and having an equal number of tail fin rays as supporting bones in the tail. *Miguashaia*, from the Late Devonian Escuminac Formation of Quebec, Canada, has many primitive coelacanth skull features and a heterocercal (asymmetrical) tail as in other sarcopterygians. It is regarded as the most primitive of all coelacanths that are reasonably well known. *Gavinia*, from the Late Devonian Mount Howitt site, Australia, is known only from a fragmentary skull and parts of the body and tail. It appears similar in many ways to *Miguashaia* but is more primitive in having a longer lower jaw.

The extraordinary coelacanths from the Early Carboniferous Bear Gulch Limestone of Montana studied by Dick Lund show the extreme adaptations of body form that coelacanths achieved. Some slender forms, like *Caridosuctor* (meaning "prawn eater") were similar to later coelacanths in the shape of their bodies, whereas *Hadronector, Lochmocercus* and *Polyosteorhynchus* were fairly squat. The most extreme body shape is seen in

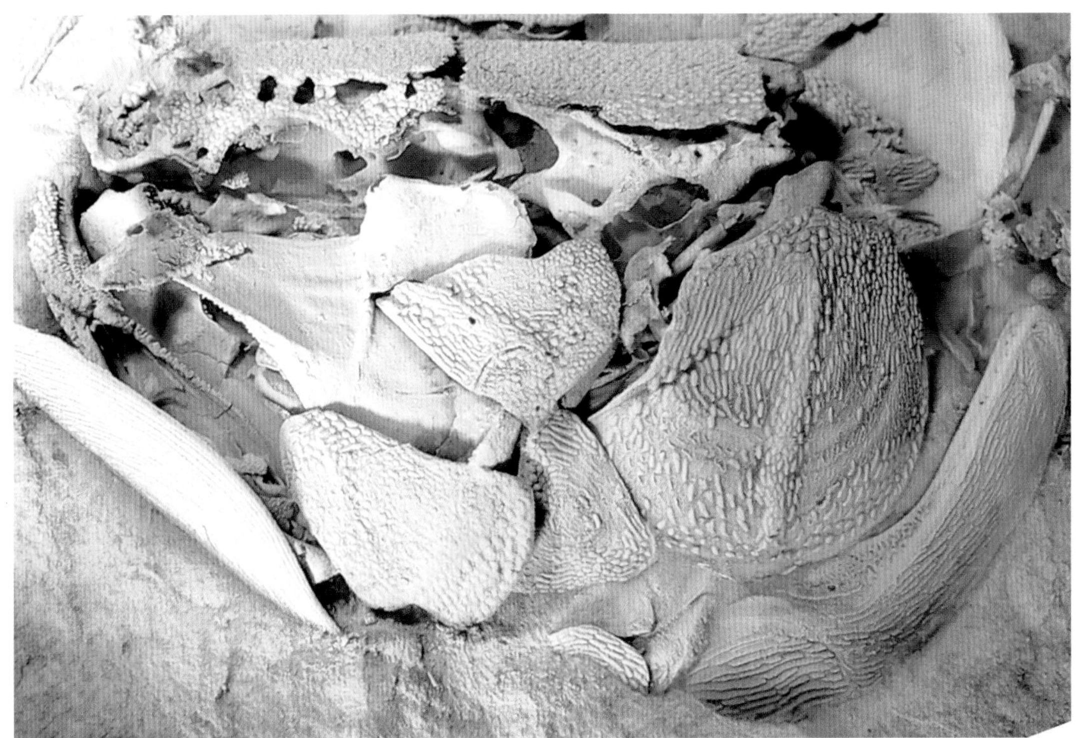

Allenypterus, a deep-bodied coelacanth with a long tail region and rather small dorsal, pelvic, and anal fins. *Allenypterus* is so unusual that it was first described as a ray-finned fish and not a coelacanth!

During the Mesozoic, the coelacanths became fairly conservative, although some managed to reach large sizes, like *Mawsonia* from the Cretaceous of South

(*Top*) Head of a late Devonian coelacanth, *Diplocercides*, from the Gogo Formation of Western Australia.

(*Bottom*) *Allenypterus*, a coelacanth from the Early Carboniferous Bear Gulch Limestone of Montana, is unusual in that its body plan could have been radically varied from that of the modern coelacanths. (Courtesy Karl Frickhinger)

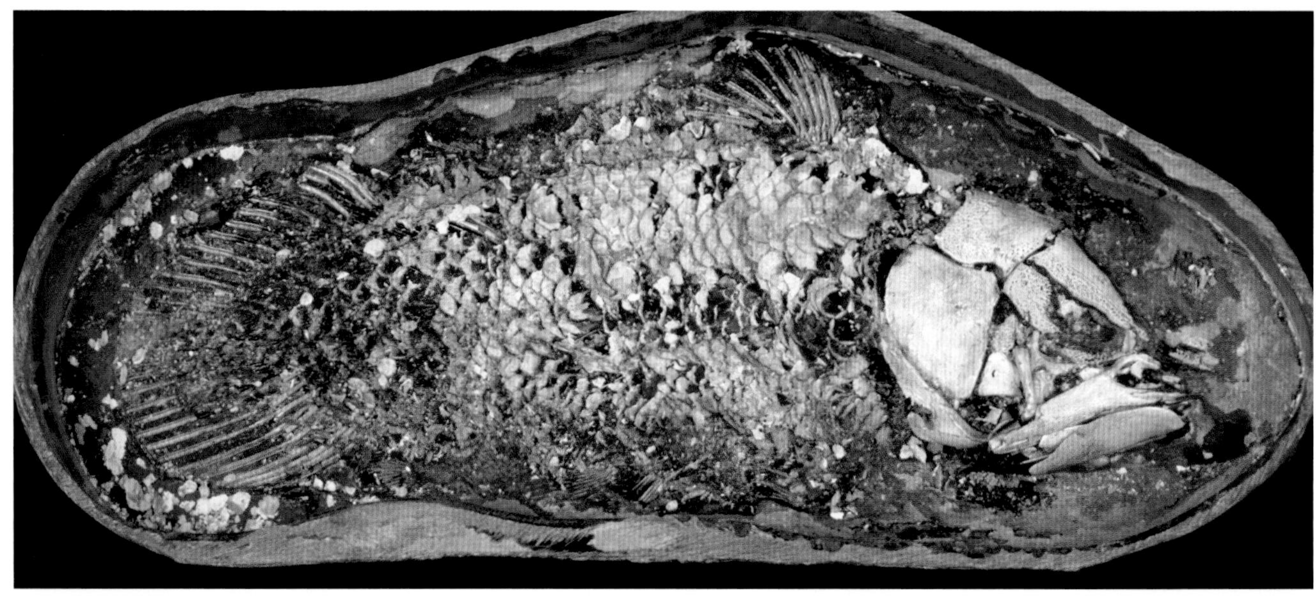

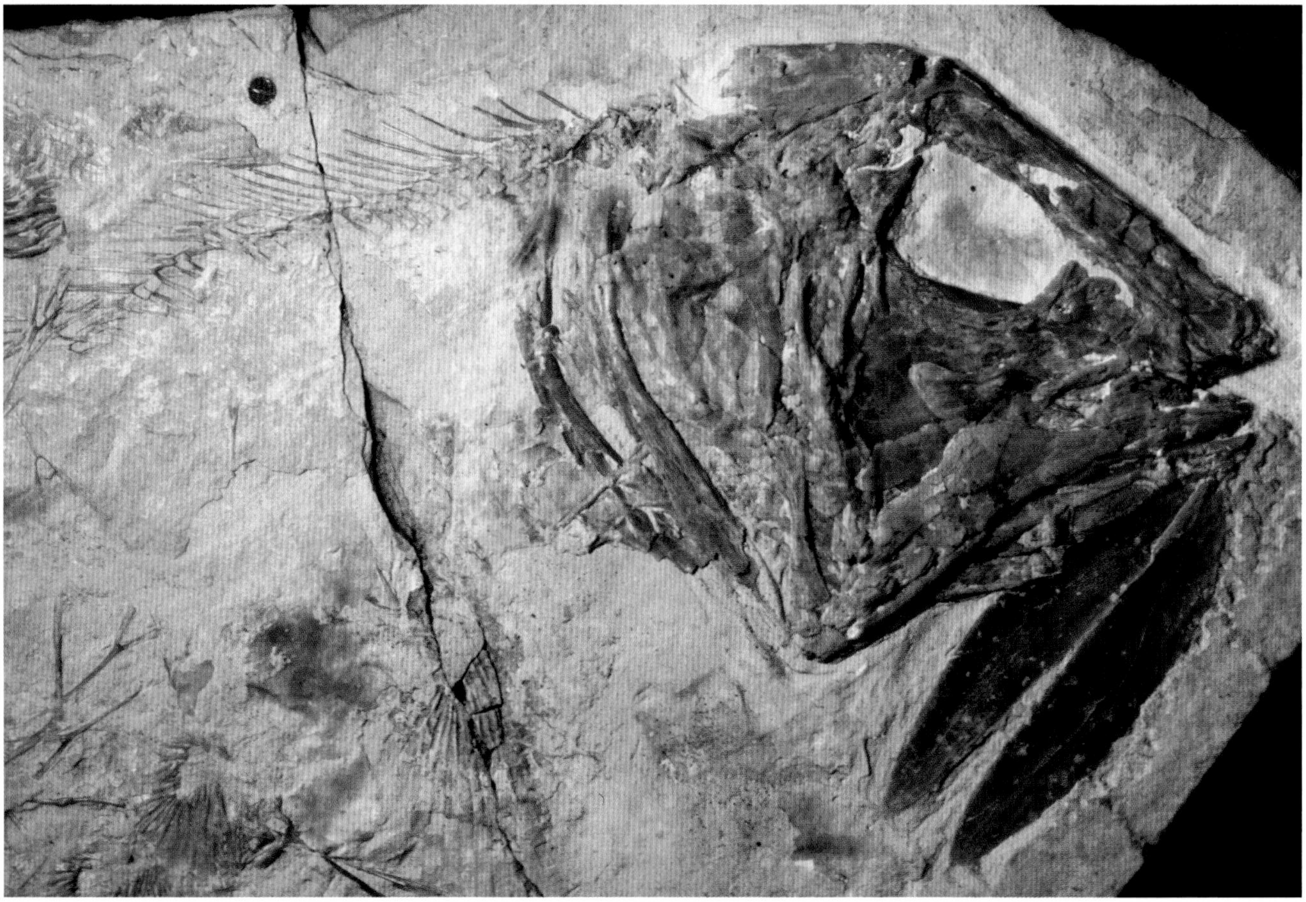

(*Top*) An Early Cretaceous coelacanth, *Axelrodichthys*, from the Santana Formation of Brazil. (Courtesy John Maisey, American Museum of Natural History)

(*Bottom*) *Macropoma*, a Late Cretaceous coelacanth from Britain and other parts of Europe.

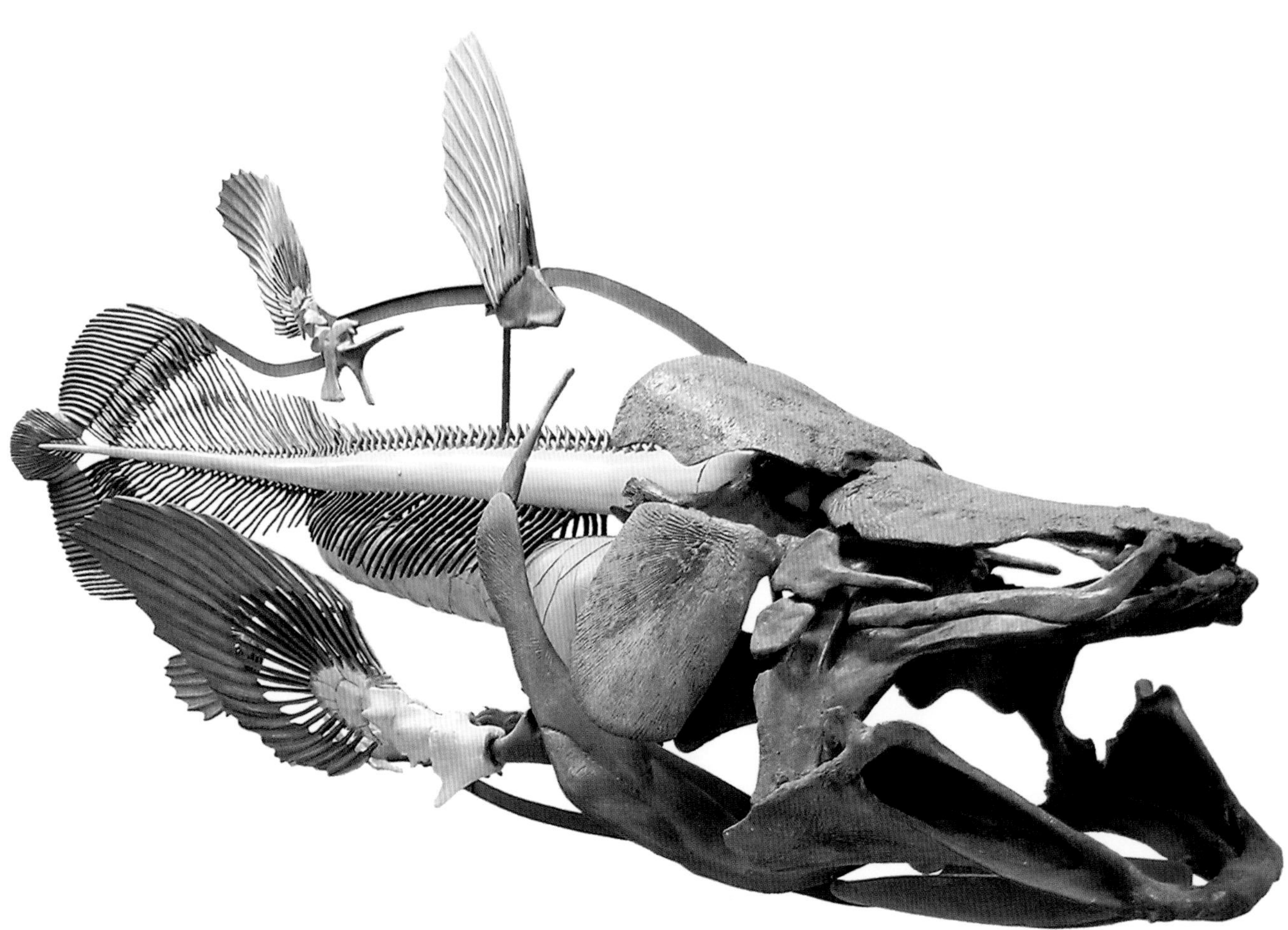

A reconstructed model of the skeleton of the giant Early Cretaceous coelacanth *Mawsonia lavocati* from Morocco, which may have reached sizes of 4 m. (Courtesy Yoshitaka Yabumoto, Kitakyushu Museum of Natural History, Japan)

America and Africa. The largest species, *Mawsonia lavocati,* may have reached lengths in excess of 4 m, and was the one of the largest predatory fish in the shallow seas of Gondwana at this time.

Relationships of the Sarcopterygian Groups

Relationships of the sarcopterygian groups have been hotly debated by vertebrate paleontologists, and consequently a number of differing opinions exist in the current literature. British paleontologists Peter Forey, Brian Gardiner, and Colin Patterson, working with American Donn Rosen, argued in 1981 that lungfishes were the closest ancestors to the land vertebrates. Modern work incorporating molecular phylogenies of vertebrates combined with new fossil discoveries have vindicated this theory, as coelacanths are now widely regarded to be outside the clade that unites lungfishes with higher tetrapods. Of course there are many extinct groups that fit in between these main lineages.

Recent work supports a split within the sarcopterygians into two main lineages: the Dipnomorpha (including lungfishes and closely related extinct forms like *Youngolepis* and porolepiforms) and the Tetrapodomorpha (extinct clades of lobe-finned fishes like rhizodontiforms and more derived groups like osteolepidids, tristichopterids, and panderichyids leading to the first tetrapods).

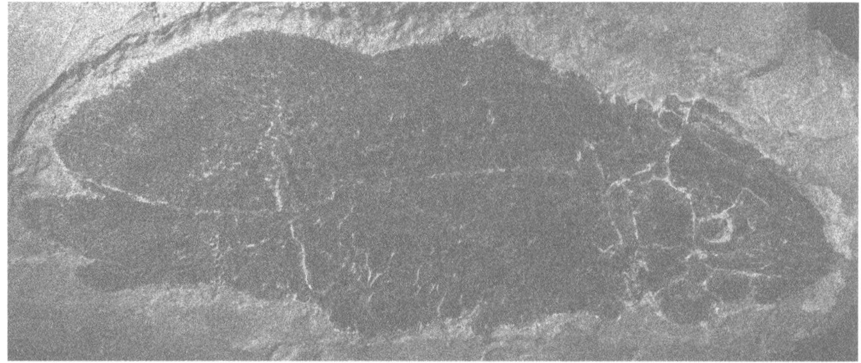

CHAPTER 11

Strangers in the Bite: Dipnomorphans

How fishes learned to breathe air

Dipnomorphans include the beady-eyed extinct porolepiforms and the lungfishes. The lungfishes have an almost complete fossil record in Australia, including superb three-dimensional skulls of Devonian age from New South Wales and Western Australia. Although the first lungfishes, dating back about 400 million years, lived in the sea, all lungfishes, since about 340 million years ago, inhabited freshwater environments. The fossil record tells us that lungfishes acquired the ability to breathe air independently of other vertebrates and are therefore not considered ancestral to the first four-legged land animals, the amphibians. Fossil burrows tell us that back in the Late Paleozoic lungfishes could estivate (lie dormant) enclosed in baked mud during the dry season, awaiting the next season's rains. The story of lungfish evolution is one of frantic and rapid change during the Devonian, the "dipnoan renaissance," with a much reduced rate of evolutionary change from the end of Carboniferous Period to recent times. The three genera of living lungfishes occur in Africa, South America, and Australia. The Queensland lungfish is the most primitive of these "living fossils"—fossils of the Australian lungfish indicate that this species has remained unchanged in Australia for at least 100 million years, making it the most enduring species of vertebrate known on Earth.

Porolepiformes: Fat-Headed, Beady-Eyed Predators

The Porolepiformes were relatively large predatory fishes that lived in the Devonian Period (Jarvik 1972). They have been found at sites all around the world but are most prevalent in Middle and Late Devonian freshwater deposits. The group takes its name from having rows of pores on their cosmine-covered scales. Other features that characterize the porolepiform fishes are a broad skull with small eyes, the presence of a prespiracular bone in the cheek, and a complex style of infolding of the enamel and dentin in the large fangs, called dendrodont tooth structure. At the front of the lower jaws is a large whorl of stabbing teeth. From the shape of their bodies and tail, most porolepiforms were thought to be ambush predators, lying in wait for passing prey that would be caught by a quick forward lunge.

The oldest example is *Porolepis* from the Early Devonian of Spitsbergen and elsewhere in Western Europe. *Porolepis* reached lengths of about 1.5 m, fairly large for its time. It had a thick cosmine cover on all its bones and scales, and its eyes were very small. *Porolepis* and a few other forms known from fragmentary remains had thick rhombic scales and are placed in the family Porolepidae. *Porolepis* was among the largest predators of the Early Devonian seas and nearshore environments. Their sluggish appearance was no doubt efficient by comparison with the heavily armored primitive placoderms and acanthodians, upon which they most likely preyed. Most Middle and Late Devonian porolepiforms are placed in the family Holoptychiidae, as these lack cosmine and have rounded scales. The earliest holoptychioid is *Nasogaluakus* from the late Early Devonian of Arctic Canada (Schultze 2008).

In the Middle Devonian, the chief group of porolepiforms emerged, the holoptychioids. These fishes grew to enormous sizes for their day, maybe 2.5–3 m, and were fearsome ambush predators. Although some were found in marginal marine deposits, most had invaded the river systems, away from the giant predatory dinichthyid placoderms, where they could be the top predators in the water. The holoptychioids are considered more evolutionarily advanced than the porolepids, in that they have lost the thick cosmine cover on the bones and scales, the scales have become rounded, the skull has a specialized set of bones around the external nostrils (including a "nariodal" bone), and the mouth

Porolepis, from the Early Devonian of Spitsbergen, Norway, was a thick-scaled, heavy-boned predator. Porolepiform fishes, once thought to be closely related to the tetrapod lineage, are now accepted as close relatives of lungfishes.

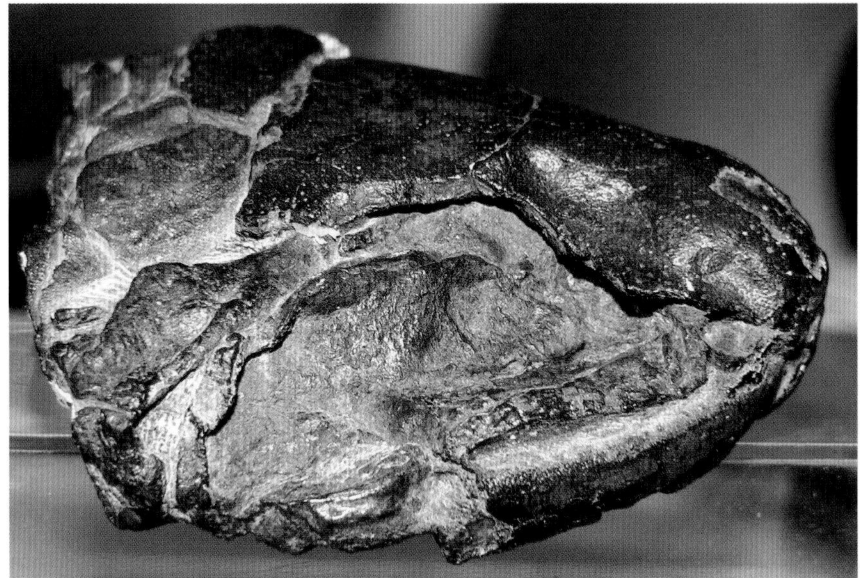

The head of *Porolepis* showing the division of the skull into two halves and robust palatoquadrate (upper jaw and roof of mouth) bones.

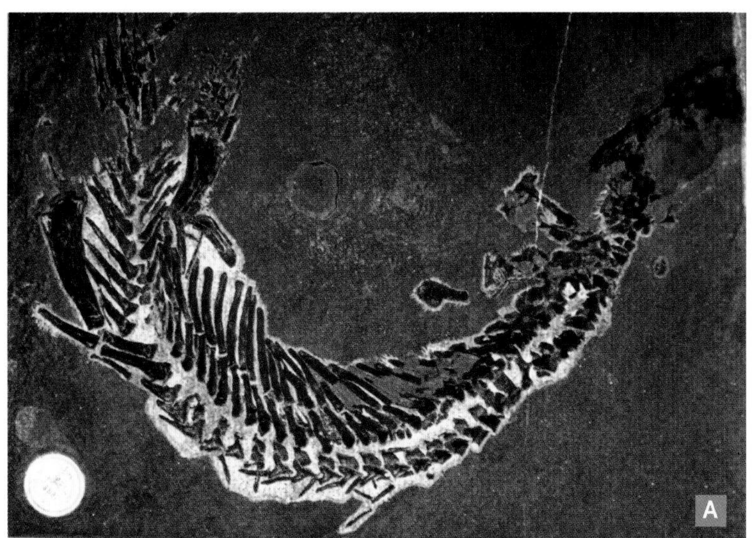

The internal skeleton of *Glyptolepis* (*A*), from the Middle Devonian of Scotland, shows the presence of strong fin support bones, a sarcopterygian feature not developed to this extent in ray-finned fishes. The Middle Devonian *Laccognathus* from the Lode Quarry in Latvia (*B*) was a flat-headed predator with very small eyes. Porolepiforms most likely relied on their sensory-line systems and ambushed their prey. (Panel B, courtesy Oleg Lebedev, Borissiak Paleontological Institute, Moscow)

bears enlarged tooth whorls at the front of the lower jaws. These tooth whorls differ from those of the onychodontids in that they have a series of large teeth in parallel rows. The whole dentition of porolepiforms reflects their predatory diet—large teeth or fangs appear regularly along the lower and upper jaws, flanked by several series of smaller gripping teeth. The robust ventral gill arch bones also had many small bones bearing toothlike denticles.

One of the more widespread Middle Devonian holoptychioids was *Glyptolepis,* known from several species found in East Greenland and Scotland and the subject of a detailed study by Erik Jarvik of Stockholm. *Glyptolepis* thrived in the Middle and early Late Devonian of the Old Red Continent and may have reached sizes close to 1 m in length. Recently some material of so-called *Glyptolepis* and *Holoptychius* species from Scotland has been redescribed by Per Ahlberg of the University of Uppsala

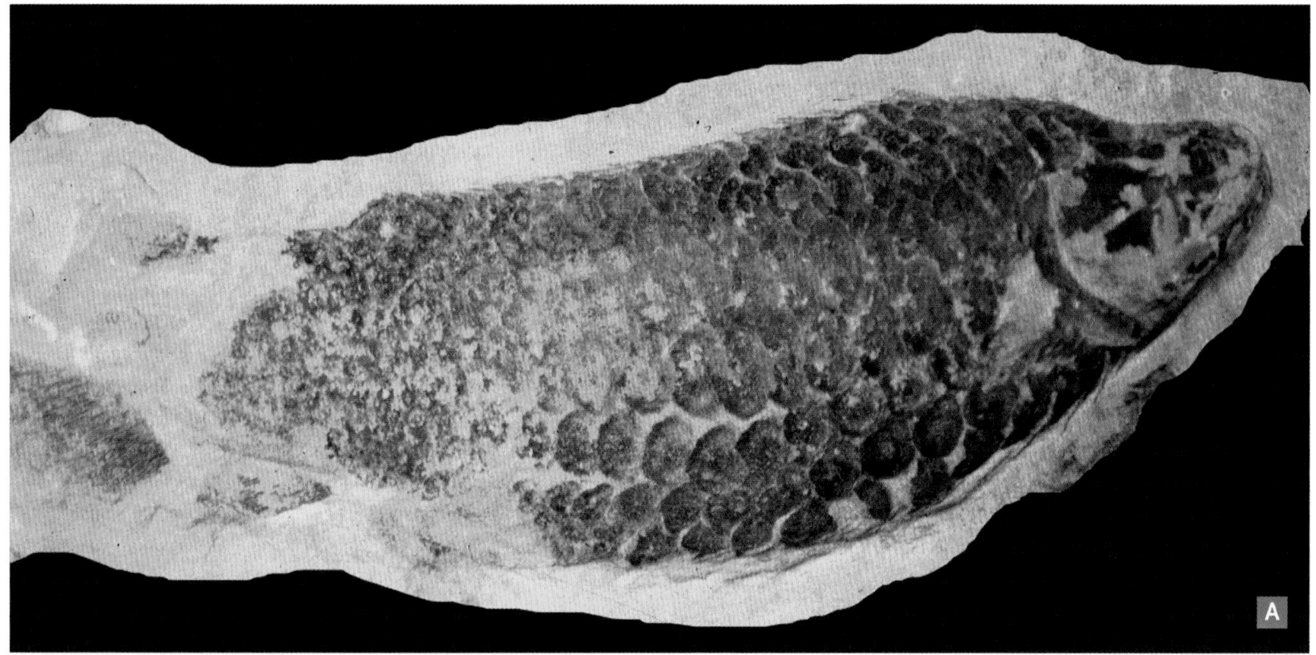

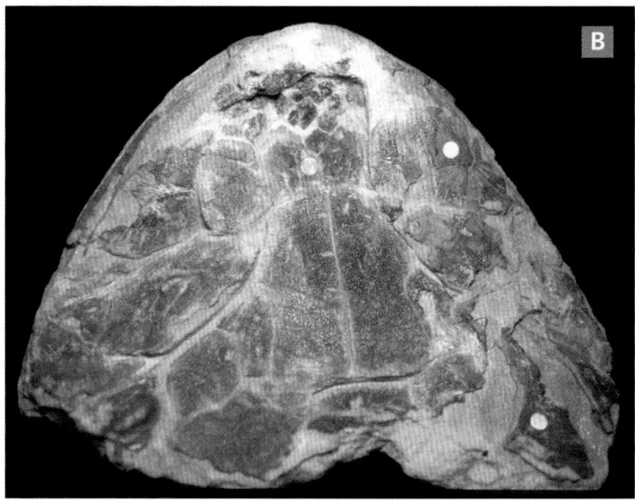

Holoptychius **was a widespread Late Devonian porolepiform. These specimens are from Dura Den, Scotland; whole fish (*A*); skull in dorsal view (*B*).** (Panel B reproduced with the permission of the Natural History Museum, London)

as new forms, such as *Duffichthys*, from the Scat Craig Beds of Elgin (Ahlberg 1994). *Duffichthys* is known only from lower jaws, which are unusual in their very large attachment area for the symphysial tooth whorl (Ahlberg 1994).

The largest and one of the more widespread members of the porolepiform group was *Holoptychius*, which lived near the end of the Devonian. The scales of the largest species, *Holoptychius nobilissimus*, indicate that the reconstructed fish may have been up to 3 m long, making it a formidable predator in the ancient river and lake systems of North America, Greenland, Europe, parts of Asia, and maybe Australia. Whole body fossils of *Holoptychius* are well known from the famous Dura Den site in Scotland, where schools of these fishes died and were rapidly buried by windblown sands. On average the Dura Den *Holoptychius* are relatively small species, less than 1 m in length. *Holoptychius*-type scales have been found around the world, indicating the genus was widely dispersed and capable of transgressing saltwater to invade new river systems. Marine porolepiforms are known from Latvia (for example, *Laccognathus* from the Lode deposit), although in general they are rarely found outside of river or lake deposits.

The Dipnoi: Twice Breathers

The name "dipnoan" for lungfishes comes from the ancient Greek meaning "two lungs" because of their ability to breathe twice, with gills in the water and with lungs to gulp air. Although fossil lungfishes have been known from the Old Red Sandstone rocks of Scotland for more than 200 years, when the three known living

Basic Structure of a Primitive Dipnoan

The one adaptation that seems to unite all the special anatomical features of early dipnoans is the ability to produce a powerful bite. The numerous tightly interconnected bones of the skull roof relate to powerful jaw muscle insertions on the inside of it. The massive size of the region where the lower jaws meet (the symphysis), the fusion of the palate to the braincase, the heavily built-up gill arch bones, and the specialized dental tissues, all reflect a skull capable of exerting great power in the bite.

The braincase of dipnoans is heavily ossified as a single piece, with the palate firmly fused to its lower surface. In primitive dipnoans, the braincase has struts supporting the skull roof, creating large chambers for the passage of jaw muscles, which attach on the inside of the skull roof and run down to the lower jaw. The parasphenoid one, which sits in the middle of the palate, is short and plowshare-shaped in primitive forms, although in later lineages it became expanded with a long posterior stalk. This elongation gave more room in the mouth for the gulping of air bubbles.

The snout of lungfishes has visible grooves for the incurrent nostril situated along the upper border of the mouth. The excurrent nostril opens from the nasal capsule directly into the palate, without any bones covering the nasal capsule. Primitive dipnoans have a complex system of minute tubules running through the bone of the snout and ends of the lower jaws. Some scientists have interpreted these as an electrosensory system, like the detecting devices used by some modern fishes for finding food in muddy environments. Other scientists believe that the complex network of tubules in the snout was part of a nutritive system that fed the skin and sensory-line canals.

Lungfishes fed by one of two main methods. They either had hardened tooth plates for crushing food or a mouth covered with small denticles that were periodically shed (called *denticle shedders* or *denticulates*). The oldest dipnoans are characterized by having a powerful crushing bite, as seen in the large area where the lower jaws meet in the midline, and the large attachment areas for the jaw

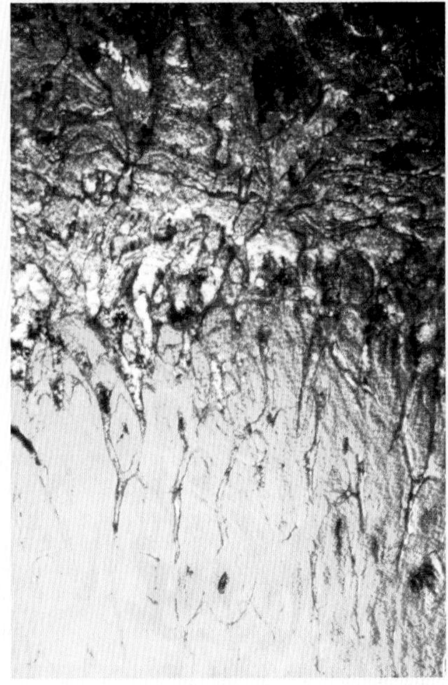

A highly magnified cross-section through the toothplate of a *Chirodipterus* highlights the dense, mineralized dentin tissue that allows such lungfish to crush up hard-shelled prey.

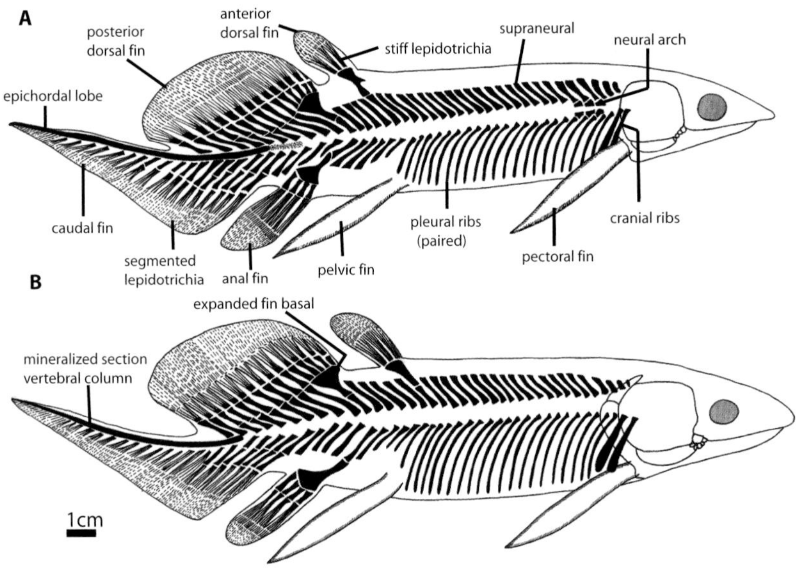

Skeletons of two closely related Middle Devonian lungfishes, *Howidipterus* and *Barwickia*. Similar postcranial skeletons indicate a relatively recent time of divergence for these separate genera.

musculature. Primitive biters include forms with palates covered by shiny dentin, a tissue found below the enamel layer in most vertebrate teeth. These dentin-plated forms gave rise to tooth plates with rows of teeth organized on tooth ridges. The denticle shedders, however, have powerful gill arch bones lined with smaller denticle-covered bones for rasping food against the denticle-covered palates. The gill arches of lungfish feature large ceratohyal elements and the hyomandibular does not take part in the jaw articulation, as in other osteichthyans. Many toothed bones accompany the gill arch series in denticle-shedding lungfishes.

The bodies of early lungfish have two equal-sized dorsal fins, a separate anal fin, and a heterocercal caudal fin, with long feathery paired pectoral and pelvic fins. Throughout the Devonian, the trend was to change to having a shorter first dorsal fin with a longer second dorsal fin, with eventual merging of the median fins and tail fin. By the end of the Devonian, 355 million years ago, lungfishes had acquired the body shape and fin plan that they were to keep for the rest of their evolution.

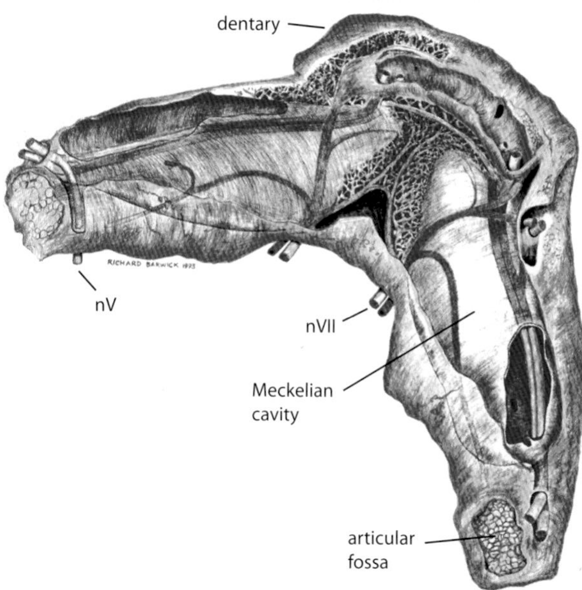

Sketch showing some key anatomical features of the lower jaw of a primitive lungfish, *Speonesydrion*. nV = nerve V in the cranial numbering system. (Courtesy Richard Barwick)

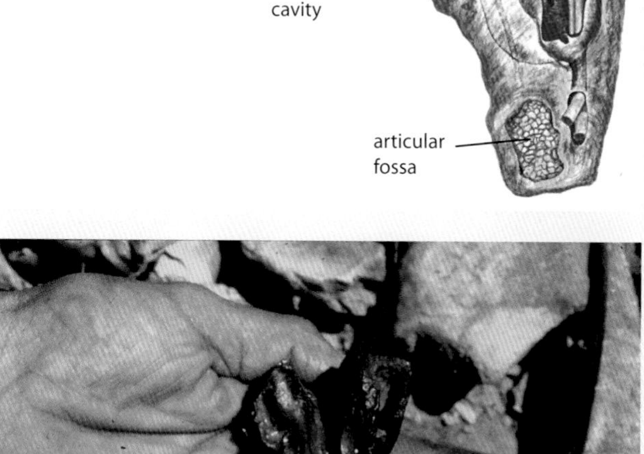

Massive lower jaws of a fossilized lungfish as found in the field at Taemas, New South Wales, Australia. (Courtesy Gavin Young)

The soft anatomy of lungfishes is characterized by a number of unique features including specializations of the nervous system such as concentrically layered olfactory bulb and Mauthner cells in the brain. Lungfishes have a three-chambered heart, although partitioning in the atrium is only partial, and in *Neoceratodus* it is barely evident. Their lungs are developed as an outpocketing of the gut, developed from the primitive osteichthyan swim-bladder. The modification of the swim-bladder to form a functional lung was simply a matter of increasing the internal surface area for improved gas-exchange ability and was thus not a complex evolutionary step.

The Early-Middle Devonian lungfishes had thick rhombic scales covered with cosmine, a shiny enameloid layer over the bones and scales, which housed a system of pores and interconnecting canals in the dentine beneath. The cosmine layer was subsequently lost in most dipnoans by the Late Devonian, and the bodies of these advanced fishes were covered by thinner, rounded scales. The snout of primitive cosmine-covered dipnoans was ossified as a stout single unit, which often broke away from the skull after death and can occasionally be found as isolated fossils (this is termed the "loose-nose problem" by Erik Jarvik). As cosmine disappeared from the skeleton, the ossified snout was replaced by one formed of soft tissue, sometimes with special small bones covering its top surface.

lungfishes were first discovered in the 1830s, some of the early zoologists could not believe that they were fishes.

The first discovery of a living lungfish was made in 1836 by Viennese naturalist Johann Natterer, who collected *Lepidosiren* specimens from the mouth of the Amazon River, Brazil. He sent his material to Leopold Fitzinger, Curator of Reptiles at the Imperial Museum in Vienna, who wrote to Count von Sternberg about the animal. The letter was later read before the Society of Natural History in Jena. Although Fitzinger's specimens were gutted, a remnant of the lung was found, and this, together with the unusual nostrils placed near the upper lip, lead him to describe the creature as "undoubtedly a reptile."

In the following year, great British anatomist Richard Owen was able to observe a specimen of an African lungfish collected from the Gambia River and presented to the Royal College of Surgeons. Owen, too, was puzzled by the unusual mosaic of fish and amphibian features seen in the specimen, but concluded, given the structure of the nasal sacs and nostrils, that it must be a fish. Although Owen had earlier proposed the genus name *Protopterus* for the specimen, he changed it back to *Lepidosiren* after receiving Natterer's paper of 1838. The first Australian lungfish, *Neoceratodus,* was found in Queensland in 1870, 32 years after the closely related genus *Ceratodus* was described by French paleontologist Louis Agassiz from fossil tooth plates. Albert Gunther, of the British Museum of Natural History, gave the first detailed anatomical description of the Queensland lungfish in 1871 and recognized it as being a fish. Because *Neoceratodus* was immediately recognized as being the most primitive of the three genera of living lungfishes, most subsequent work focused on the anatomy and embryology of this genus.

Fritz Muller coined the name Dipnoi in 1844, but the confusion over how dipnoans, both living and fossil, should be classified resulted in a string of newly proposed names over the next two decades, none of which has stood the test of time: order Ichthyosirenes (Castelnau 1855), family Pneumoichthyes (Hyrtl 1845), Ichthyosirens (M'Donnel 1860), order Protopteri (Owen 1853), order Pseudoichthyes (Owen 1859). Hogg (1841)

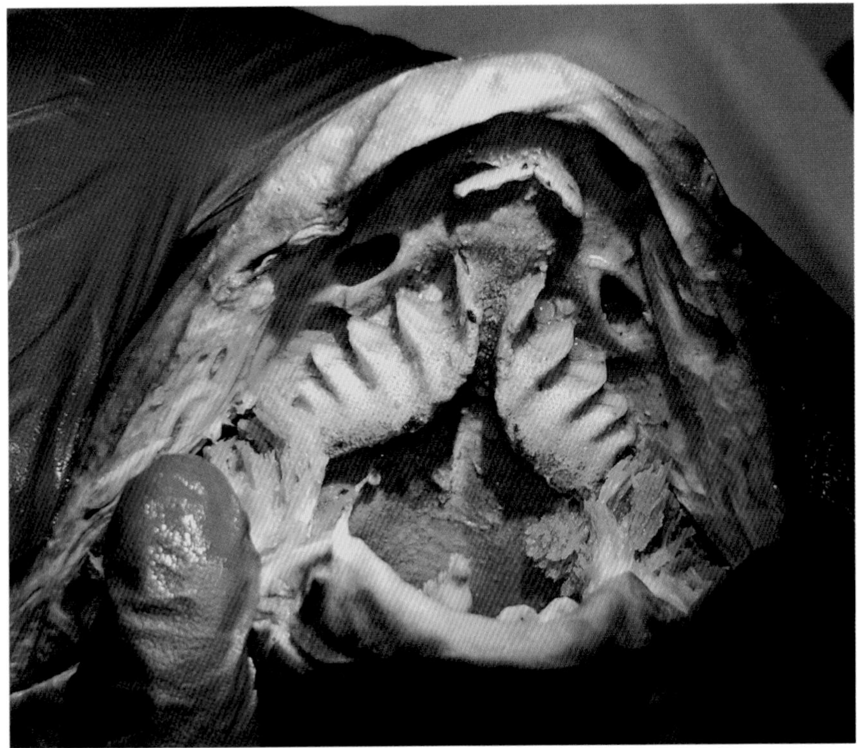

The Queensland lungfish, *Neoceratodus forsteri*, with mouth open to show tooth plates and palatal nostrils.

probably takes the cake for placing them in his tribe Fimbribranchia and family Amphibichthyidae! Lungfishes are now placed in the subclass Dipnoi, and three living genera (making up six living species) are placed in two families, the Lepidosirenidae (*Lepidosiren, Protopterus*) and the Ceratodontidae (*Neoceratodus*).

Today we have an excellent fossil record of lungfishes, especially so from Australia, where remains of these fishes have been found from almost every period of geological time since their first appearance at the beginning of the Devonian.

The remains of the first lungfishes come from limestones and shales containing abundant fossils of marine invertebrates, indicating that the lungfishes lived in shallow sea environments. Since the end of the Carboniferous Period dipnoans have inhabited only freshwater environments, and in the Permian Period some had acquired the ability to estivate. The story of lungfish evolution can be summarized as one of frantic, marvelous change in the Devonian Period, followed by slow steady change during the later part of the Paleozoic Era and into the Mesozoic. Since then they have remained almost unchanged. Our story is largely concerned with that initial burst of evolutionary radiation back in the Devonian, when most of the radical changes took place.

Lungfish Origins

From their first appearance, the lungfishes were a highly unique group, easily recognized by a number of anatomical features in their skull roof and dentition. *Diabolepis,* meaning "devil scale" (first named *Diabolichthys* in 1984), is the first member of the group leading to lungfishes. It was found back in the early 1980s from the Early Devonian of Yunnan, China, and described by Chang Meeman. This fish has a skull roof pattern with

(*Left*) The palate of *Diabolepis speratus* from the Early Devonian of China, the oldest and most basal member of the lungfish lineage. Note that the denticles cover the parasphenoid and vomer regions, showing an adaptation for crushing food. (Courtesy Meeman Chang, Institute of Vertebrate Paleontology and Paleoanthropology, China)

(*Right*) The skull roof of *Diabolepis*, showing the remnant suture between two parts of the braincase, which has fused in all later lungfishes. (Courtesy Meeman Chang, Institute of Vertebrate Paleontology and Paleoanthropology, China)

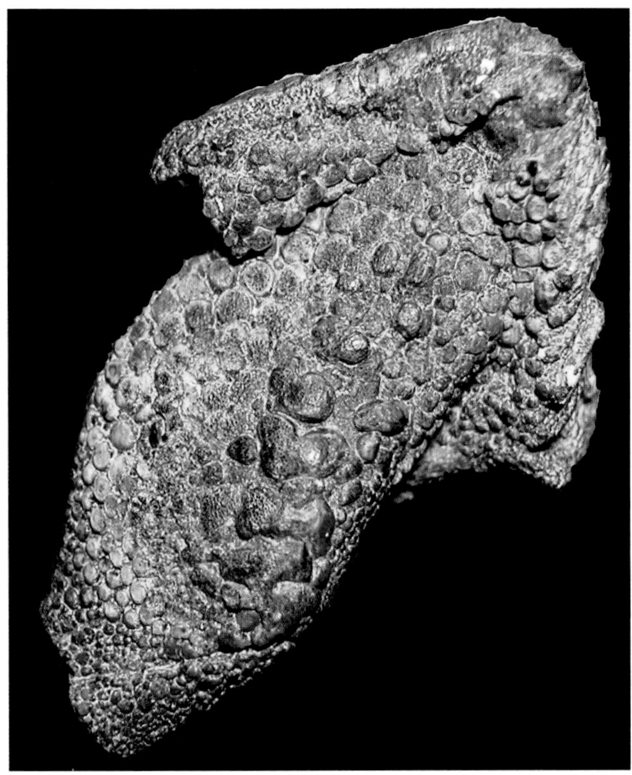

(*Top*) **A lower jaw of *Diabolepis*, showing crushing prearticular tooth plate with robust rounded teeth.** (Courtesy Institute of Vertebrate Paleontology and Paleoanthropology, China)

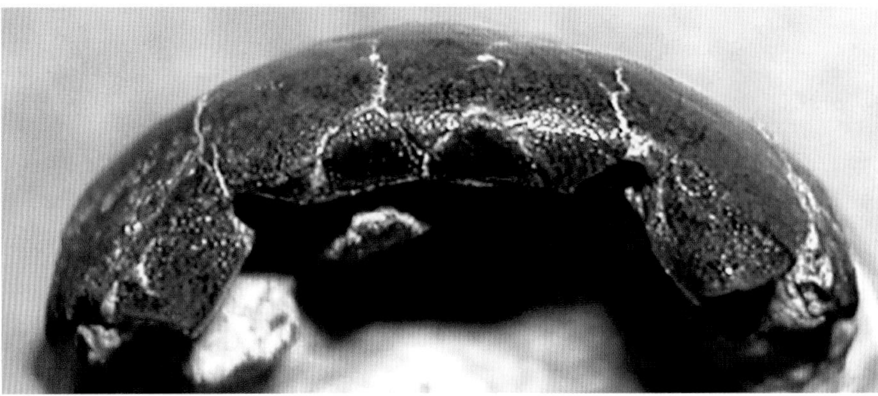

(*Bottom*) **The front of the snout of *Diabolepis* shows the narial grooves for the palatal nostrils.** (Courtesy Institute of Vertebrate Paleontology and Paleoanthropology, China)

a median "B" bone in front of the contacting "I" bones (as in nearly all other sarcopterygians) and the two sides of the lower jaws meet in an extensive, strong symphysis. The anterior external nostril is situated on the upper mouth margin as in dipnoans, and broad crushing tooth plates are present. Despite the ongoing debate over the relationships of *Diabolepis,* most now agree that it is undoubtedly a close link between the first true dipnoans and the stem sarcoptetrygian fishes. Although the oldest true dipnoan, *Uranolophus,* comes from North America, the Early Devonian limestone rocks of southeastern Australia hold the greatest diversity of primitive dipnoans and possibly the key to understanding the early evolution of the group.

Dipnoan Diversity

Dipnoans exhibit a wide range of dentitions, from thick dentin sheets covering the plate and jaws through to tooth-plated forms and those with a shagreen of denticles for rasping hard kinds of prey. An amazing specialization of some lungfishes, like the holodontids, is the ability to remodel and reshape the dentition throughout their life. Dipnoan tooth plates may comprise a great diversity of tissue types, many of which have been lost in later dipnoans. Heavily mineralized tissues, such as petrodentin, gave certain dipnoans the ability to develop extremely powerful crushing bites and to crack and grind up hard-shelled prey items, such as primitive clams, lamp-shells, and corals.

The earliest well-preserved lungfish is *Uranolophus,* from North America, an Early Devonian form with a denticle-covered palate. Its primitive nature is seen by the two equal-sized dorsal fins, thick cosmine-covered rhombic scales and a skull roof pattern with large "I" bones meeting each other behind the "B" bone, and numerous small bones forming the snout. The lower jaws meet in a strong contact zone, suggesting that *Uranolophus* had powerful jaw muscles for exerting much pressure when it bit. Other Early Devonian fishes like the dipnorhynchids had powerful sheets of dentin covering the plate for crushing up hard-shelled prey such as clams. Several species of these large lungfishes have been described from the Taemas limestones by Ken Campbell and Richard Barwick. Some, like *Speonesydrion* and *Ichnomylax,* show the beginnings of tooth rows and distinct heels on the lower jaw tooth plates.

In the Middle and Late Devonian, denticulate lungfishes existed side by side with tooth-plated ones, often together in the same environments. Many of the fossil fish faunas of this age contain examples of both groups,

(*Left*) *Uranolophus*, an Early Devonian lungfish from Wyoming, showing the denticle-covered palate. Here tooth plates covered in denticles are merged with palatal bones to form an entire palatal surface for grinding food.

(*Right*) The palate of *Dipnorhynchus sussmilchi* from the Early Devonian of Australia shows the heavy layers of dentin and roughened tuberosities used to crush hard-shelled marine organisms.

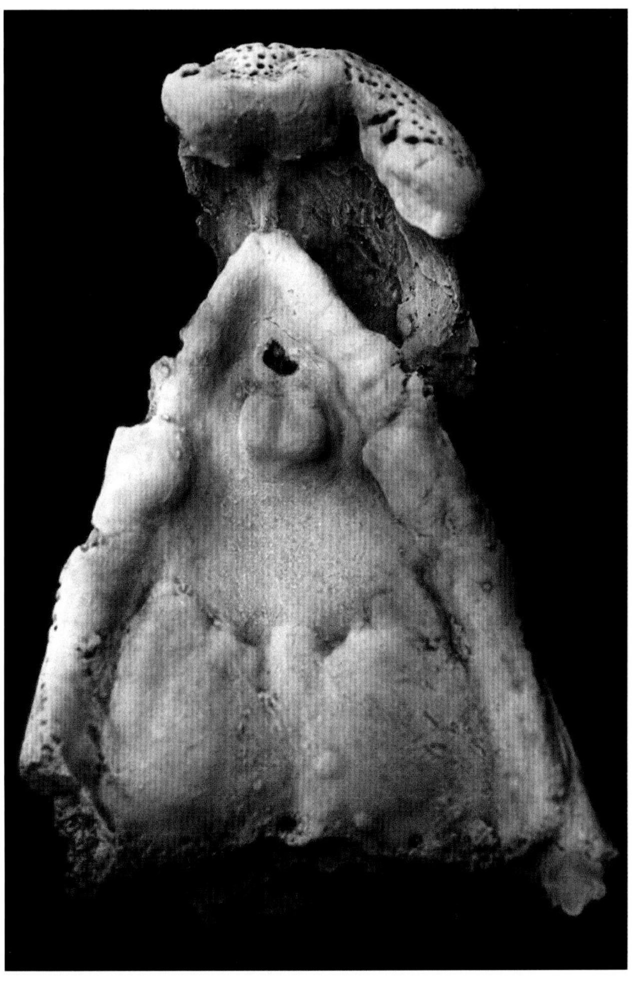

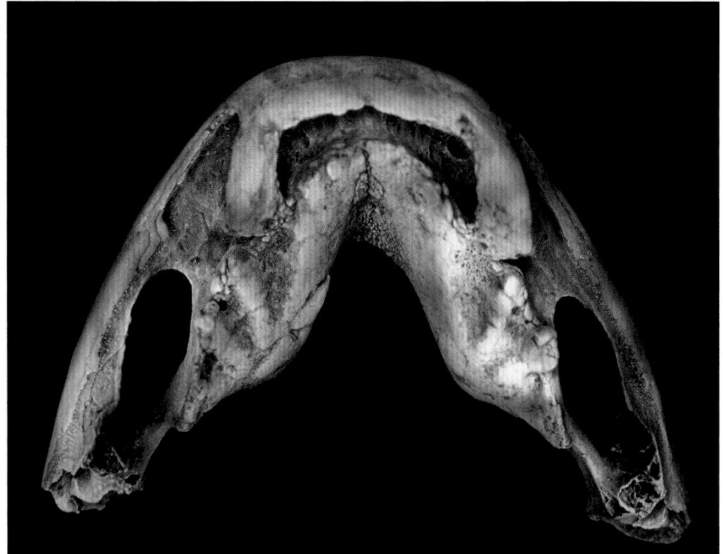

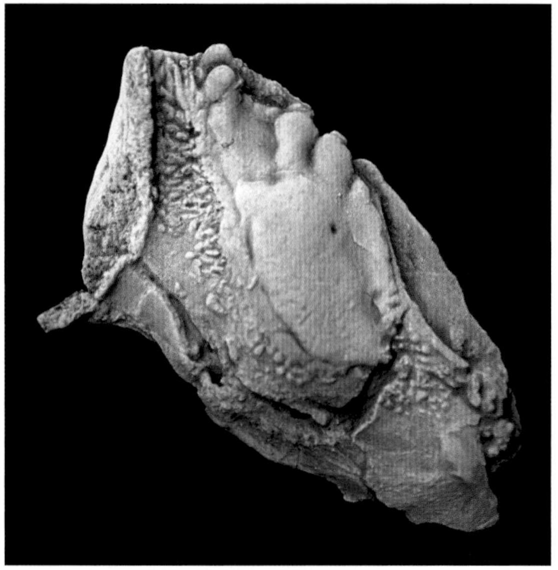

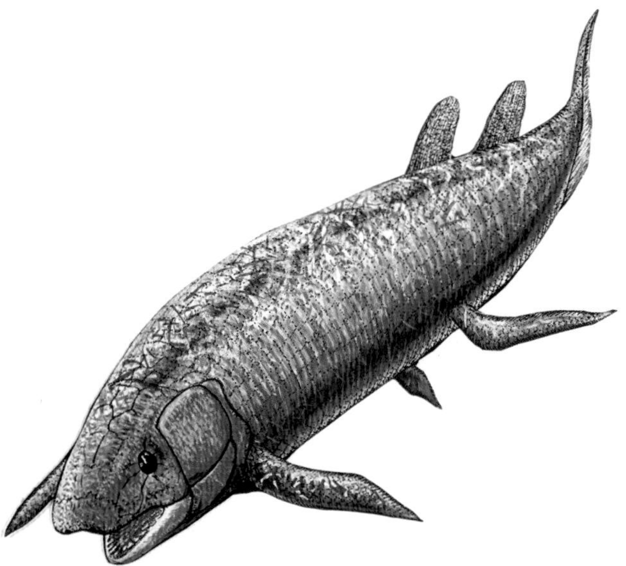

(*Top Left*) ***Speonesydrion iani***, **an Early Devonian lungfish found near Taemas, New South Wales, Australia, represents the first stage of lungfish that acquired tooth plates. The lower jaw shown here has developed rudimentary tooth rows.**

(*Top Right*) ***Ichnomylax kurnai*** **is one of the oldest lungfishes known. It is represented solely by the damaged half of a lower jaw found in limestones exposed on the beach near Walkerville, Victoria, Australia.**

(*Bottom*) **Restoration of** ***Dipnorhynchus***, **a large Early Devonian lungfish from Australia, with some lower jaw specimens indicating body lengths of up to 1.5 m.**

indicating that various dipnoans with different feeding strategies could exist within the same environment. One of the most successful of the tooth-plated forms was *Dipterus*, from the Old Red Sandstone of Scotland and other Euramerican localities. It had an ability to invade freshwater habitats of the Orcadian Basins first after water level changes. At first it was thought that when lungfishes went from marine to freshwater they began to gulp air. Recent work has shown that a marine lungfish, *Rhinodipterus,* had acquired specializations for breathing air, so the main driver for this event could well have been lower global oxygen levels (Clement and Long 2010). The denticulate dipnoans include the long-snouted *Griphognathus* and *Soederberghia* that inhabited both marine and freshwater environments. Australia has several good examples of these Devonian dipnoans, such as the duck-billed lungfish *Griphognathus whitei,* a common fish at Gogo in Western Australia, and the holodontid lungfishes, also known from Gogo, which include several species of the short-snouted, massively built *Holodipterus* as well as new forms only recently described like the massive *Robinsondipterus*. *Soederberghia,* first described from the Late Devonian of East Greenland, is also known in Australia from one skull roof found in red mudstones from a quarry at Jemalong Gap, near Forbes in New South Wales, indicating widespread dispersal of some dipnoans by this time.

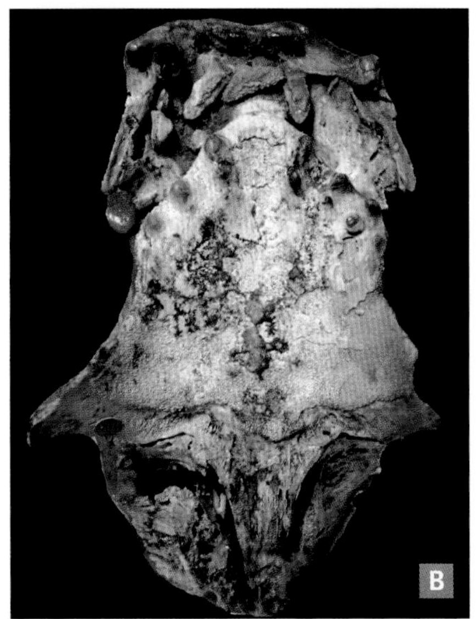

***Dipterus*, a Middle Devonian lungfish (*A*) from Scotland, is one of the oldest lungfishes known to have had paired cranial ribs that attached to the skull. This suggested that these fishes were capable of air-gulping. *Dipterus* was also the first fossil lungfish to be scientifically described, by Sedgewick and Murchison in 1828. Skull of the large lungfish, *Holodipterus* (*B*), one of the rarer fossil lungfishes found in the Gogo Formation of Western Australia. Note the bulbous crushing surfaces of the palate. Skull of *Robinsondipterus* (*C*), a large holodontid from the Late Devonian Gogo Formation of Western Australia. This skull measures 20 cm in length, indicating it was one of the largest of the Gogo lungfishes.**

Griphognathus is an unusual-looking fish with a long, flat ducklike bill. Its strong gill arch skeleton has many muscle attachment scars indicating strong muscles to move the ventral gill arch bone (basibranchial) sideways and up and down, exactly like a file rasping away at a hard surface. This technique, coupled with the pliers-like duckbill, may have enabled *Griphognathus* to snap off long pieces of branching coral and then grind them up with the denticle-covered gill arch bones, palate, and lower jaws. Alternatively, it may have used its long snout for nuzzling along the muddy seafloor, sensing for soft-bodied worms and other creatures. The success of *Griphognathus* is seen by the fact

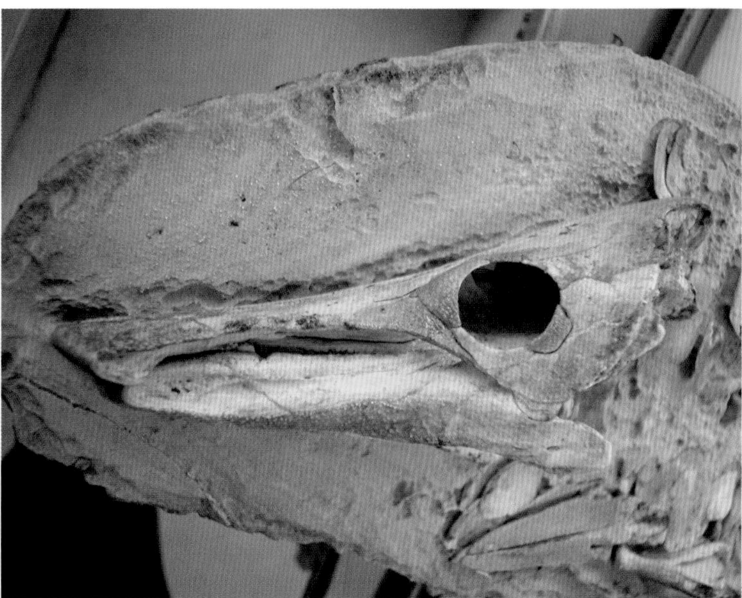

The snout of a *Griphognathus* just poking out of the rock (*A*) as found in the field by the author at Gogo in 1986. A superbly preserved skull of the long-snouted lungfish, *Griphognathus whitei* (*B*), from the Late Devonian Gogo Formation, Western Australia. Its long snout may have been used for nuzzling around the muddy seafloor while using its electrosensory system to detect invertebrates.

that it is one of the most widespread of all dipnoans in the Late Devonian, being found in North America, Europe, and Australia.

I'll never forget the day in August 1986 when I found a complete *Griphognathus* on my first season working at Gogo in Australia. First I picked up a large block showing only the shiny cosmine tip of the snout emerging (shown above, top left). I knew immediately that I had a complete undamaged skull inside the rock. Then, only a meter or so away from this, I found another three blocks, each showing a cross-section of the body in the round. Incredibly the blocks all fitted together to make a large sausage-shaped nodule. Inside was one of the world's most complete Devonian lungfishes.

The last known denticle-shedding lungfish is *Conchopoma*, from the Permian of Germany and North America, which had a tail fin like modern lungfishes and broad median palate bone (parasphenoid) covered in numerous denticles. In appearance *Conchopoma* resembles many of its contemporary tooth-plated lungfishes. The reason for these assumed parallel series of changes in differing lungfish lineages may be that both had a similar developmental plan. If juvenile fish of both groups follow the same pattern of growth and development (called *ontogeny*), then changes in the timing of this development can produce parallel changes in different groups. This form of evolutionary change is called *heterochrony* and is now used to explain how much of evolution was not necessarily driven by external environmental pressures but controlled instead by internal developmental factors within the organism. A clear evolutionary trend in some Devonian lungfishes is the retention of juvenile features into the adult stage of later species. This is called *paedomorphosis* and can explain why some features, such as large eyes and shorter cheeks, can evolve rapidly just by earlier sexual maturity within a lineage of fishes (Bemis 1984).

The transition from denticulate palates to tooth plates may not have been that great an evolutionary step. There are some denticulate lungfish that show massive dentin-covered palates with large toothlike cusps (for example, *Holodipterus gogoensis*), while other species found in the same site are slender-jawed with fine denticle shagreen covering the palate and biting areas of the lower jaw (*Robinsondipterus*).

Dentin-Covered Palates: The Early Crushers

Three of the world's few known Early Devonian lungfish have been found in southeastern Australia. All of these are primitive lungfishes with powerful crushing dentitions formed of thick dentin sheets covering the palate and biting surface of the lower jaws. The skulls are known in two of these genera and show the primitive condition of the bone pattern, in having many small bones around the snout and front of the skull roof. The best known of these, *Dipnorhynchus* (meaning "two lungs snout"), is represented by three species, known from the Taemas-Wee Jasper and Cooma regions of New South Wales and Buchan district of Eastern Victoria. All three species feature a heavily built palate with bulbous tuberosities used for exerting great pressure on their prey, functioning much like a nutcracker, to smash clams and other hard-shelled delectables.

Speonesydrion (from the Greek, meaning "cave island") is named after the site where the fossil was found, on Cave Island in Burrunjuck Dam, near Taemas. This fish has a skull roof pattern similar to *Dipnorhynchus*, but the palate and lower jaws have rudimentary rows of tuberosities, forming the primitive plan for a crushing tooth plate. Similarly, *Ichnomylax*, known only from one side of the lower jaw found near Bell's Point (Waratah

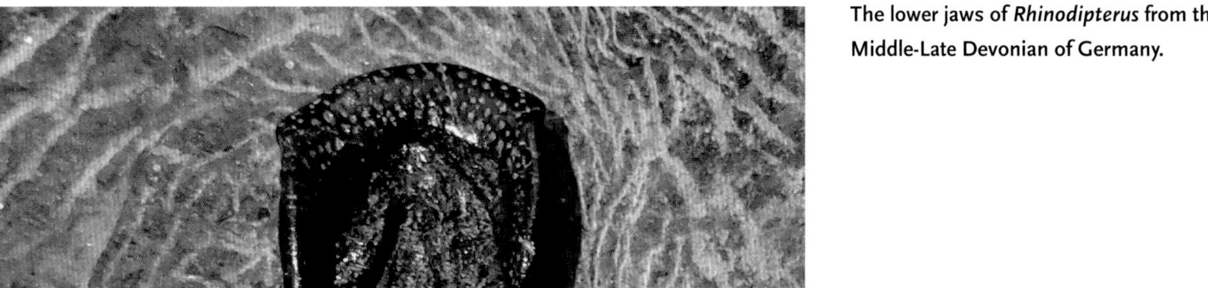

The lower jaws of *Rhinodipterus* from the Middle-Late Devonian of Germany.

Bay), Victoria, is older than the previous forms and has primitive tooth ridges and a bulbous dentin-covered crushing heel on the inside of the lower jaw.

Refined Feeding Mechanisms

Anyone who has watched a Queensland lungfish feeding will never forget the experience. Food is taken into the mouth, chewed up using the tooth plates, and extruded out again as a pulpy long tube. Then it is chewed again and again, until all the material has been reduced to a readily digestible mass. The ability to feed in this manner must have been in use in the Devonian when the first true tooth-plated dipnoans evolved. One group, the chirodipterid family, has tooth plates that lack teeth, most lacking any cusps. These are termed *dental plates,* as opposed to true tooth plates that possess teeth (individual cusps that are added on at the margins of the plates with continued growth). In Australia, the chirodipterids are well represented by three species, all from the Late Devonian Gogo Formation of Western Australia.

Chirodipterus, first described from material in Europe and North America, is a commonly found genus in the Gogo Formation. *Chirodipterus australis* has broad crushing tooth plates with weak tooth ridges on the dental plates, whereas *Gogodipterus paddyensis* has strongly developed tooth ridges with deep grooves between them on each dental plate. The third Gogo chirodipterid species, *Pillararhynchus longi,* has a deeper skull than either of the other forms and possesses long, narrow tooth plates with concave crushing surfaces. The palate bone (parasphenoid) has a patch of dentin on its front surface, indicating that the fish crushed food with that surface. The largest of the chirodipterids was a monster called *Palaedaphus insignis* from the Late Devonian marine deposits of Belgium. Its huge tooth plates measure

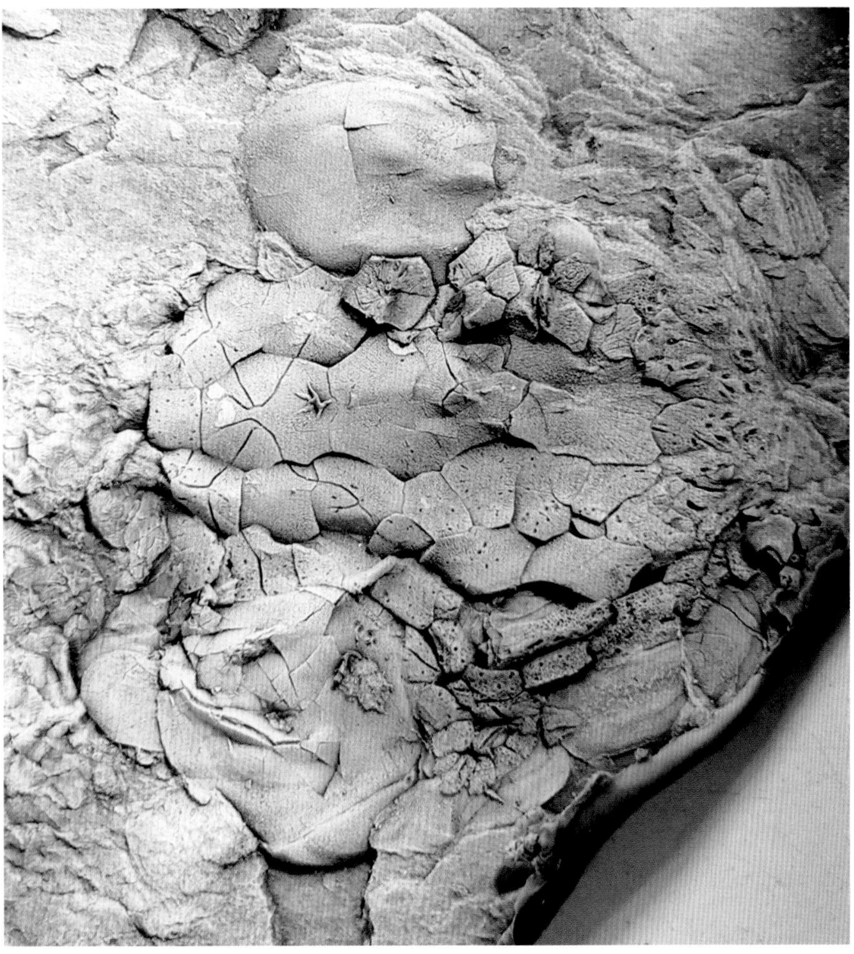

Skull of *Howidipterus*, showing much variation in the shape of bones and the degree of fusion between bones. This is a latex cast of the cleaned fossil that has been whitened with ammonium chloride to highlight the surface details. *Howidipterus* is a Late Devonian lungfish from Mount Howitt, Victoria, Australia.

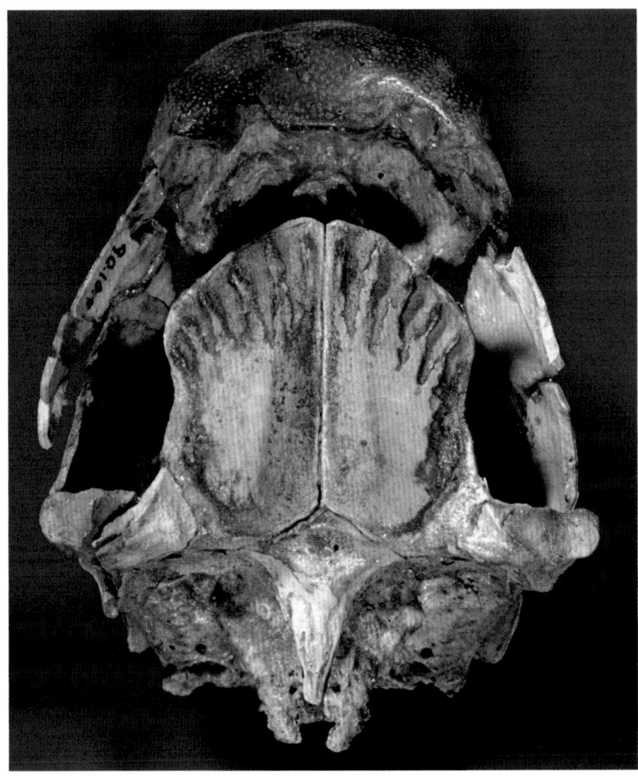

Palate and tooth plates of *Chirodipterus australis* from the Late Devonian Gogo Formation of Western Australia.

almost 14 cm long and 10 cm wide, suggesting that the fish may have grown to about 2 m in length.

Most of the known families of lungfishes have tooth plates. The Devonian dipterids represent the most primitive grade, as these fishes have two dorsal fins and possess shiny cosmine on the dermal bones. The fleurantiids and phaneropleurids are more derived than the dipterids, in that they have the first dorsal fin much reduced and the second dorsal fin enlarged, and they possess simplified skull roof and cheek patterns. Within the general evolution of Devonian lungfishes, we see the reduction of dorsal fins and merging of the anal fin with the caudal fin to give the same appearance as the modern Queensland lungfish. Steps in achieving this plan are seen in the transformation series going from *Uranolophus* to *Dipterus* to *Howidipterus* to *Scaumenacia* to *Phaneropleuron*. The final successful body plan was achieved by the end of the Devonian, as seen in the Scottish genus *Phaneropleuron* and retained in all later lungfish.

Howidipterus, from the Late Devonian of Victoria, had unusual tooth plates with well-developed teeth along the margins of each plate and smooth crushing surfaces toward the center of the plates. The skull had many primitive features such as the retention of the large "D" and "K" bones. The parasphenoid had a well-developed stalk, unlike the primitive small diamond-shaped type seen in *Dipterus*. A recent study of two lungfishes from the Late Devonian Mount Howitt site, Victoria, shows that although the two have superficially differing dentitions—*Howidipterus* has tooth plates, *Barwickia* appears to be a denticle-shedder—both have identical body shapes, a similar numbers of ribs, and a similar plan of fin-support bones. And when the dentitions are studied more closely, it appears that they have similar types of tooth plates, one being dominated by rows of teeth with few denticles (*Howidipterus*), the other having few rows of teeth but many denticles (*Barwickia*). They are regarded as both members of the fleurantiid group.

Other fleurantiids, such as *Andreyevichthys* from Russia, show that both kinds of dentition can exist in the same species as a matter of growth variations. Such precise similarities in the postcranial skeletons of the two Mount Howitt forms are unknown in other fossil lungfishes and strongly suggest that they evolved from a common ancestor with a similar body plan, into two distinct lineages having different feeding strategies. The Mount Howitt lungfishes lived in a large lake environment. Modern fish communities living in such lake environments are often based on a common ancestor entering the lake and then speciating into many similar forms, each with a slightly different feeding strategy (the lake-dwelling cichlids of Africa are an example).

All dipnoans of the Carboniferous Period had only one continuous median fin that merged with the tail fin, while their tooth plates could be quite specialized with numerous tooth rows and closely packed cusps (for example *Ctenodus*). The only Carboniferous dipnoan described from Australia is *Delatitia breviceps*, first named as a new species of the European genus *Ctenodus* by Ar-

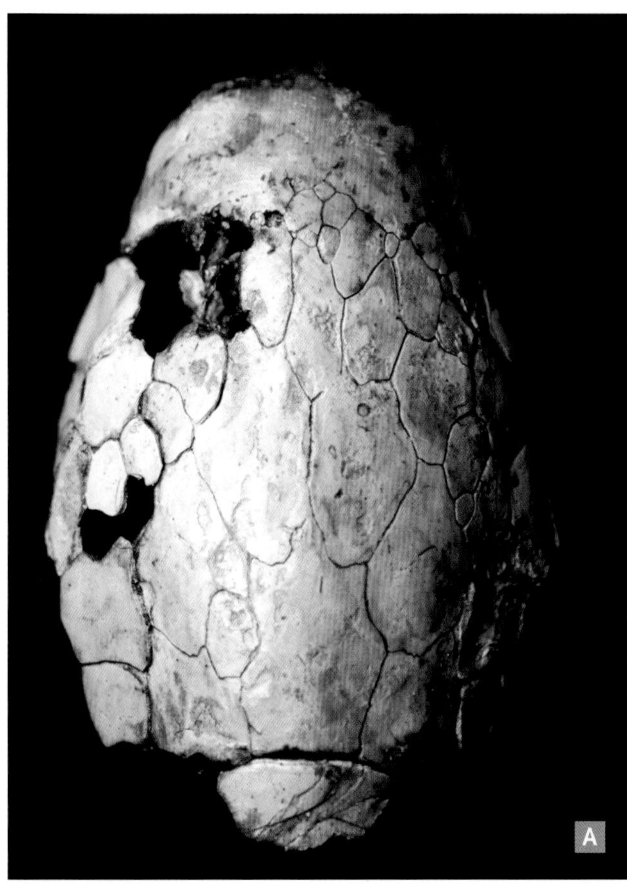

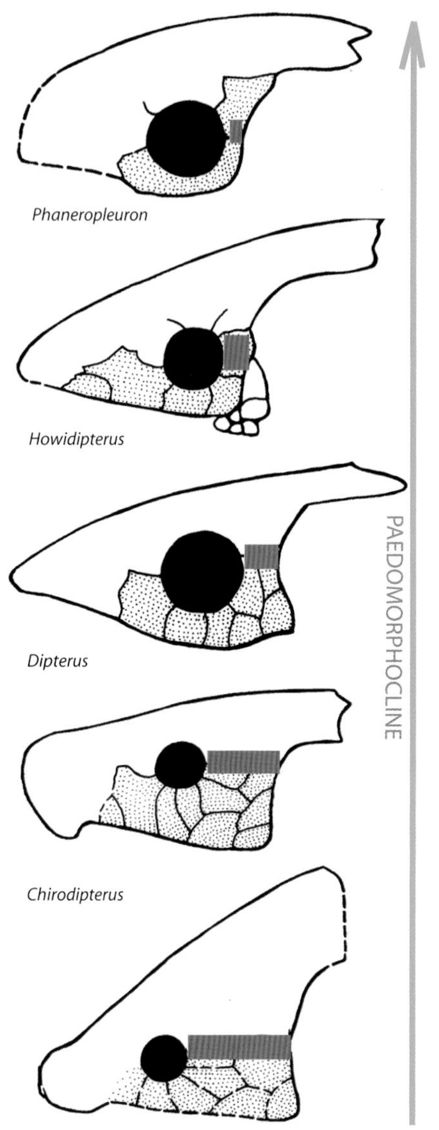

The skull roof of *Chirodipterus australis* from the Late Devonian of Australia (*A*). Evolution through heterochrony—literally "different timing," meaning when species evolve either juvenile or adult characteristics of their ancestors—as shown by the trend in Devonian lungfishes of shortening the cheek size and enlargement of the orbits (*B*), both features seen in juveniles of other species. This form of evolution is called paedomorphosis.

thur Smith-Woodward in 1906. Well-preserved lungfishes of the Carboniferous and Permian Periods of Europe and North America include *Conchopoma*, a denticulate form, and *Uronemus*, with modified narrow tooth plates. These and all other known subsequent lungfishes had achieved the body and fin pattern that still exists in the present day. The North American Permian *Gnathorhiza* is famous for being found preserved within its fossil burrows, testifying to the ability of some lungfishes of Paleozoic times to overcome droughts by their ability to estivate.

By the Mesozoic Era, the majority of lungfishes were ceratodontids (the group that includes the living Australian lungfish, *Neoceratodus*) or lepidosirenids (the group including the living African and South American lungfishes).

One of Australia's best-preserved Triassic lungfishes is *Gosfordia truncata*, from the Hawkesbury Sandstone near Gosford, New South Wales. It was first described

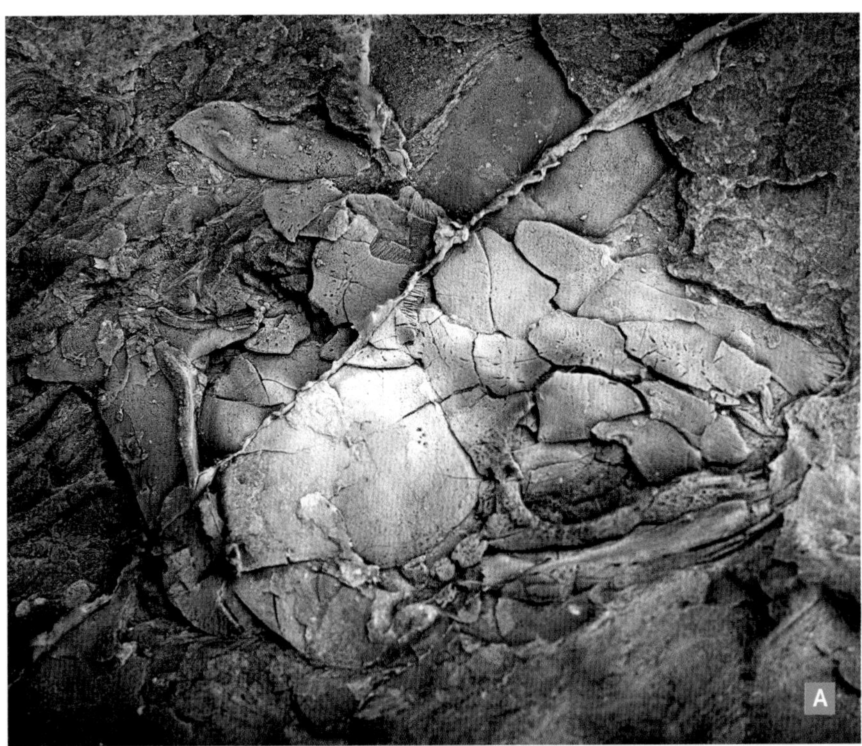

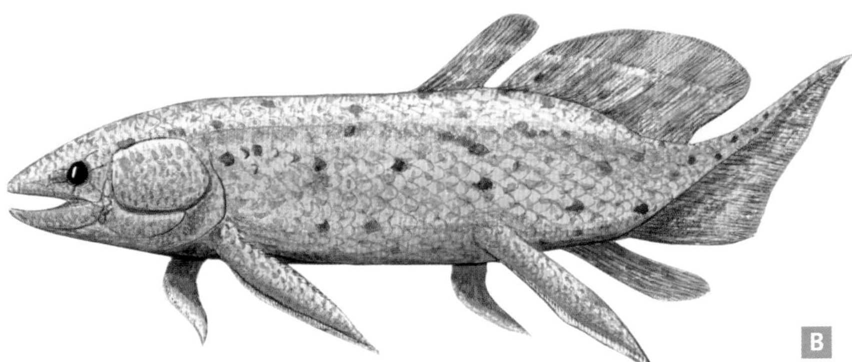

(*Top*) Skull of *Barwickia downunda* (*A*) from the Middle Devonian of Mount Howitt, Australia. Reconstruction of the Middle Devonian lungfish *Barwickia* (*B*).

(*Bottom*) *Scaumenacia*, a Late Devonian tooth-plated lungfish from Miguasha, Canada, has resemblances to the Australian *Howidipterus* but is more advanced in having an elongated first dorsal fin and in several features of the skull.

Gosfordia truncata, a complete fossil lungfish found in Triassic rocks near Gosford, New South Wales. _Gosfordia_ is a close relative of the lungfish lineage leading to _Neoceratodus forsteri_, the modern Queensland lungfish. (Courtesy Alex Ritchie, Australia National University)

Fossil tooth plate of the lungfish _Arganodus_ from the Early Cretaceous of Niger Republic, Africa, showing the distinctive tooth ridges used to triturate food.

and named by British paleontologist Arthur Smith-Woodward from about five incomplete specimens and was later redescribed by Alex Ritchie of the Australian Museum from a superb new specimen found in 1980 by quarryman John Costigan. *Gosfordia* has a deep, plump body and broad tail, and in overall length was about 50 cm. The body of *Gosfordia* suggests that it was a strong swimmer and not adapted for estivation like the South American and African lungfishes. *Gosfordia* was probably typical of the shape and size of most Triassic lungfishes.

Most other Mesozoic lungfishes from Australia are represented by only isolated tooth plates, such as those of *Ceratodus* found in the Triassic Blina Shale of North Western Australia, the Rewan Formation (of Southern Queensland), and the Wianamatta Group (of New South Wales). Tooth plates found in Early Cretaceous rocks indicate that the extant Queensland lungfish, *Neoceratodus forsteri,* was then living near Lightning Ridge,

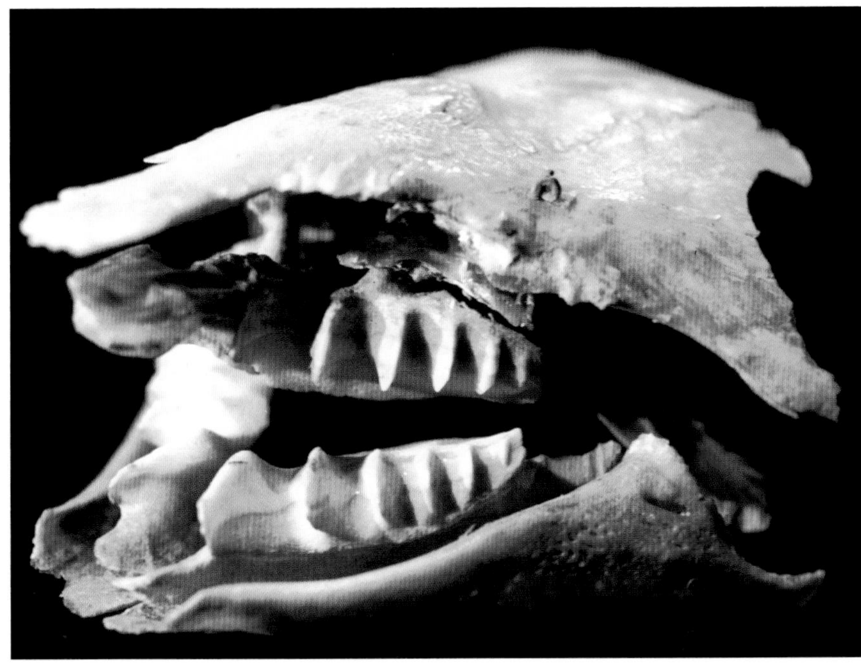

Skull of the extant lungfish *Neoceratodus forsteri*, showing the strong tooth plates for grinding up food.

while in Victoria *Ceratodus nargun* survived in rivers of the cold rift valley that formed as Antarctica and Australia were beginning their separation. The biggest lungfish of all time was *Neoceratodus tuberculatus*, from the Late Cretaceous of Egypt. Based on a single toothplate measuring 10 cm, it must have been around 3–4 m long (Churcher 1995).

During the Mesozoic Era, lungfish skull roof bones become simpler, and there is great variation in toothplate morphology, even within an individual genus such as *Neoceratodus*. In the Tertiary, fishes such as *Neoceratodus gregoryi* may have reached lengths in excess of 3 m, based on the large tooth plates and skull roof bones found from the Miocene river and lake deposits of central South Australia—and least four other *Neoceratodus* species were widely distributed in the lakes and rivers of central Australia and Queensland. Although many species are known only from tooth plates, the Redbank Plains area of Southern Queensland has yielded semiarticulated skull and body remains of *Neoceratodus denticulatus*, which are probably Eocene in age. These were first described in 1941 by Edwin Sherbon Hills of the University of Melbourne.

Today the Dipnoi are known from three surviving genera: *Proptopterus* in Africa, *Lepidosiren* in South America, and *Neoceratodus* in Queensland, Australia. Of these, the Queensland lungfish is by far the most primitive; it has changed little, if in fact at all, in more than 100 million years. The lepidosirenids are longer, more slender fish with greatly reduced pectoral and pelvic fins for sensory functioning. During the dry season *Protopterus* can burrow into the ground and await the next rainy season and is therefore adapted to survive in harsh climatic conditions where the Queensland lungfish would not survive. However, despite this, the Queensland lungfish has amazingly withstood Australia's harsh climatic changes and is still Australia's only primary freshwater fish—a species that has evolved here rather than being an immigrant. Despite the building of dams that threaten its breeding grounds in Queensland, one can only hope that its future is not critically endangered by humankind or our pollution of the planet.

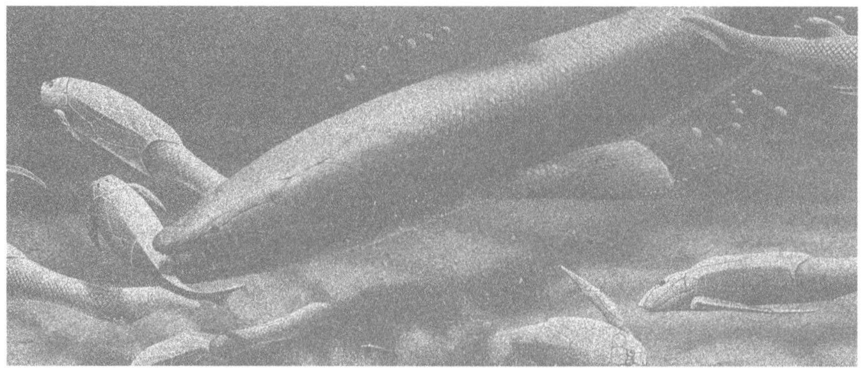

CHAPTER 12

Big Teeth, Strong Fins

Fishes with fin bone patterns like human arms and legs

The tetrapodomorphan, or four-foot shaped, fishes first appear in the late Early Devonian, represented by *Kenichthys* from China. They were characterized by their single pair of external nostrils and also possessed a choana, or palatal nostril, formed by migration of the rear nostril to the palate. But more important for future steps in evolution, the group had robust powerful limb skeletons that would serve them well to later develop arms and legs and to transition to living on land. The most successful members of this tetrapodomorphan group were a mixed bag of taxa known collectively as osteolepiforms, a branch of the sarcopterygians discussed in Chapter 10. These include the well-known *Eusthenopteron* from the Late Devonian of Canada and *Gogonasus* from Australia. Large members of the tristichopterid group terrorized the rivers and lake systems of the Late Devonian, with forms like *Hyneria* reaching 4 m in length. They were succeeded by even larger rhizodontiform fishes growing to lengths of 6 m or more. These slow-moving river and lake dwelling monsters were specialized to ambush amphibian or large fish prey from below and twist them into a savage death roll.

The Beginnings of the Tetrapod-like Fishes

The osteolepiforms were once regarded as a natural or monophyletic group, but recent work is showing them to be a grade group—species that share common characteristics but are not descended from the same ancestors—without any robust specialized features. Nonetheless, they were a diverse array of sarcopterygians that first appeared in the late Early Devonian with forms like *Kenichthys* from China, named by Zhu Min in honor of Ken Campbell. *Kenichthys* is unusual in having one pair of external nostrils (the anterior or incurrent nares) with a second or posterior pair situated along the upper jaw margin, straddling part of the palate, as demonstrated by Zhu Min and Per Ahlberg in their *Nature* paper (2004). The fishes thus show an intermediate stage in the evolution of the choana. Their cheeks show that some of the bones of early tetrapodomorphans might have primitively been fused into larger units, a condition seen also in very basal sarcopterygians like *Psarolepis*.

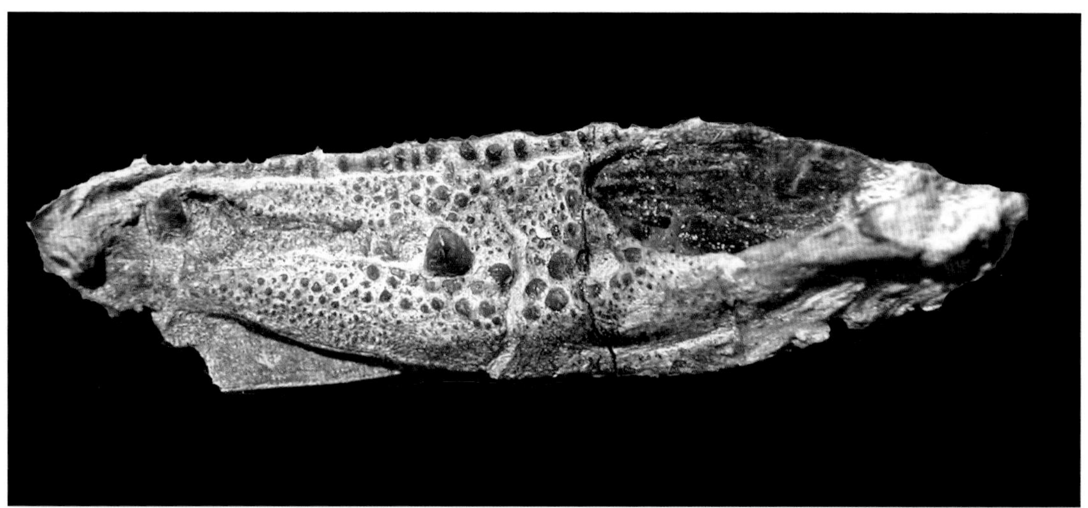

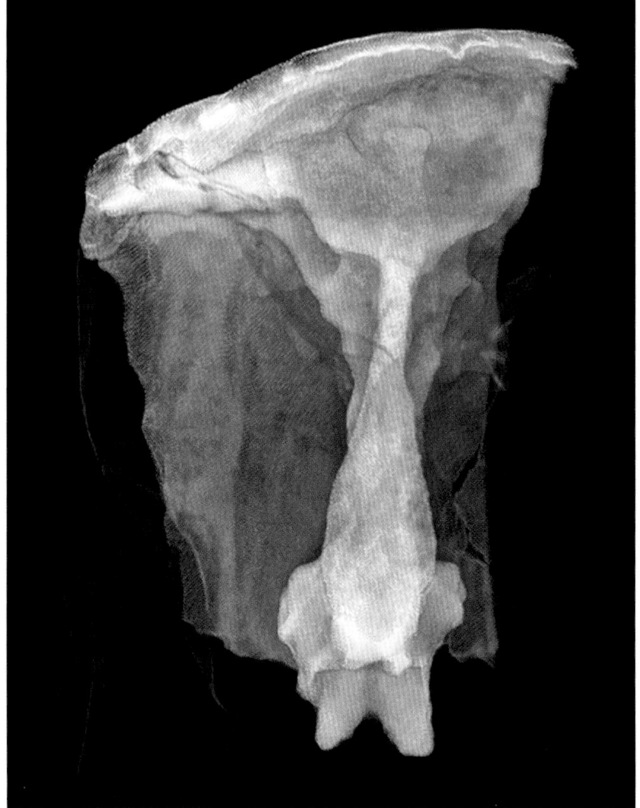

(*Top*) Lower jaw of *Kenichthys* from the Early-Middle Devonian of Yunnan, China, the most basal of all the tetrapodomorphan fishes. (Courtesy Zhu Min, IVPP, China).

(*Bottom*) Micro CT scan image of the front part of the skull of *Kenichthys* from China. (Courtesy Tim Sendin, Australian National University)

The next offshoot of the line leading to tetrapods is represented by the Rhizodontida, a group of mostly large fishes that appear late in the geological record (late Middle Devonian) but are fairly generalized in their anatomy.

Giant Killers of the Carboniferous: Rhizodontiformes

The rhizodontiforms (meaning "root tooth," so-named because of long fangs that extend deeply into the jaws) were the largest and most voracious of the sarcopterygian fishes, reaching estimated sizes of 6 to 7 m. As they are generally known only from large pieces, we had little information about them until the first complete material was described in 1986 and the first detailed study of a complete skull published in 1989. In general, the group is characterized by stiff fins, which have long, unbranched bony rods (lepidotrichia) supporting the main part of the fin. The pectoral fins were very strong, supported internally by a robust humerus and strong ulna and radius bones. This pattern of arm bones is also seen in osteolepiforms and all higher land vertebrates.

The function of the shoulder joint in rhizodontiforms may have been capable of powerful rotational movements, so that a fish could use its large, stiff fins to twist around in the water, similar to the way in which crocodiles wrestle and tear flesh off their prey. The teeth of some large rhizodontiforms are laterally compressed to form a razor-sharp blade edge, also a characteristic feature of the group but also found in some tristichopterids.

The rhizodontiforms first appeared in the late Middle Devonian, represented by *Aztecia*, known from a lower jaw and part of the shoulder girdle discovered near the top of Mount Ritchie in South Victoria Land, Antarctica, by Alex Ritchie in the early 1970s. A large form named *Notorhizodon* (meaning "southern root tooth") was found at the same site nearby and first described as a large rhizodont but is now regarded as a tristichopterid, the group to which *Eusthenopteron* belonged. Other Devonian rhizodonts include *Sauripterus* from the Late Devonian of North America, known principally from a large fossil pectoral fin skeleton and some other bits and pieces, but with a fin skeleton displaying eight sets of fin radials, coincidentally similar to the eight digits seen on the limbs of the early amphibian tetrapod *Acanthostega*.

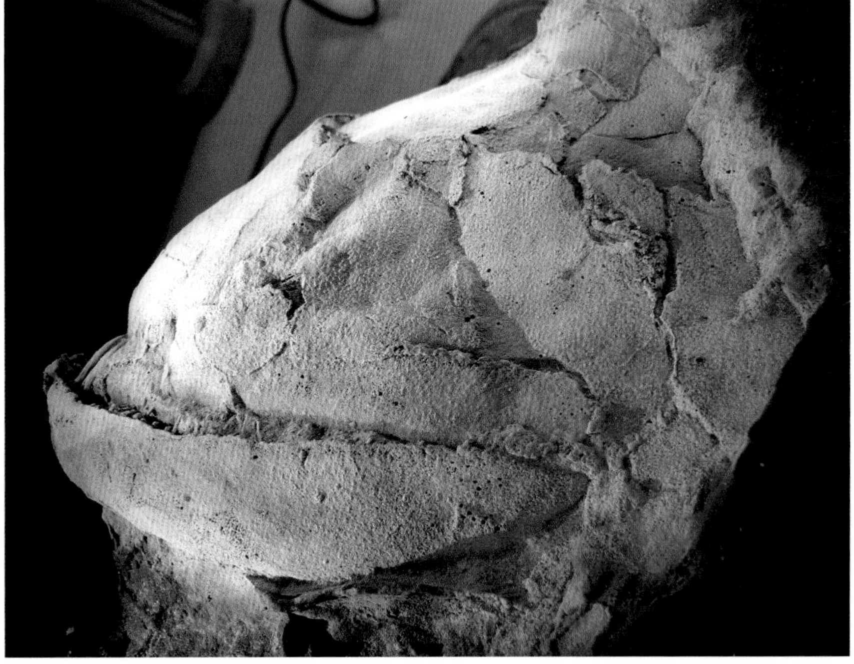

Skull of *Goologongia*, the most basal of all rhizodontid fishes from the Late Devonian Canowindra site in Australia (cast whitened).

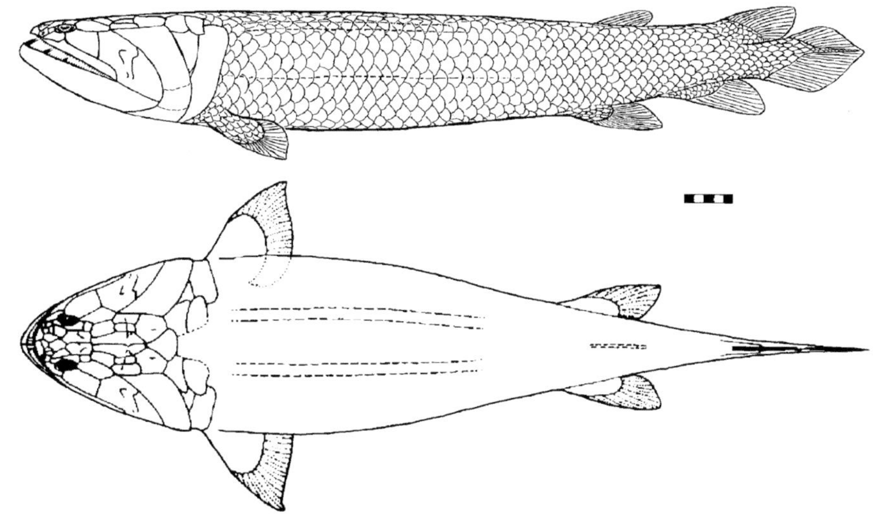

(*Top*) Restored body of *Goologongia* from the Late Devonian of Australia. Bar scale in centimeters. (Courtesy Per Ahlberg and Zerina Johanson).

(*Bottom*) A pectoral fin and shoulder girdle of the rhizodont *Sauripterus halli*, from the Late Devonian near Blossberg, Pennsylvania. Note the humerus articulating with the ulna and radius, the same pattern of bones seen in all tetrapods. (With the permission of the American Museum of Natural History)

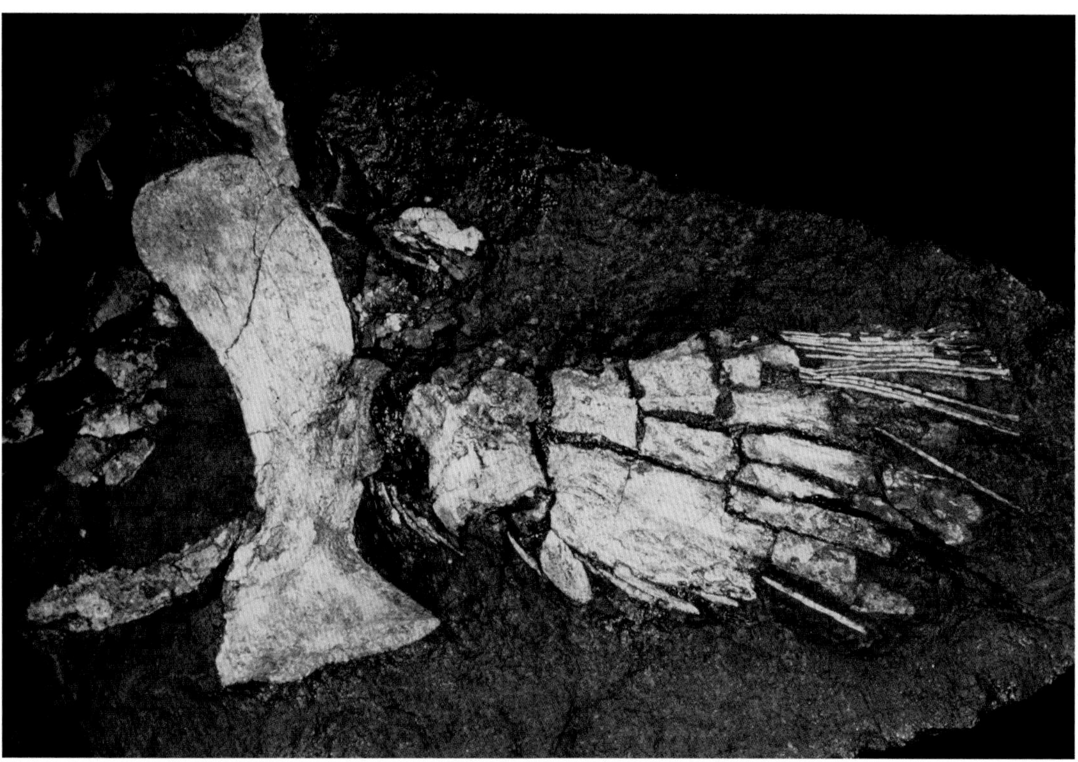

One of the best-known early rhizodonts is *Goologongia* from the Frasnian Canowindra fish mortality site in New South Wales, Australia, which was first described by Zerina Johanson and Per Ahlberg in 1999. *Goologongia*'s skull is quite well preserved and shows the lower jaws protruding beyond the upper, suggesting it hunted prey by sneaking up from murky waters underneath.

During the Carboniferous, the rhizodontiforms reached a peak of diversity and size. The largest known was *Rhizodus* from Scotland. An isolated lower jaw in the National Museum of Scotland is almost 1 m long, suggesting a maximum size of 6 to 7 m for its owner. The largest known teeth are the fangs at the front of the mouth, some of which are 22 cm long. Other forms such as *Strepsodus, Barameda,* and *Screbinodus* also reached large sizes and had similar large fangs at the front of the lower jaws. *Barameda*, known from two species found at Mansfield, southeastern Australia, and *Goologongia* show that the head of rhizodontiforms is of a similar pattern to that in some osteolepiforms and that the cranial joint enabled great frontal lift of the snout when opening the mouth. The rhizodontiforms were probably hunters of the large amphibians and fishes that lived in the murky coal swamps and lakes of their time.

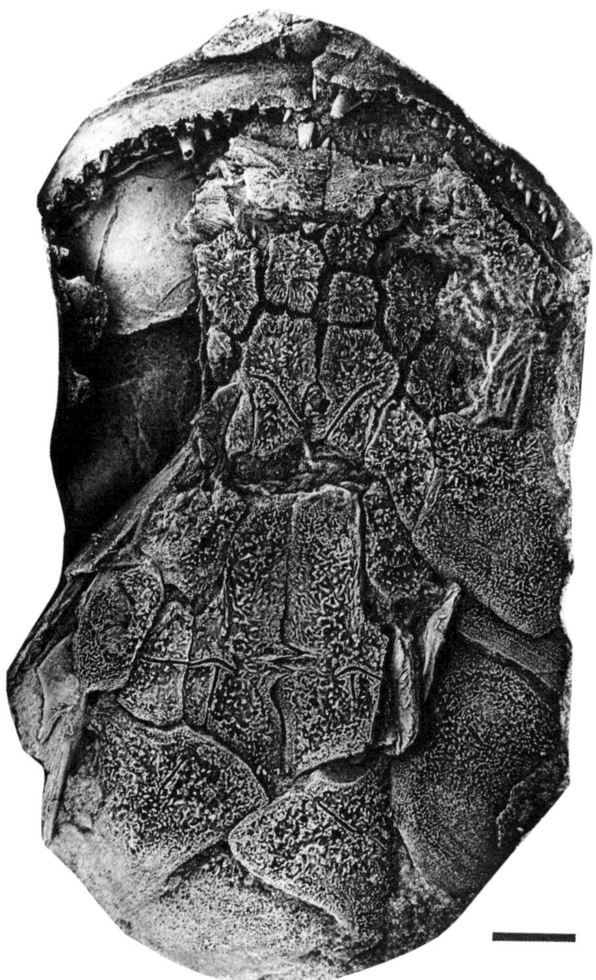

The only complete fossil of a rhizodontiform is a small specimen of *Strepsodus*, whose body was relatively elongated with small pelvic, dorsal, and anal fins and large paddle-like pectoral fins. This body shape is ideal for a slow-swimming stalker, capable of occasional fast bursts of activity as unsuspecting prey swim nearby. The last rhizodontiforms died out by the start of the Permian Period, probably outcompeted by the rapidly growing number of large aquatic amphibians and large freshwater sharks sharing their habitat.

A Step Toward Land: Stem Tetrapodomorphans

By the Middle Devonian, the tetrapodomorphans were established around the globe, reaching a peak of diversity in the Late Devonian before diminishing in diversity and becoming extinct during the Permian. The only exception to this was the one group that went on to become the successful tetrapods. They are the only groups of sarcopterygians that possess a single external pair of nasal openings and a choana and generally have a set pattern of seven bones forming the cheek unit. Like rhizodontiforms, they have strongly ossified paired fins, the pectoral fin having a solid humerus, ulna, and radius. The most primitive members of the

(*Left*) The skull of *Barameda mitchelli*, from the Early Carboniferous of Victoria, Australia, was the first well-preserved rhizodont skull ever described.

(*Bottom*) Sketch showing the main features of the lower jaw of a rhizodontid, *Barameda decipiens*. ifd = infradentary formina. (Courtesy of Tim Holland)

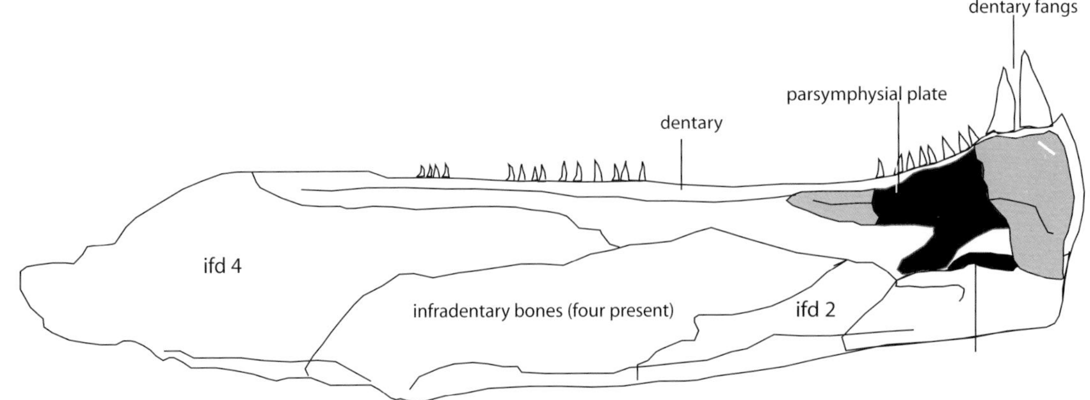

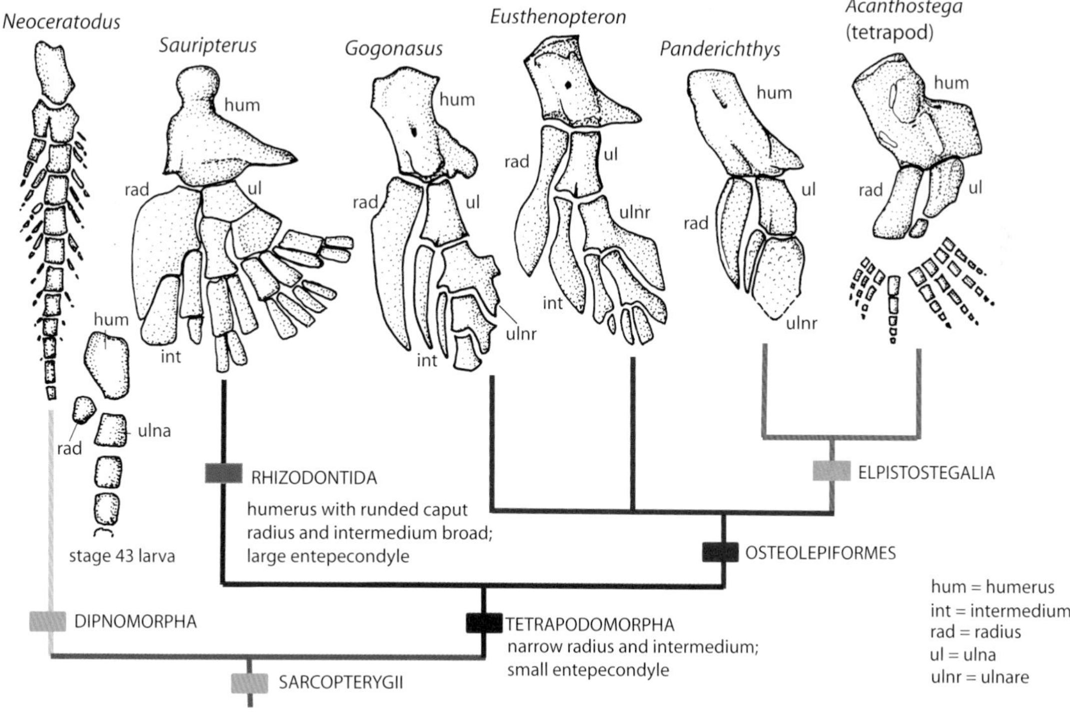

group have thick rhombic scales and all dermal bones have a cosmine layer, although in several later lineages the cosmine is lost and the scales become thinner and rounded.

There are several other groups of basal tetrapodomorphans that can be mentioned, although most dif-

(*Top*) Restoration of the rhizodontid *Barameda mitchelli*, Early Carboniferous of Victoria, Australia (Courtesy Peter Schouten)

(*Bottom*) Sarcopterygian fin bone patterns and hypothesis of relationships of tetrapodomorphan fishes. Note the characteristic tetrapodomorph pattern with robust humerus, ulna, radius, and intermedium present.

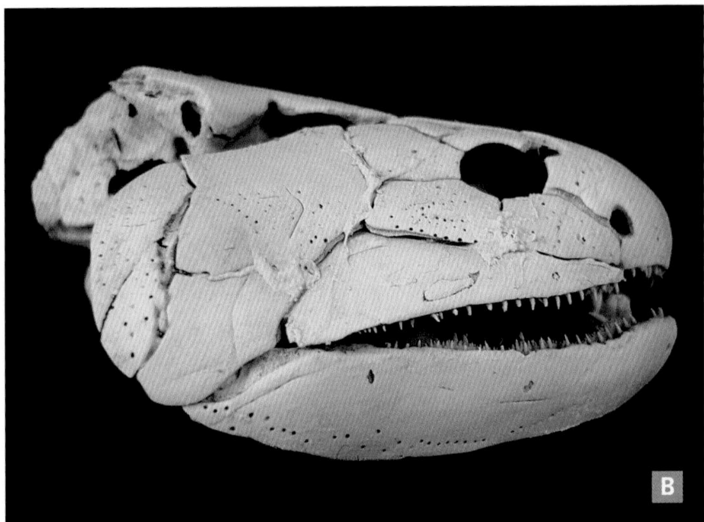

fer in only minor technical ways from the general plan of the group. Among the recently discovered new types is the family Canowindridae, a clade of three genera unique to the East Gondwanan region (Australia and Antarctica). Characteristics of this primitive group of osteolepiforms include large deep opercular bones, wide flat skulls with very small eyes, and possibly additional bones occupying the position of the postorbital bone in the cheek units. The first described genus, *Canowindra grossi* (named after the New South Wales town of

(*Top*) Snout of the Late Devonian *Gogonasus* emerging from the limestone rock (*A*). Note the single external nostril that typifies all tetrapodomorphan fishes and tetrapods. Head of *Gogonasus* (*B*), an acid-prepared Late Devonian tetrapodomorph fish from the Gogo Formation, Western Australia, which has unusually large spiracular openings on top of the head.

(*Bottom*) Restoration of *Gogonasus* living on an ancient Devonian reef. (Courtesy Brian Choo)

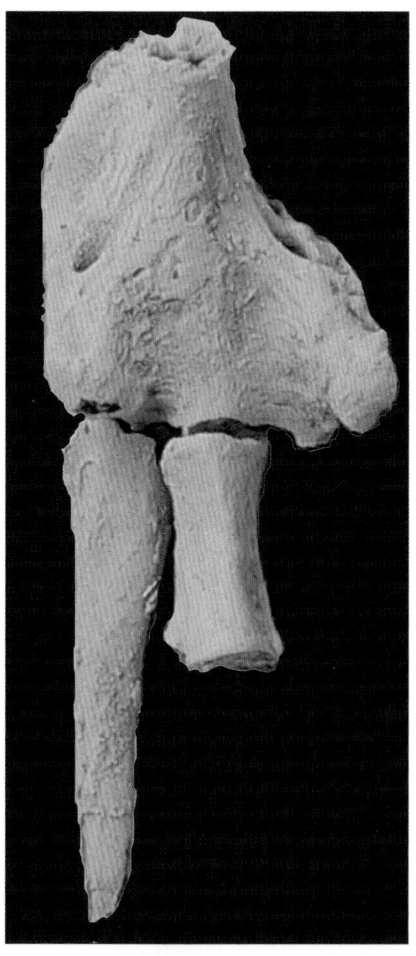

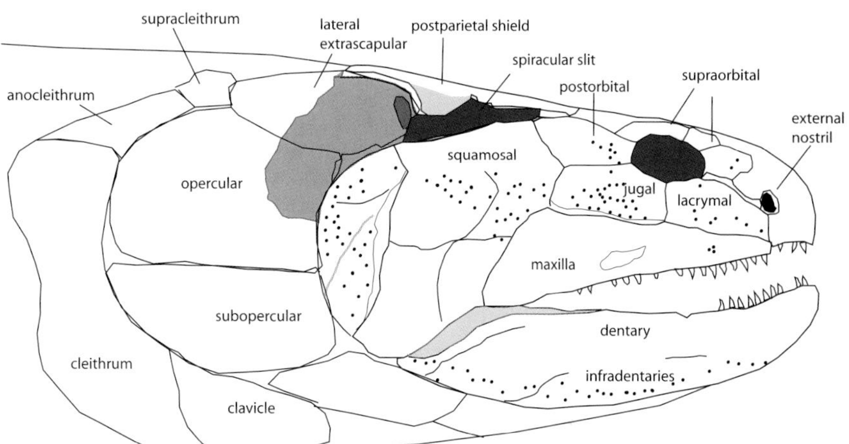

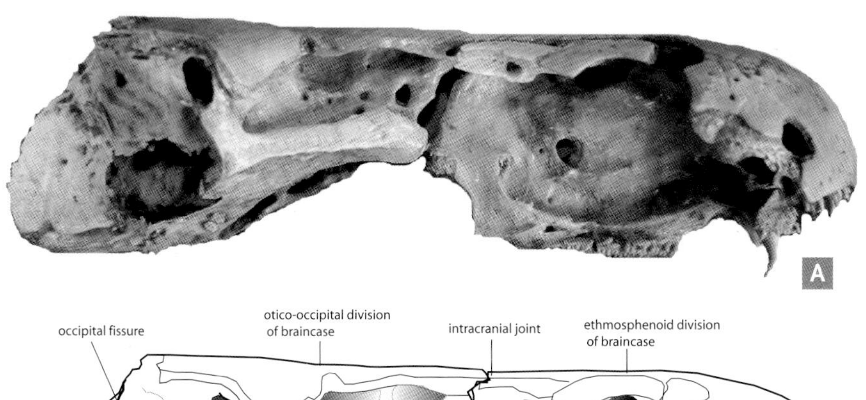

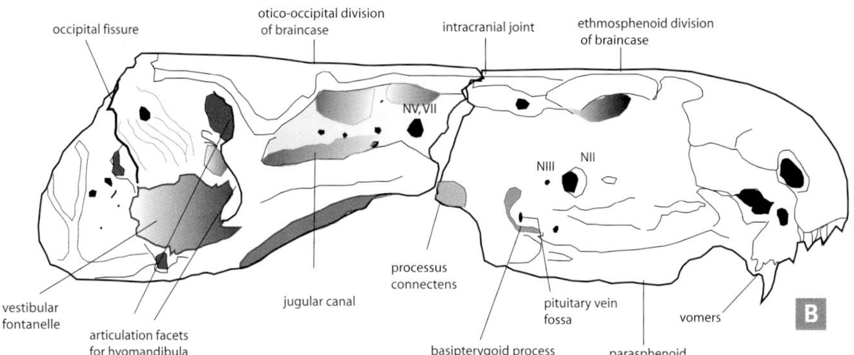

(*Top Left*) Proximal three bones of the pectoral fin of *Gogonasus*, showing humerus (*above*), radius (*left*), and ulna (*right*).

(*Top Right*) Dermal bones of the head of *Gogonasus*.

(*Left*) Braincase of *Gogonasus* (*A*), with anatomical features indicated (*B*).

Canowindra), was studied by Keith Thompson in 1973 but could not be attributed to any particular group of sarcopterygians at that time. In recent years, additional genera have been described from Victoria (*Beelarongia*) and Antarctica (*Koharolepis*), filling in the evolutionary sequence. The most primitive members have rhombic cosmine scales and bones, whereas the most derived genus, *Canowindra,* lacks cosmine and has rounded scales with a ventral boss, or protuberance, on them, as in tristichopterids.

The Osteolepididae are represented by well-known forms such as *Osteolepis, Gyroptychius,* and *Thursius* from the Old Red Sandstone beds of Scotland. These are generally less than 50 cm long and relatively conservative in their patterns of dermal bones. All have simple heterocercal tails, two dorsal fins, and thick rhombic scales.

Megalichthyinids were a group of advanced osteolepidids that survived until the Middle Permian, long after all other families had died out. They retained their cosmine cover but were specialized in having tectal bones wrapping around the nostril, wide palatal cavities, and a long median process extending into the mouth from the premaxillary bones. Some megalichthyinids also possessed a special articulation on the dermal bones linking the two halves of the skull roof. The group were highly successful and inhabited the coal swamps and lakes around Gondwana and also in Euramerica, represented by examples such as *Megalichthys* and *Megapomus,* the largest species reaching sizes of about 1 to 2 m or more.

The Tristichopteridae include forms with reduced or no cosmine cover, and most have lost the extratemporal bone from the skull roof, except for one primitive genus, *Marsdenichthys,* from Australia (which might fall outside the group, but is closely allied; Holland et al.

Cast-whitened skull of *Canowindra grossi*, from the Late Devonian of New South Wales, Australia. It is a member of a family of tetrapodomorphans endemic to Australia and Antarctica (East Gondwana Province).

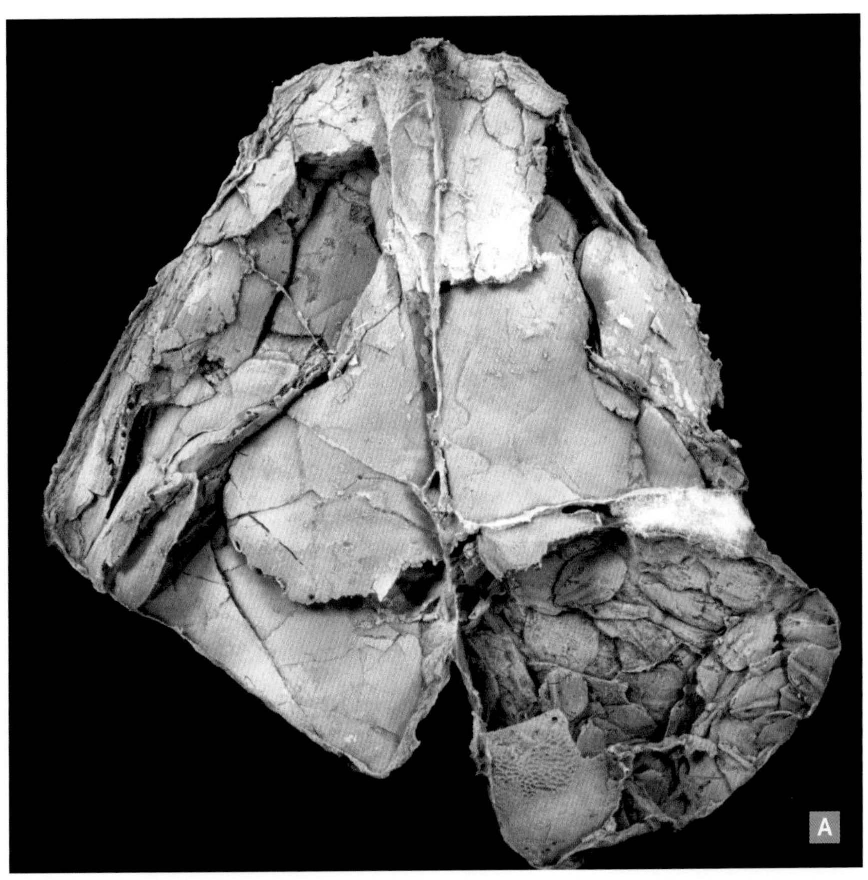

Cast-whitened skull of *Beelarongia* (*A*) from the Middle Devonian of Australia. Middle Devonian Mount Howitt site in the mountains of central Victoria has produced many complete fish fossils (*B*).

***Koharolepis*, a Middle Devonian canowindrid osteolepiform from Mount Crean, Victoria Land, Antarctica. These fishes are thought to be the most primitive of all known osteolepiforms.**

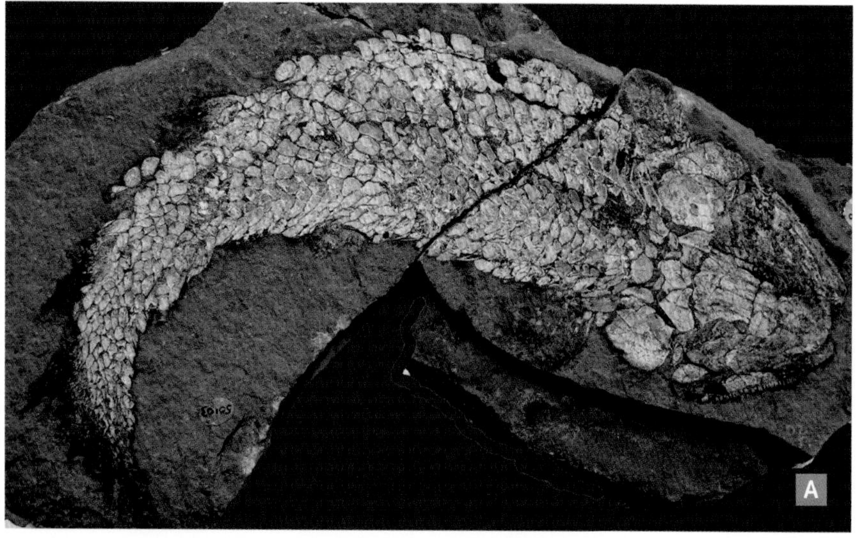

***Osteolepis*, a cosmine-covered tetrapodomorph (*A*) from the Middle Devonian Old Red Sandstone of Scotland. Skull of *Osteolepis* in dorsal view (*B*).** (With the permission of the Natural History Museum, London)

2010). The best-known member of the family is *Eusthenopteron foordi* from the Late Devonian Escuminac Bay fauna of Canada, a medium-sized fish about 1 m long. *Eusthenopteron* was the subject of 25 years of detailed study by Erik Jarvik of Stockholm, who made a large wax model of its braincase and gill arches from serial sections of a skull. The skull was embedded in resin, then ground down by a tenth of a millimeter each time. The section was then photographed and drawn, magnified by 10, and a wax layer was cut out to show the position of bone and cartilage. By slowly building up the wax layers, Jarvik constructed his model and then pub-

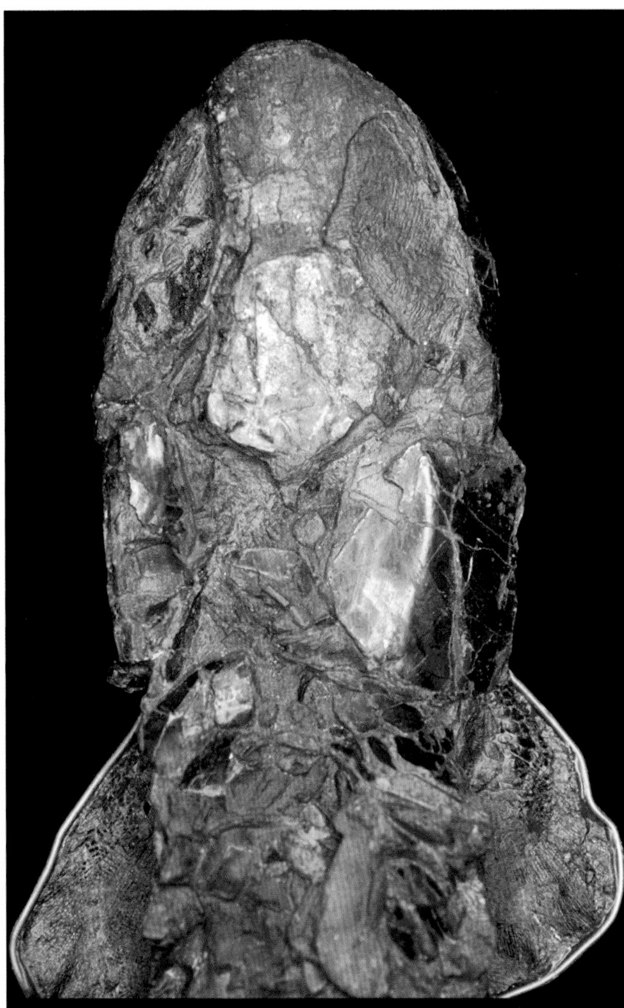

(*Top Left*) Skull roof of *Megalichthys laticeps*, from the Carboniferous of Scotland. Note the shiny cosmine surface. (With the permission of the Natural History Museum, London)

(*Top Right*) *Claradosymblema*, a megalichthyinid fish from the Early Carboniferous of central Queensland, Australia, viewed from above.

(*Bottom*) Skull of *Megapomus*, a Carboniferous megalichthyinid fish from Russia. (Courtesy Oleg Lebedev, Borissiak Paleontological Institute, Moscow)

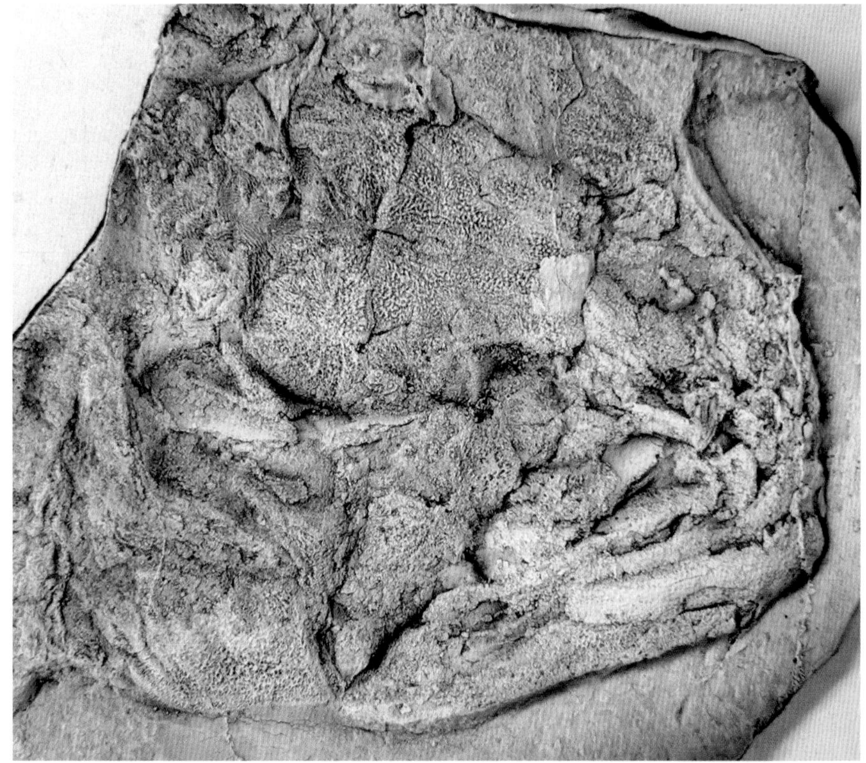

(*Top*) *Marsdenichthys longioccipitus*, a Middle Devonian tetrapodomorphan fish, from Mount Howitt, Victoria, Australia, that originated near the beginning of the tristichopterids. This is a latex peel of the skull.

(*Bottom*) *Eusthenopteron*, from the Late Devonian Escuminac Formation, Quebec. *Eusthenopteron* is perhaps one of the most intensely studied of all fossil fishes. Its anatomy was reconstructed by Swedish paleontologist Erik Jarvik by making painstaking wax models built from a specimen ground away in layers of one-tenth of a millimeter thick. The work took nearly 30 years.

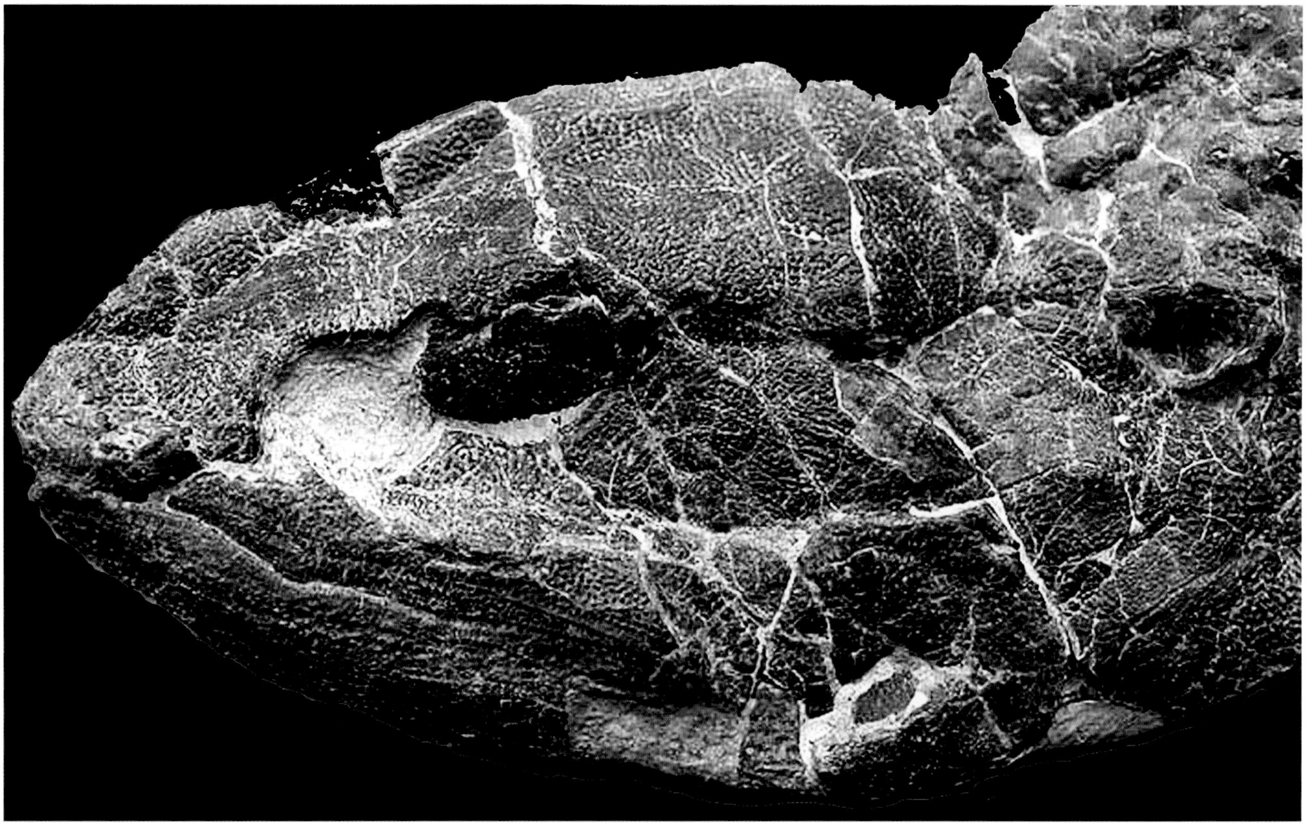

Skull roof and some of the cheek and jawbones of *Jarvikia*, a tristichopterid from the Late Devonian of Russia. (Courtesy Oleg Lebedev, Borissiak Paleontological Institute, Moscow)

lished several large papers elucidating the fine anatomical structure of *Eusthenopteron*. Today this genus is still one of the best known of all Palaeozoic fishes. Complete fossils of *Eusthenopteron* are frequently found in the Escuminac Formation of Quebec, and other species of the genus have been described from Russia and Europe. *Jarvikia* is a large form found in Russia, and the biggest members of the group include *Eusthenodon* from East Greenland and *Hyneria* from North America.

In recent years some very well-preserved tristichopterids have been described from the Late Devonian Canowindra site in New South Wales, Australia, by Per Ahlberg and Zerina Johanson. These include the large *Mandageria* and the smaller *Cabonnichthys*. *Mandageria* has an acutely pointed snout and small eyes. It had a well-developed neck joint that enabled faster opening of its huge fang-filled mouth. It was a fast-swimming predator with a streamlined body shape. Recent work by Gavin Young (2008) suggests that Australian tristichopterids could represent an endemic group that shares unique features of the palate and cheekbone pattern.

Although representing several grades of organization—or steps in evolution—from the first choanate fishes like *Kenichthys* to the tetrapods, the osteolepiforms are perhaps the most important group of fishes in studying the evolutionary transition from fish to land animal, particularly the derived end members of the clade, the elpistostegalians. These fishes have more in common with primitive amphibians than with any other fish. This great step in vertebrate evolution is discussed in more detail in the last chapter of this book.

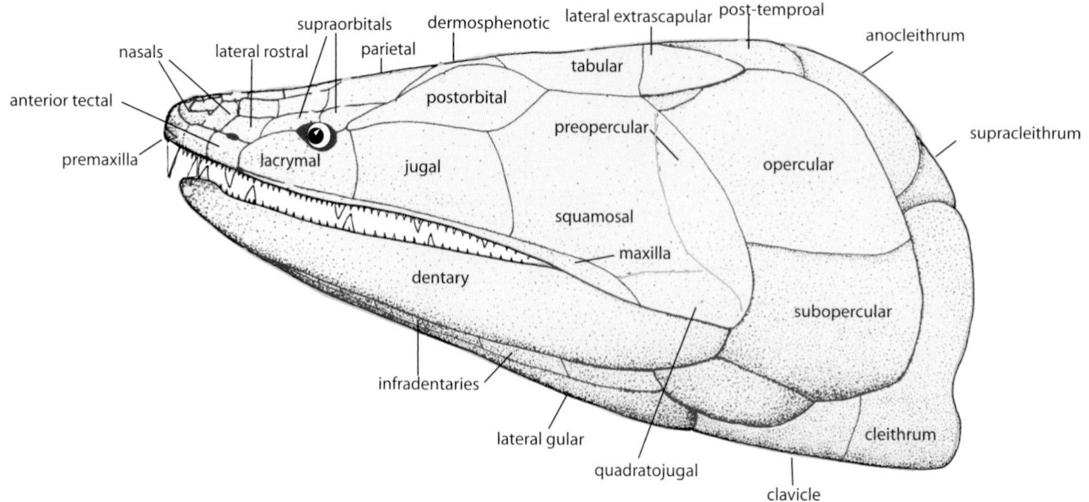

(*Top*) Head of the streamlined large tristichopterid *Mandageria* from the Late Devonian Canowindra site, New South Wales, Australia.

(*Bottom*) Reconstructed scene of the late Devonian Canowindra fauna (Australia) showing the predatory *Mandageria* hunting placoderms (*Remigolepis, Bothriolepis*, and a *Groenlandaspis*) with *Canowindra* lurking in the background. (Courtesy Age of Fishes Museum, Canowindra, Australia).

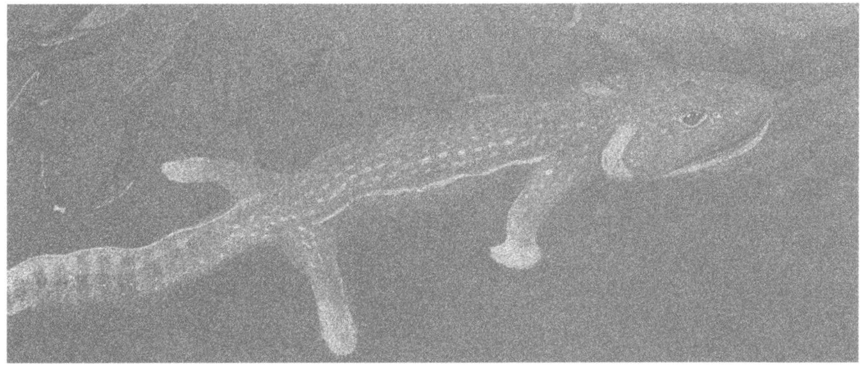

CHAPTER 13

The Greatest Step in Evolution

From swimming fishes to walking land animals

The greatest step in vertebrate evolution is undoubtedly the transition from aqueous gill-respiring fishes to air-breathing, walking, land animals. Despite the numerous complex anatomical changes that this required, the transition from the elpistostegalian branch of sarcopterygian fishes, discussed in detail below, to primitive tetrapods (amphibians) was no great deal, as the prerequisites for the change were already inherent in the group. The limbs and shoulder and hip girdles of these fishes, exemplified by *Panderichthys* and *Tiktaalik,* were strong. Their arms and legs (as fins) already had the same bone pattern that all land animals would retain. The pattern of their skull and cheekbones did not have to change at all. From these fishes arose primitive amphibians.

The first evidence of tetrapods comes from 395-million-year-old trackways found in shallow marine sediments in Poland (Niedźwiedzki et al. 2010), suggesting there is a ghost record of missing forms, as these trackways predate the oldest known elpistostegalian fishes by 10 million years. By the Late Devonian, tetrapods had begun to radiate into different parts of the globe. These early examples did not conform to the standard tetrapod pattern and retained many fishlike characteristics, such as gill breathing, a fishlike tail, and scales covering much of the body. It is now accepted that the earliest well-known forms, like *Acanthostega* and *Ventastega,* were fully aquatic. Within a relatively short time after the first amphibians began to diversify, the first protoreptile had appeared, in the Early Carboniferous. The rest of vertebrate evolution is simple by comparison, being merely a tale of what lineages stemmed from the basic reptilian pattern.

The transition from fishes to four-legged land vertebrates (tetrapods) has always been a difficult area of evolution to grasp and has, until quite recently, been a hotly disputed area of paleontology, particularly with regard to which fishes may have given rise to the first amphibians. The following passage taken from an early twentieth-century popular geology book nicely invokes some of the mystery surrounding this major step in evolution:

> These facts, coupled with the fish-like structure of certain genera [of fossil amphibians], tempt one to imagine them as having slowly evolved, in the midst of the Carboniferous marshes, from true fishes; first wriggling helplessly among the slime, and afterwards generally acquiring lungs for breathing air and limbs for locomotive purposes, and lastly, their strong and peculiar teeth for masticating the vegetation on which they may be presumed to have principally lived. (from B. Webster Smith, *The World in The Past*, Warne & Co., London, 1926)

Thus the early twentieth-century scientists saw land animals as springing from fishes wriggling in the mud, which, being out of water long enough, somehow acquired lungs and limbs more or less at the same time. Yet, we now know that fishes already had lungs well before they ventured from the water and that certain fishes had the same sets of arm and leg bones as land animals. In fact, most of the major transitions needed for the invasion of land had already occurred within fish evolution. Today, we have a far more complete record of early fossil amphibians and more detailed knowledge of the physiology of living air-breathing fishes, such as the dipnoans and many groups of actinopterygians that use accessory air-breathing organs (Graham 1995). More significantly, we can study the nature of growth and change in organisms and document the major morphological transformations that result from simple changes in the timing of a creature's embryonic or growth development (termed *heterochrony*). By combining all of

A frog is a living representative of the most primitive of all living land animals, the amphibians. Amphibians, like reptiles, birds, and mammals, are all tetrapods, animals having a four-legged body plan.

Snout of *Panderichthys*, an Early-Late Devonian elpistostegalian fish from Latvia. (Courtesy Oleg Lebedev, Borissiak Paleontological Institute, Moscow)

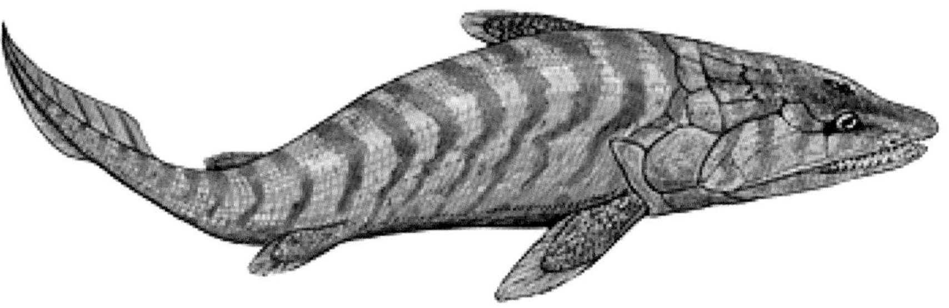

Reconstruction of *Panderichthys*. Note the large head, eyes situated at the top of the skull and located close together, and the robust paired limbs (still retaining fins).

these new results and discoveries, an accurate picture of the origins of land vertebrates is beginning to emerge.

The discovery of fossil amphibians that lived in the Carboniferous Period goes back a long way in paleontology. However, it was only in the twentieth century that the first remains of amphibians from Devonian age rocks were made. Furthermore, only in the past decade have there been many new finds of well-preserved Devonian amphibians and their trackways that have shed much new light on the remarkable transition between fishes and the tetrapods. Such discoveries show that the first tetrapods were not very different from their fishy ancestors. They may have had limbs with digits, as opposed to fins, but some remained fully aquatic animals, capable of underwater gill respiration, and probably spent most of their time living like fishes.

From Water to Land: How to Survive

Life in the water is quite different from life on land. Fishes have a different body shape from land animals because the water holds them up and because the forces acting on their bodies are different. Thus gravity is one force, friction from the water is another, and the buoyancy of their bodies within the water column is yet another force. By balancing its buoyancy, a fish can achieve neutral weight in the water (that is, weigh nothing) and so can direct its energy from the tail and hydrodynamic lift from fins to take it up or down in its environment. In reality, however, it is not as simple. Many fishes (osteichthyans) use a swim-bladder to regulate gases inside the body to achieve lift or fall in the water through subtle changes in buoyancy. Sharks and other fishes without swim-bladders use different methods to achieve neutral buoyancy, like having a large oil-filled liver and having winglike pectoral fins to give greater lift in the water.

But there is more to life in water that just supporting your body and going up and down. There is also breathing, sensing, excreting, and eating and trying to avoid being eaten, not to mention the most important of all functions, reproduction. All of these functions needed to be modified for a successful conquering of the terrestrial environment.

Breathing through gills requires that there is always a certain sustainable level of oxygen dissolved in the water so that the oxygen can be taken up through the fine membranes of the gills into the bloodstream. To keep a constant flow of water over the gills, fishes use different pumping methods or, as in sharks, tend to keep moving or rest in active currents of water. Thus water comes in through the mouth and either flows over the gills and out via the operculum or is pumped over the gills by moving the mouth and altering the volumes inside the gill chamber. The transition to air breathing was not difficult in this regard. Many modern gill-breathing fishes also have an ability to breathe air for limited times, using the swim-bladder as a gas-exchange organ to obtain small amounts of oxygen (for example, mudskippers).

Eating may seem to be a basic requirement of all creatures and may not necessarily have been a driving force for fishes to leave the water. Indeed, the first land vertebrates were of rather clumsy design and probably ventured away from the water for only short periods. Their crossopterygian-like bodies and skulls were much better suited to catching prey in the water. Eating food on land and acquiring the body design to do so efficiently came much later in the evolution of the tetrapods. For a slow-moving amphibian to catch land-living insects as a source of food, one of two requirements is needed. Either the animal must be capable of short fast lunges or bouts of running to catch the unsuspecting prey; or as in many modern frogs, they must develop a long prehensile tongue that can simply dart out and catch a flying insect. We have no evidence that the long tongue existed in early tetrapods and is more likely a specialization of later amphibians; so we must, therefore, rely on evidence from their body shapes and limb joints that they were not efficient hunters on land.

Sensing the environment around you is another important part of invading a new habitat. In the water, fishes rely heavily on their lateral-line sensory system for detecting movements in the water, either to find prey or to detect a larger predator approaching them. They use their eyes and hearing to lesser extent. Fishes that leave water for short periods of time rely even more on their eyes for detecting food items or approaching danger. One of the major differences between the bones of fishes and those of fossil tetrapods is that fishes tend to have the lateral-line canals enclosed with rows of pores open to the external environment, whereas the tetrapods have wide, open grooves in the dermal bones for their sensory lines. Eventually, as these animals became more adapted to life on land, the other senses took over from the lateral line system, which is really only useful when the creature is in water. One of the first major transitions to occur in this respect was the evolution of the eardrum in early amphibians. The long hyomandibula bone that braced the jaw joint of higher osteolepiform fishes is, in tetrapods, modified to brace the otic membrane above the inner ear. Thus it becomes a stapes, a bone that evolved as a means to convey airborne vibrations into the inner ear. The evolution of more complex

The tetrapod stapes (*above*) evolved from the hyomandibula or first gill arch bone series in tetrapodomorphan fishes (*below*).

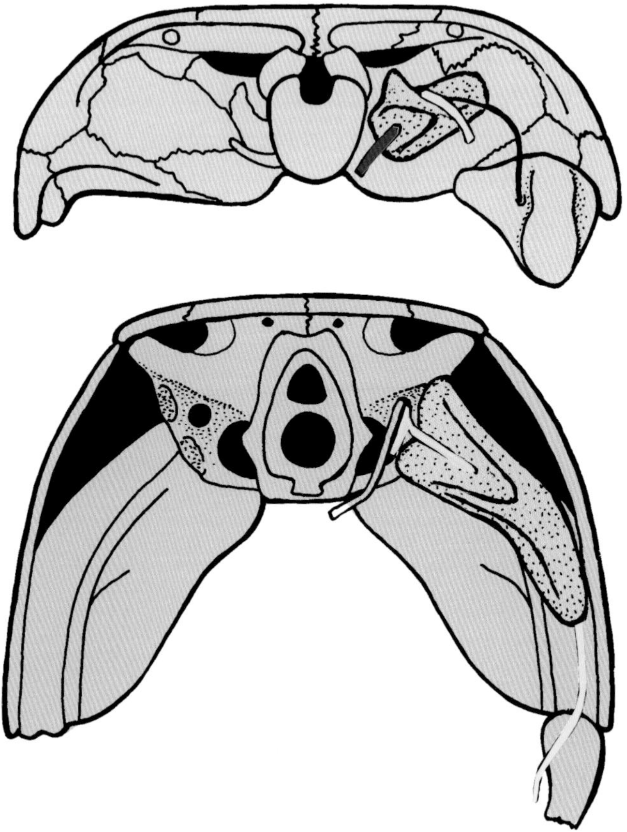

middle ear-bones, such as the incus and malleus, came later with the reptiles and mammals. Although the eyes and nostrils became increasingly more important for the land animal to sense its surrounding habitat, there is virtually no initial change in these structures from the fish to the first tetrapod, apart from the origin of the lacrimal gland at some stage to keep the eye moist.

Excreting is another vital bodily function and is linked with the overall problem of preventing the animal from dehydrating once out of its aqueous environment. Fishes and water-dwelling animals that excrete into the water do not risk dehydration, as they are always immersed in water. However, a land animal that excretes urine and moist feces is continually lowering its level of body moisture and must replenish that supply by drinking more. Early tetrapods had skins that were covered with fishlike scales to protect the skin and prevent some dehydration.

The first real adaptation for living away from the water was to come with the reptiles, only a short time after the first amphibians appeared. This great innovation was the hard-shelled egg, and this testifies that the egg-layers were the first creatures capable of living away from the water for any significant length of time. Yes, this does indicate the true answer to the age-old riddle—the egg came well before the chicken.

Elpistostegalians—Fishes or Amphibians with Fins?

The last few chapters in this book have given a broad overview of the lobe-finned fishes and their diversification through the Paleozoic Era. However, one very special group, the elpistostegalians (which include panderichthyid fishes and early tetrapods) have been saved until last. Until quite recently they were always classified as a group within the osteolepiforms, although a paper published in 1991 by Hans-Peter Schultze working with Russian paleontologist Emilia Vorobyeva redescribed the group and placed these fishes in a new order, the Panderichthyida. The panderichthyids, it seems, share more anatomical features in common with early tetrapods than with any other fish group. So what exactly makes a panderichthyid? To answer this question we will treat the panderichthyids alongside the first tetrapods, as these all belong in the group Elpistostegalia.

The Elpistostegalia is now known to contain about a dozen species, most of which have been recognized, discovered, or described only in the past decade. Of these, only two, *Tiktaalik* (Nunavit for "big fish") and *Panderichthys* (meaning "Pander's fish"), are fishes that are known in any detail. *Panderichthys* was the first elpistostegalian fish to be known from more than just fragmentary remains. It was described from isolated snout and skull material by German scientist Walter Gross in 1941 from remains found in the famous Latvian site at Lode. Fossil material from this site is beautifully preserved in three-dimensional form, as the surrounding sediment was a soft clay. New finds of more-complete specimens of *Panderichthys* have enabled description of nearly all aspects of its anatomy mainly by Schultze and Vorobyeva, along with Per Ahlberg, Catherine Boisvert, and others.

Recently, Boisvert identified that the pectoral fin of *Panderichthys* contains digit-like elements that appear homologous to the digits in early tetrapods. The pelvic fin is very small and must have not been very powerful in moving the animal out of water.

Panderichthyid fishes are characterized by the following few features: the median rostral bone of the snout does not contact the premaxilla; they have a very large median gular bone under the head; the mouth is underneath the protruding snout; and they have a lateral recess in the nasal capsules.

Tiktaalik is known from several well-preserved individuals found in the far north of Arctic Canada over a series of field trips, led by Neil Shubin's team from 1999–2004 (and outlined in his book, Shubin 2008). The cranial anatomy and pectoral fin skeleton been described by Neil Shubin, Ted Daeschler, Farish Jenkins Jr., and Jason Downes in a series of recent papers (Shubin et al. 2006, Daeschler et al. 2006, Downes et al. 2008). The main differences between *Tiktaalik* and *Panderichthys* is

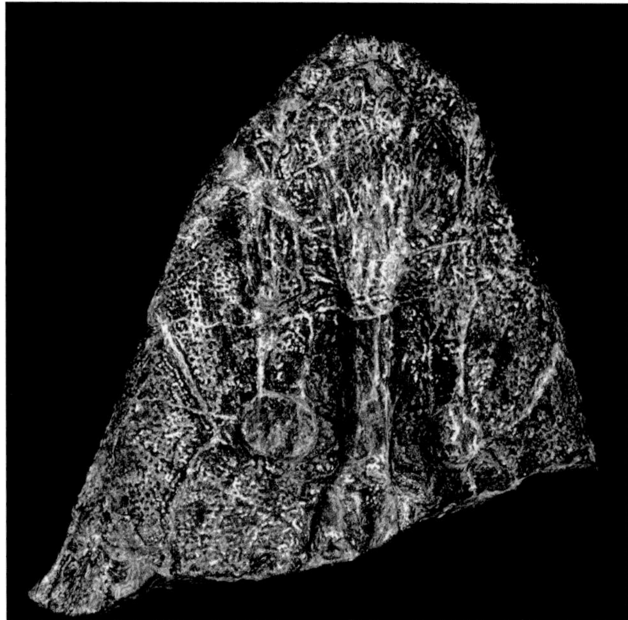

(*Left*) Partial skull of *Elpistostege* from the Late Devonian Escuminac Formation of Canada in top view. When first described, it was thought to be a tetrapod but was later shown to be a fish.

(*Bottom*) Front half of the fish most closely related to land animals, *Tiktaalik rosae*. Discovered by Neil Shubin and his team from Frasnian beds in Arctic Canada and announced to the world in 2006, *Tiktaalik* perfectly bridges the gap between fishes and amphibians. (© T. Daeschler)

that *Tiktaalik* lacks opercular bones in the skull and has a longer snout and more robust pectoral fin bones than *Panderichthys* does. It has been suggested that *Tiktaalik* may have had an incipient wrist joint to help push the head up out of water. The skull roof of *Tiktaalik* has a very large spiracular slit, also associated with accessory air-breathing.

These fishes share a number of specialized features with amphibians, as the following list demonstrates: They are long-bodied with a large head that is nearly one-quarter their total length. They have large pectoral fins and smaller pelvic fins but no dorsal or anal fins. The skull is broad and flat, with eyes on the top of the head and rather close together with distinct brow ridges. The external nostrils are placed ventrally, close to the margin of the mouth. A cross-section of the large teeth shows a complex form of labyrinthine infolding of the enamel and dentin, as occurs in many amphibians. The bones of the skull roof have three pairs of median bones from the back of the skull to the eyes, rather than two as in other tetrapodomorphans. The cheekbones are large, and the jugal bone separates the squamosal from the maxilla, as occurs in amphibians. The cheekbones do not rigidly meet the skull roof for some distance, leaving a large spiracular slit along each side of the skull table. The external pattern of skull roof bones shows that the intracranial joint has fused, so that the skull was not kinetic as in many other sarcopterygians. The pectoral fins have strongly ossified humerus, ulna, and radius, with the humerus having a longer shaft than for any other fish. The vertebrae are also unusual in having only the ventral component (intercentrum) present, with large neural arches straddling the notochord. Ribs are attached to the neural arch and intercentrum, exactly as in tetrapods. The body is covered in rhombic bony scales, but cosmine is absent.

Elpistostege, from the Escuminac Formation of Que-

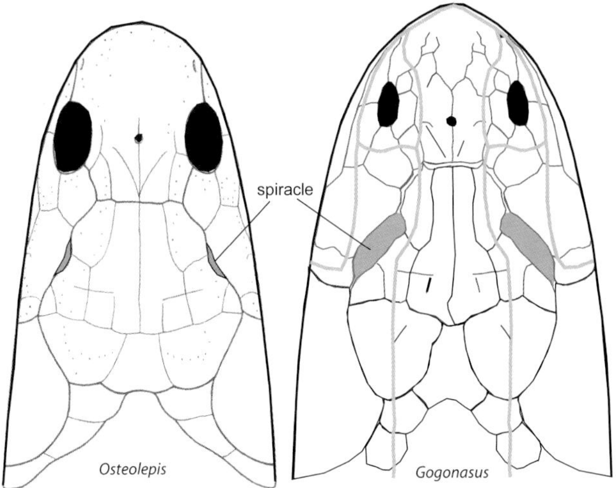

(*Top*) Site where *Tiktaalik* was discovered, at Nunavit, Arctic Canada. (© T. Daeschler)

(*Bottom*) Diagram showing the size of the spiracular opening on the heads of *Gogonasus* and *Osteolepis*. The larger spiracle in *Gogonasus* may be an adaptation for taking in air as an auxiliary method of breathing during times of low global oxygen levels.

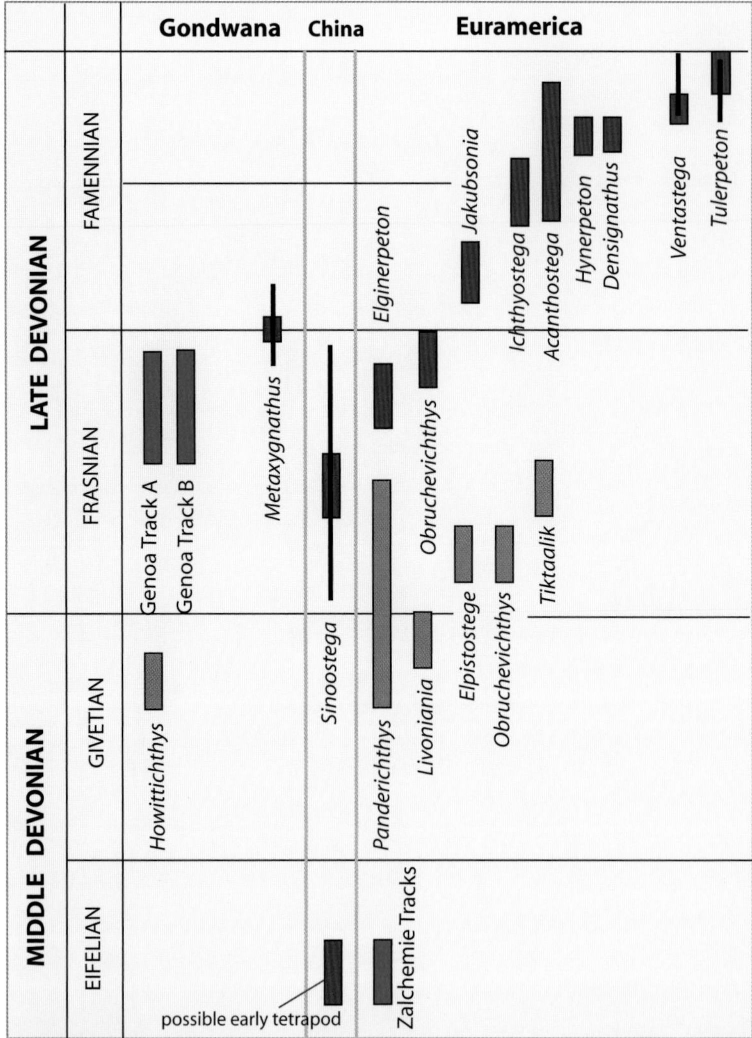

(*Top*) Ranges and relationships of the earliest tetrapods and elpistostegalian fishes in time. By the Early Middle Devonian, tetrapods were established on the Euramerican continent as shown by trackways from Poland.

(*Bottom*) *Obruchevichthys* is an elpisosteglian fish known only from a few skeletal remains, such as this lower jaw from the Frasnian deposits of Russia. (Courtesy Oleg Lebedev, Borissiak Paleontological Institute, Moscow)

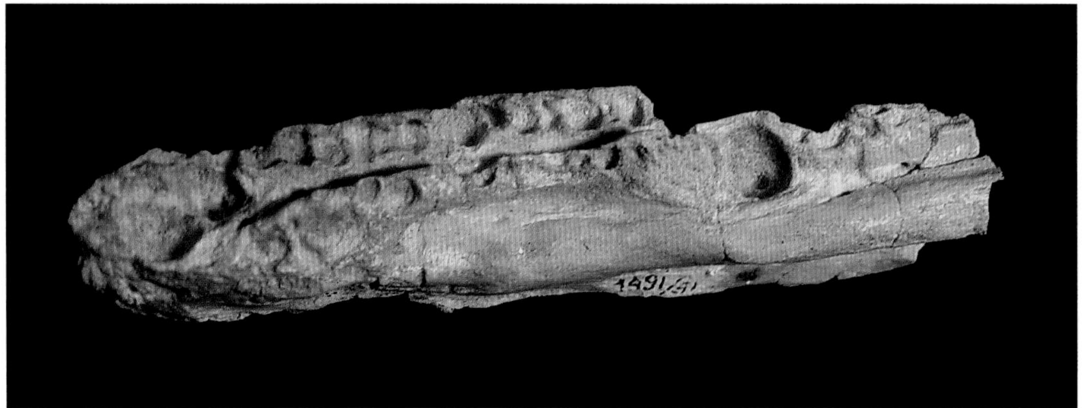

bec is another example of this group, but it is poorly known. When British paleontologist Stanley Westoll first described an incomplete skull roof of this creature in 1938, he was certain it was a primitive amphibian. It was not until new material was recovered in the 1980s that it was shown to be a panderichthyid fish, and its redescription led Hans-Peter Schultze to pursue the close link between panderichthyids and tetrapods. *Elpistostege* is known from only a few partial skulls and is, in general terms, very close to *Panderichthys*. *Obruchevichthys* is another Russian form assigned to this group. This determination is based only on a lower jaw that shows many features similar to early tetrapods.

Thus the panderichthyids, when preserved as partial skull material, are almost impossible to distinguish from early amphibians, and this has led to some confusion in the past. Before we look at the sequence of events leading to the fish-amphibian transition, it is first necessary to see exactly what the earliest true tetrapods were like.

Stem-Group Tetrapods

The oldest definite evidence for tetrapod origins comes from 395-million-year-old trackways recently discovered preserved in shallow marine carbonates from the Holy Cross Mountains of Poland. Described by Grzegorz Niedźwiedzki (2010) of the University of Warsaw and colleagues, they show unambiguous digit impressions and indicate a series of track makers (at least two forms) were present at this time, one of which

The discovery of these 395-million-year-old trackways made a by an early tetrapod from Zalchemie quarry in Poland was announced in early 2010, forcing scientists to rethink their ideas about when fishes left the water to invade land. It now appears that stem tetrapods had evolved by the start of the Middle Devonian. Recent finds of tetrapod bones of this age from China also confirm this view. (Courtesy Piotr Szrek, University of Warsaw)

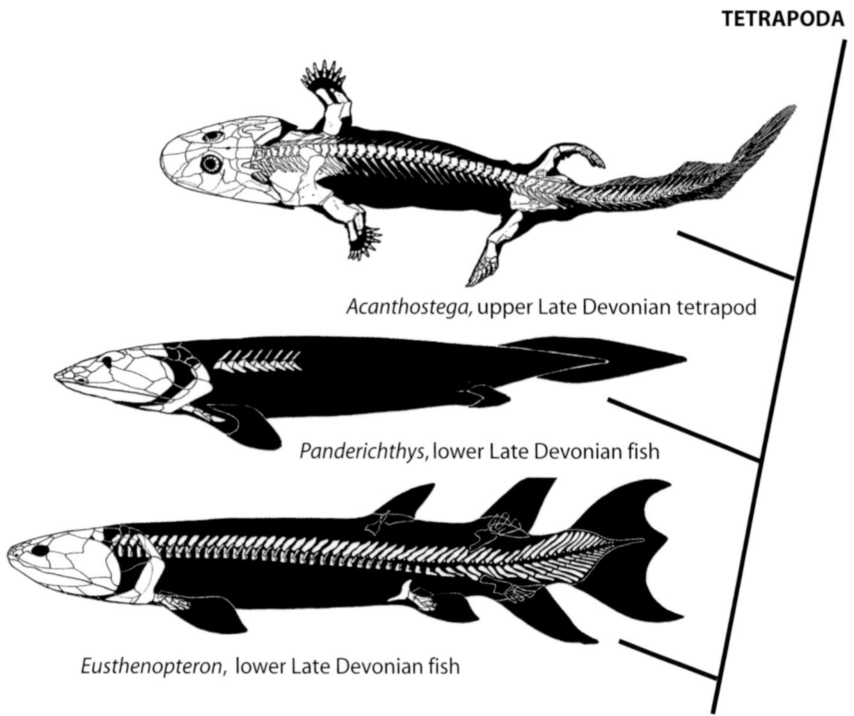

Skeletal diagram showing the main steps between tetrapodomorph fishes like *Eusthenopteron* and *Panderichthys* and a basal tetrapod, *Acanthostega.* (Courtesy Michael Coates, University of Chicago)

was quite large, around 2.5 m in length. This find predates all other tetrapod tracks or skeletal remains by about 10 to 12 million years and occurs well before the first elpistostegalian fishes, the accepted sister group to early tetrapods.

All of this indicates that tetrapods may well have originated by the end of the Early Devonian. Some enigmatic trackways from the Glenilsa site, Grampians in Victoria, Australia, were described as possible early tetrapods by Anne Warren and colleagues back in 1986 but dismissed by other researchers as being too old (Silurian-Devonian boundary) and lacking clear digit impressions. The new discovery has opened our minds that traces of tetrapods could exist much farther back in time than previously expected and that paleontologists should be searching other environmental settings, such as shallow marine and lagoonal settings, for their fossils. As this book goes to press I have just heard of Middle Devonian (Eifelian) tetrapod bones being discovered in China, so this discovery, when published, may cement the fact that tetrapods had appeared and diversified by this time period.

The oldest identifiable tetrapod remains belong to *Elginerpeton,* described from bits of jaws, of the skull and shoulder girdle, and of the postcranial skeleton by Per Ahlberg from a site the Scat Craig in Scotland. They date from the upper part of the Frasnian (lower Late Devonian).

Trackways made by an early tetrapod were found from the Genoa River site in Victoria, Australia, in the early 1970s. These trackways were restudied by Jenny Clack in 1997, who identified that at least two kinds of track makers left their impressions in the red sandstone, deposited by a slow-flowing stream, some 365 million years ago. The age of the site has been dated as Early-Middle Frasnian by Gavin Young, making these equally as old, if not older, than *Elginerpeton.* They indicate that by the start of the Late Devonian tetrapods had diversified and radiated across the seas separating Gondwana from Euramerica.

Fragmentary tetrapod remains have been discovered and described in recent years from the Famennian Red Hill site in Pennsylvania by Ted Daeschler and Neil Shubin. These include jaws and shoulder girdle bones attributed to *Densignathus* and *Hynerpeton,* plus an isolated humerus from an unnamed form that appears to

Trackways made by two different tetrapod track makers are present on this slab of sandstone from the Genoa River Beds, Victoria, Australia. Dated as Early-Middle Frasnian, they represent early evidence of tetrapods in Gondwana. Older trackways are known from the Grampians in western Victoria.

straddle the condition seen in the humeri of panderichthyid fishes, while lacking advanced features seen in *Acanthostega* and *Ichthyostega* limbs.

One of the most primitive tetrapods known from well-preserved material is *Ventastaga curonica*, from the late Famennian Ketleri Formation of the Venta River site in western Latvia. First discovered in 1994, continued excavations yielded a near-complete skull along with well-preserved shoulder girdle and axial skeleton bones recently described by Per Alhberg and colleagues (2008). *Ventastega*, along with *Densignathus* and *Elginerpeton*, appears to show an advanced tetrapod-like lower jaw structure, while those with skull remains show the panderichthyid fishlike pattern of cranial bones.

The first tetrapods known in considerable detail are the Late Devonian ichthyostegalids from East Greenland, *Ichthyostega* and *Acanthostega*. Early descriptive work by the Swedish scientists Gunnar Save-Soderbergh and Erik Jarvik revealed much about the basic structure of these amphibians, although in the past couple of decades research by Clack, Ahlberg, and Mike Coates has revealed many new aspects of the anatomy of these forms, based on new discoveries from East Greenland in the 1980s.

Humerus (upper arm bone) of an early tetrapod or near-tetrapod-like fish, from the Red Hill site near Hyner, Pennsylvania. (© T. Daeschler)

(*Left*) Lower jaw of *Densignathus*, an early tetrapod from the Red Hill site near Hyner, Pennsylvania. (© T. Daeschler)

(*Below*) Restoration of the head and neck of *Ventastega*, a Late Devonian tetrapod discovered by Per Ahlberg, Ervins Lucevics, and colleagues from the Venta River site in Latvia.

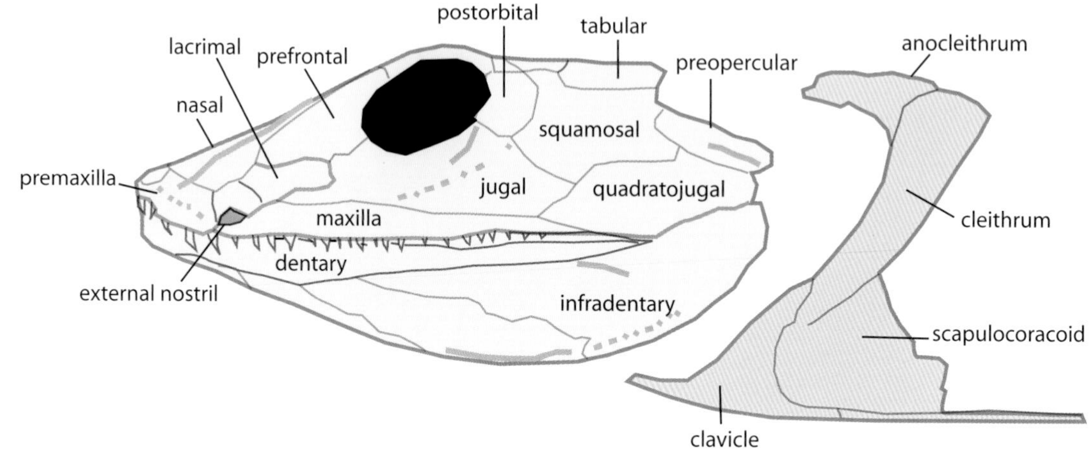

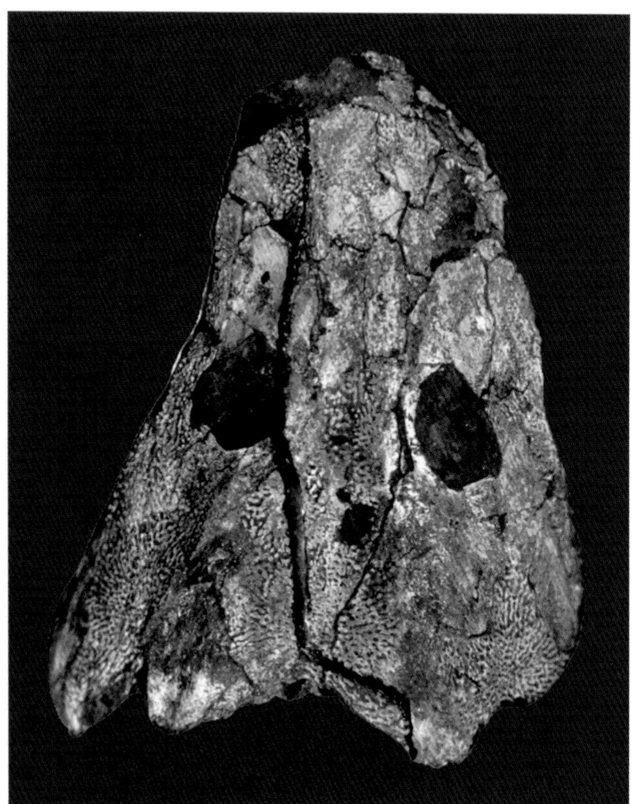

Skull of *Ichthyostega*, one of the first discovered Late Devonian tetrapods, found in East Greenland in the 1930s and described in detail in later years by Erik Jarvik of Sweden.

The skull is well ossified, and there is no sign of an intracranial joint as seen in tetrapodomorphan fishes. Otherwise the skull roof and cheek pattern is much like that of a panderichthyid fish, with the exception of *Ichthyostega,* which has an unusual fused median bone in rear of the skull roof. The head is large relative to the overall body size, and the eyes are placed in the middle and on top of the skull. The single pair of external nostrils opens close to the mouth and faces downward. Inside the mouth, there is a large palatal nostril (choana) present, bordered by the vomer and other toothed palatal bones. The braincase is not hinged and is much reduced in size relative to overall size of the skull. The gill arch elements of *Acanthostega* revealed that the animal was capable of aquatic respiration in the adult phase. Thus these early amphibians were thus still highly dependent on living in the water.

The bodies of these early amphibians show long tails that bear a well-developed tail fin supported by rods of bone (lepidotrichs). Scales looking like thin slivers of dermal bone may cover the ventral or belly surface. The limbs feature many digits (seven or eight) on the front

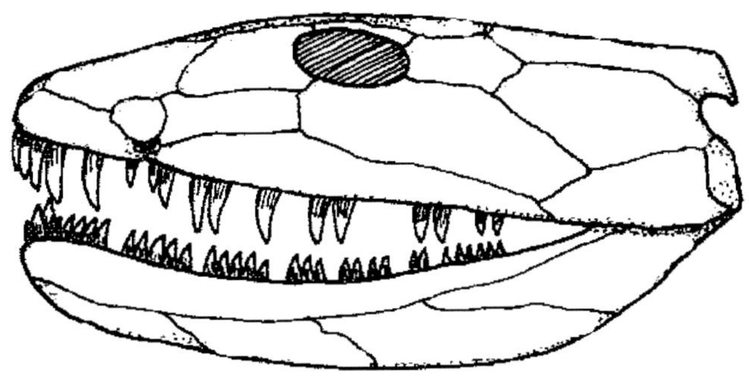

Head of *Ichthyostega* restored in side view. The teeth really are large, indicating it was a predator that hunted fishes and maybe other tetrapods.

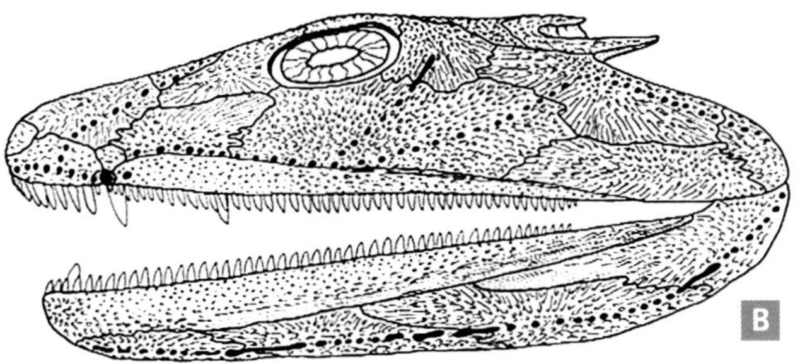

Skull and partial skeleton of *Acanthostega* (*A*), a Late Devonian tetrapod from east Greenland that bore eight digits on the hands and feet and had gills for aqueous respiration, like a fish. Skull of *Acanthostega*, showing major bones of the skull (*B*). (With the permission of the Zoology Museum, Cambridge University)

and hind limbs, and these may be divided on each hand or foot into a series of large digits and series of much smaller elements. Another Devonian amphibian, *Tulerpeton*, described from Russia, has six digits on its front limbs. There are no external shoulder girdle bones seen on the outside of these animals, although the cleithrum is a large, high bone. Coates and Oleg Lebedev regard *Tulerpeton* as being the closest to reptiles of all the Devonian tetrapods. The fact that it shows the beginnings of an asymmetrical foot suggests it may have been partially terrestrial.

In general, the structure of these earliest amphibians is much like that already described for panderichthyids. Only the tetrapods have limbs with digits present, while

(*Top*) **A scene from the Late Devonian of East Greenland, about 360 million years ago. The tetrapod *Acanthostega* is hunting a group of *Remigolepis* placoderms, while the porolepiform fish *Holoptychius* moves quietly in the background.** (Courtesy Brian Choo)

(*Left*) **The six-toed foot of the late Devonian tetrapod *Tulerpeton*, from the Andreyevka site in Russia.** (Courtesy Oleg Lebedev, Borissiak Paleontological Institute, Moscow)

the fish have fins and most of them, except for *Tiktaalik*, have complete series of operculogular bones that cover the gill chamber.

From the geographical range of the fossils discussed above, including the Polish trackways dated at around 395 million years old, it would be logical to assume that Euramerica was the most likely place for tetrapods to have evolved. However, recent finds from China and Eastern Gondwana appear to compete as an alternative center for much of the tetrapodomorphan diversifica-

Eucritta melanolimnetes **(meaning the "Creature from the Black Lagoon") was a small early tetrapod from the Carboniferous of Scotland. By this time tetrapods had begun their invasion of land.**

tion. East Gondwana's unique faunas containing stem-group tetrapodomorphs, the canowindrids, the earliest and most generalized rhizodontiforms and possibly the most primitive known tristichopterids (although this is still debated) would suggest that the evolution of the higher clade, the elpistostegalians, could well have first taken place there. These data, considered with the facts that the Genoa River trackways, from Gondwana, lead me to consider Gondwana could alternatively be a likely place for this great step in evolution. This model would assume that the elpistostegalians radiated out to invade the northern hemisphere after first evolving, although this is purely a speculative idea on my part. Evidence clearly shows that elpistostegalian fishes are currently only known from the northern hemisphere continents (Laurasia). New fossil fish and tetrapod sites in Antarctica and parts of Australia might one day hold the key to this great mystery.

The first truly terrestrial tetrapods, which might be classified as amphibians, are represented by forms like *Pederpes* from the Early Carboniferous of Scotland, (Clack and Finney 2005). *Pederpes* shows the familiar pattern of having five digits on the hands and feet, and the fact that the digits are asymmetrical indicates adaptation for walking on land, rather than swimming in water. A similarly aged enigmatic form from Ducabrook, Queensland, Australia is the newly discovered *Ossinodus*, another primitive tetrapod, whose skull features place it very close to *Pederpes*. A short time after the appearance of these forms, more advanced amphibians began to appear and diversify throughout the Carboniferous Period (Clack 2002).

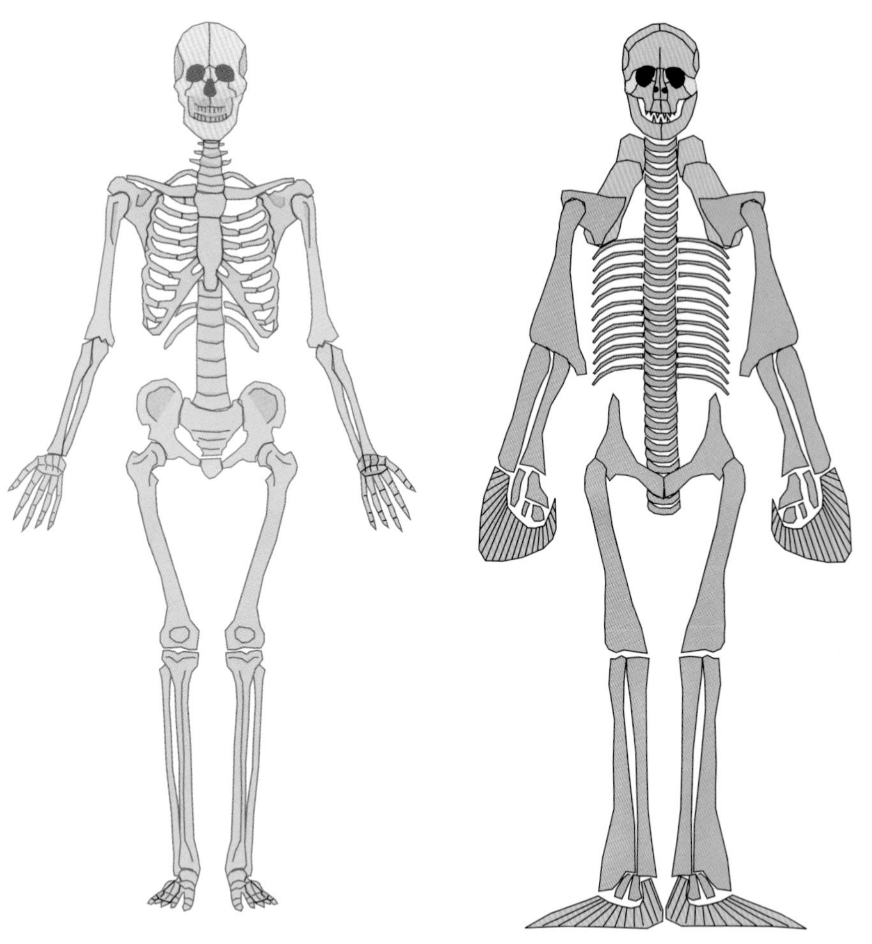

Skeletal body plan of a Devonian tetrapodomorph fish, when scaled up to match human proportions, is powerful visual evidence of just how much of the human body plan had evolved in fishes. Once tetrapods left the water to invade land, fine tuning of the existing pattern was all that was needed to develop reptiles, birds, dinosaurs, mammals, and, ultimately, humans.

A Small Step for Fishkind but a Giant Step for Man

We have now seen the contenders—the fishes—and the end result—the amphibians—and have looked at where this giant step may have occurred. But why did they take it? What factors drove fishes to eventually become landlubbers? Was it simply a matter of escaping the piscine rat race of the Devonian seas, rivers, and lakes, or were there more fundamental reasons for invading a new, but hostile, habitat? Well before this time, land had been colonized, not only by plants (with their great diversity), but also by many forms of arthropods and other invertebrates. No doubt there were new food sources to be exploited, but this was probably not reason enough to evolve digits and air-breathing capability. Like most innovations in evolution, the precursor organs for the invasion of land were already installed in the fishes.

We have seen that the lungs of amphibians are just modified versions of the swim-bladder of an osteichthyan fish. The digits are just the ends of the fins, homologues to the distal radials seen in *Panderichthys* (Boisvert and Ahlberg 2009); and already those fins in osteolepiforms and rhizodontiform fish had robust humerus, ulna, and radius bones. The skulls of early tetrapods and panderichthyid fishes are almost identical, and little change was needed in this department. Recent work on the Late Devonian East Greenland and Latvian amphibians shows that they were really just slightly modified fishes—they still inhabited the water and respired by gills with an assumed air-breathing capability in place, based on the presence of their internal palatal nostril or choana. The numerous digits on the hands and feet are closely akin to the paddle-shaped fins of their closest piscine ancestors. In other words, all of the adaptations evolved in some early amphibians were probably developed as specializations enabling them more efficiency in the water for their particular lifestyle.

Many of the lobe-finned fishes may have been capable of venturing out of the water for short crawls or wriggles on the banks of the river. The first amphibians possibly evolved almost nonchalantly from advanced panderichthyid fishes, which were either sexually maturing at earlier generations (at younger stages of development) or growing slower and retaining more juvenile features into maturity, for some unexplained environmentally or dietary-driven reason. These early tetrapods continued in their piscine lifestyle unperturbed by the fact that they actually modified the ends of their fins to form primitive fingers and toes. Perhaps in their early phases, they may have possibly had short fin rays emerging to support the web between the digits. The expanded digits on the limbs may have become useful adaptations for changing direction more efficiently when maneuvering in the water or for climbing up muddy riverbanks.

The occasional ventures onto land were then facilitated as the wrist and ankle joints evolved by simple modification of the limb structure. Almost serendipitously, the limbs became suddenly more practical, and longer jaunts away from the water may have resulted as new food sources were discovered or simply because such journeys enabled the animals to seek out quieter, isolated pools of water for safer breeding. Reproduction is often a powerful driving force in evolution, as obviously it is paramount to the continuation of the species.

The real innovation was yet to come, sometime within the next 50 million years or so, when amphibians became more proficient at terrestrial locomotion. Their next phases of evolution centered on the limbs becoming adapted to lift the animal off the ground and bear the weight under gravity, as opposed to the buoyancy provided in water. The axial skeleton underwent much modification as the backbone became strengthened to support the arms and legs for efficient movement on land. But despite these adaptations and the ensuing great radiation of amphibian forms in the Carboniferous Period, it was water that they always had to return to for living and breeding.

Our studies of the anatomy of fishes like *Gogonasus* or *Tiktaalik* show that most of the bones found in our own human skeletons first appeared in these Devonian fishes. Indeed, if the elements present in *Gogonasus* were simply scaled up to the same proportions as that in a human we can see a powerful example of how evolution plays tricks on our perceptions. We find that much of the human body plan was already established by the Devonian Period, and after that, when fishes invaded land as tetrapods, the rest of vertebrate evolution was mostly about fine tuning and tweaking a successful body plan.

In my opinion, it was the evolution of the hard-shelled, or amniote, egg (containing the embryonic amnion) that was the greatest single advancement in the evolution of vertebrates from fish to human. This enabled the early tetrapods to venture away from the seas, rivers, and lakes and to seek new inland habitats. Their ancestral ties with the water had been finally broken. The invasion of continents by vertebrates was then just beginning. The rest is history.

A Classification of Fishes

This classification is primarily for referring to fossil forms and is adapted from J. Nelson (2006) to uniformly show modern fish taxonomic endings where possible. Names in parentheses indicate alternatives still in wide usage; those in brackets indicate the widely used taxonomic name but one that doesn't conform to modern ichthyological format following Nelson (2006). Incertae sedis indicates taxonomic groups that are difficult to place with any degree of certainty.

SUBPHYLUM CRANIATA

Superclass Myxinomorphi
CLASS MYXINI
Order Myxiniformes (Hyperotreti)
Superclass Petromyzontomorphi
CLASS PETROMYZONTIDA
Order Petromyzontiformes (Hyperoartii)
Superclass Conodontomorpha
CLASS CONODONTA (CONODONTS, KNOWN MOSTLY FROM MICROFOSSILS)
Superclass Pteraspidomorphi
CLASS PTERASPIDOMORPHI (DIPLORHINA)
Subclass Astraspida
Order Astraspidiformes (e.g., *Astraspis*)
Subclass Arandaspida
Order Arandaspidiformes (e.g., *Arandaspis*)
Subclass Heterostraci
Incertae Sedis: Cardipeltida, Corvaspidida, Lepidaspidida, Tesseraspidida, Traquairaspidiformes, Tolypelepidida
Subclass Cyathaspidiformes
Order Pteraspidiformes
Superclass Anaspida
CLASS ANASPIDA
Order Birkeniida
Order Jamoytiiformes
Superclass Thelodontida
CLASS THELODONTI
Order Sandiviformes
Order Loganelliformes
Order Shieliiformes
Order Phlebolepidiformes
Order Thelodontiformes
Order Furcacaudiformes
Superclass Osteostracomorphii
CLASS CEPHALASPIDOMORPHI (MONORHINA)
Order Galeaspidiformes
Order Pituriaspidiformes

Order Cephalaspidiformes (Osteostraci)
Incertae Sedis: *Ateleaspis*
Clade Cornuata
Superclass Gnathostomata
Grade Placodermiomorphi
CLASS PLACODERMI
Incertae Sedis: *Stensioella, Pseudopetalichthys*
Order Antiarchiformes
Suborder Yunnanolepidoidei [Yunnanolepida]
Suborder Bothriolepoidei
Order Acanthothoraciformes [Acanthothoraci]
Order Rhenaniformes [Rhenanida]
Order Petalichthyiformes [Petalichthyida]
Order Ptyctodontiformes [Ptyctodontida]
Order Arthrodiriformes [Arthrodira]
Suborder Actinolepidoidei
Suborder Phyllolepida
Suborder Phylctaeniida
Suborder Brachythoraci

Grade Chondrichthiomorphi
CLASS CHONDRICHTHYES
Incertae Sedis: Family Mongolepidida (e.g., *Mongolepis*), *Doliodus, Pucapumpella*
Subclass Holocephali
Order Orodontiformes
Order Petalodontiformes
Order Helodontiformes
Order Iniopterygiformes
Order Debeeriformes
Order Eugeneodontiformes
SUPERORDER HOLOCEPHALOMORPHA
Order Psammodontiformes
Order Copodontiformes
Order Squalorajiformes
Order Chondrenchelyiformes
Order Menaspiformes
Order Cochliodontiformes
Order Chimaeriformes
Suborder Echinochimaeroidei
Suborder Myracanthiodei
Suborder Chimaeroidei
Subclass Elasmobranchii
Incertae Sedis: *Plesioselachus*
Infraclass Cladoselachimorpha
Order Cladoselachiformes
Infraclass Xenacanthimorpha
Order Xenacanthiformes
Infraclass Euselachii
Order Ctenacanthiformes
Division Hybodontida
Order Hybodontiformes
Division Neoselachii
Subdivision Selachii (sharks)
SUPERORDER GALEOMORPHI
Order Heterodontiformes
Order Orectolobiformes
Order Lamniformes
Order Carchariniformes
SUPERORDER SQUALIMORPHI
Order Hexanchiformes
Order Echinorhiniformes
Order Squaliformes
Order Squatiniformes
Order Pristiphoriformes
Subdivision Batoidea (rays)
Order Torpediniformes
Order Pristiformes
Order Rajiformes
Order Myliobatiformes

Grade Teleostomi
CLASS ACANTHODII
Incertae Sedis: *Yealepis, Seretolepis*
Order Climatiiformes
Order Ischnacanthiformes
Order Acanthodiformes

Grade Osteichthyes (= Euteleostomi)

Incertae Sedis: *Lophosteus, Andreolepis*

CLASS ACTINOPTERYGII

SUBCLASS CLADISTIA

ORDER POLYPTERIFORMES

SUBCLASS CHONDROSTEI

ORDER CHEIROLEPIFORMES

ORDER PALAEONISCIFORMES

ORDER TARRASIIFORMES

ORDER GUIDAYICHTHYIFORMES

ORDER PHANERORHYNCHIFORMES

ORDER SAURICHTHYIFORMES

ORDER ACIPENSERIFORMES

ORDER PTYCHOLEPIFORMES

ORDER PHOLIDOPLEURIFORMES

ORDER PERLEIDIFORMES

ORDER LUGANOIIFORMES

SUBCLASS NEOPTERYGII

ORDER MACROSEMIIFORMES

ORDER SEMIONOTIFORMES

ORDER LEPISOSTEIFORMES

ORDER PYCNODONTIFORMES

ORDER AMIIFORMES

ORDER ASPIDORHYNCHIFORMES

ORDER PACHYCORMIFORMES

Division Teleostei

ORDER PHOLIDOPHORIFORMES

ORDER LEPTOLEPIFORMES

ORDER TSELFATIIFORMES

Subdivision Osteossomorpha

ORDER ICHTHYODECTIFORMES

ORDER LYCOPTERIFORMES

ORDER HIODONTIFORMES

ORDER OSTEOGLOSSIFORMES

Subdivision Elopomorpha

ORDER ELOPIFORMES

ORDER ALBULIFORMES

ORDER ANGUILLIFORMES

ORDER SACCOPHARYGIFORMES

ORDER CROSSOGNATHIFORMES

Subdivision Ostarioclupeomorpha (Otocephala)

Superorder Clupeomorpha

ORDER ELLIMMICHTHYIFORMES

ORDER CLUPEIFORMES

SUPERORDER OSTARIOPHYSI

Series Anotophysi

ORDER GONORHYNCHIFORMES

Series Otophysi

ORDER CYPRINIFORMES

ORDER CHARACIFORMES

ORDER SILURIFORMES

ORDER GYMNOTIFORMES

Subdivision Euteleosteii

SUPERORDER PROTACANTHOPTERYGII

ORDER ARGENTINIFORMES

ORDER OSMERIFORMES

ORDER SALMONIFORMES

ORDER ESOCIFORMES

SUPERORDER STENOPTERYGII

ORDER STOMIIFORMES

SUPERORDER ATELEOPODOMORPHA

ORDER ATELEOPODIFORMES

SUPERORDER CYCLOSQAMATA

ORDER AULOPIFORMES

SUPERORDER SCOPELOMORPHA

ORDER MYCTOPHIFORMES

SUPERORDER LAMPRIOMORPHA

ORDER LAMPRIFORMES

SUPERORDER POLYMIXIOMORPHA

ORDER POLYMIXIFORMES

ORDER CTENOTHRISSIFORMES

SUPERORDER PARACANTHOPTERYGII

ORDER PERCOPSIFORMES

ORDER SPHENOCEPHALIFORMES

ORDER GADIFORMES

ORDER OPHIDIFORMES

ORDER BATRACHOIDIFORMES

ORDER LOPHIIFORMES

SUPERORDER ACANTHOPTERYGII

Series Mugilomorpha

Order Mugiliformes

Series Atherinomorpha

SUPERORDER ATHERINEA

Order Atheriniformes

SUPERORDER CYPRINODONTEA

Order Beloniformes

Order Cyprinodontiformes

Series Percomorpha

Order Stephanoberyciformes

Order Beryciformes

Order Zeiformes

Order Gasterosteiformes

Order Synbranchiformes

Order Scorpaeniformes

Order Perciformes

Order Pleuronectiformes

Order Tetraodontiformes

CLASS SARCOPTERYGII

Incertae Sedis: *Ligulalepis, Guiyu, Meemania*

Order Onychodontiformes (= Onychodontida, Struniiformes)

Subclass Coelacanthimorpha

Order Coelacanthiformes (Actinistia)

Subclass Dipnotetrapodomorpha

Incertae Sedis: *Diabolepis, Styloichthys,* Dipnomorpha

Order Porolepiformes (= Holoptychiiformes)

SUPERORDER DIPTERIMORPHA

Order Dipnoiformes

Order Ceratodontiformes

Incertae Sedis: Tetrapodomorpha, Rhizodontimorpha

Order Rhizodontiformes

Incertae Sedis: Osteolepidimorpha

Order Osteolepidiformes

Unranked Clade: Elpistostegalia + Tetrapoda

Infraclass Elpistostegalia

Order Elpistostegaliformes

Infraclass Tetrapoda

Glossary

Words in parentheses are the vernacular terms derived from the formal scientific terms or are alternatives in common usage. Geological time periods and their dates can be found on the chart in Chapter 1.

Acanthodiformes (acanthodiform): One of the three main orders of acanthodian fishes, characterized by one dorsal fin and lack of shoulder armor and with well-developed gill rakers for a filter-feeding lifestyle.

Acanthodii (acanthodian): A class of jawed fishes having bony fin spines preceding all fins. They lived between the Silurian and Permian Periods and belonged the three main orders: Climatiiformes, Ischnacanthiformes, and Acanthodiformes. Not monophyletic.

Acanthothoraci (acanthothoracid = palaeacanthaspid): One of the seven major groups of placoderm fishes that thrived during the Early Devonian. They have heavily ossified armors and skulls, often with elaborate surface ornaments on the bones. Example: *Brindabellaspis.*

Acrodin: A dense dentinious tissue found on the tips of the teeth of all ray-finned fishes (actinopterygians), excluding a few primitive Devonian forms such as *Cheirolepis.*

Actinistia (actinistian, coelacanth): One of the major groups of lobe-finned fishes (sarcopterygians) characterized by a tassel-finned tail and a lower jaw with short dentary and long angular bone, among other features. Known popularly as coelacanths, the only living genus *Latimeria.*

Actinopterygyii (actinopterygian): One of three major subclasses of bony fishes (osteichthyans), containing the vast majority of all living fishes today. Characterized principally by having fins supported mainly by lepidotrichia; also ganoine layers in scales. Examples: trout, goldfish, salmon.

Agnatha: Meaning "without jaws," in this book the term refers to jawless fishes.

Anaspida (anaspids): Group of elongated, jawless fishes that lacked bony armor and paired fins and lived in the Silurian and Devonian Periods of Euramerica. Examples: *Birkenia, Jamoytius.*

Antiarchi (antiarch): One of seven major orders of placoderm fishes, characterized by having a long trunk shield, short head shield with eyes and nostrils centrally placed, and pectoral fins modified as bony props. Examples: *Bothriolepis, Asterolepis.*

Arthrodira (arthrodire, euathrodire): One of seven major orders of placoderm fishes that is characterized by having two pairs of upper-jaw tooth plates (superognathals) and includes more than 60% of all known placoderm species. Arthrodires (meaning jointed neck) generally have well-developed neck joints between the head shield and trunk shield. Includes the largest placoderms, the gigantic predatory dinichthyids. Examples: *Coccosteus, Eastmanosteus.*

Aspidin: Special type of bone found in some extinct jawless fishes, with the surface covered by ornamented dentinous sculpture, a spongiose middle layer, and acellular base.

Asterolepidoidei (Asterolepididae, asterolepid): One of the major groups of antiarch placoderm fishes, characterized by having a short, robust pectoral fin appendage that does not extend beyond the long trunk shield; short head shield with very large orbital fenestra (central hole in head shield for eyes and nostrils). Examples: *Asterolepis, Pterichthyodes.*

Basibranchial: The large ventral bone or series of bones in the gill arches, present in all gnathostome fishes. Osteichthyans have small toothed bones covering the basibranchials.

Batoidei (batoids, rays): Rays are flattened chondrichthyans with large winglike pectoral fins. Example: *Myliobatis.*

Benthic: Bottom-dwelling (living near the floor or bed of the sea, river, or lake).

Biostratigraphy: Using fossils to correlate strata or determine the relative age of sedimentary layers.

Bothriolepidoidei (Bothriolepididae, bothriolepid): One of the major suborders of antiarch placoderm fishes, characterized by having a long, segmented pectoral fin appendage that generally extends beyond the trunk shield; large head shield with small orbital fenstra (central hole for eyes and nostrils), and often has postpineal plate recessed within the nuchal. Example: *Bothriolepis.*

Camuropiscidae (camuropiscid): Family of arthrodires (placoderms) known only from the Late Devonian Gogo Formation, Western Australia. They were streamlined and elongated fishes, with large eyes and durophagous gnathal plates, some evolving long tubular snouts (rostral plates). Examples: *Camuropiscis, Rolfosteus.*

Carcharhiniformes: One of the major orders of sharks, containing the whaler sharks and black-tipped reef sharks. Example: *Carcharhinus.*

Ceratodontidae (ceratodontids): Family of tooth-plated lungfishes with much reduced skull-roof pattern and having a continuous median fin (anal and dorsal fin merged together with caudal fin). Mesozoic to recent. Examples: *Ceradotus, Neoceradotus.*

Ceratohyal: A large bone in the first gill arch (hyoid arch) of jawed fishes situated between the hypohyal and the hyomandibular (or epihyal) elements. The ceratohyal is especially large in lungfishes and some crossopterygians.

Cheirolepidae (chirolepid): Middle-Late Devonian family of primitive ray-finned fishes (actinopterygians) having tiny scales and long jaws and lacking many advanced features such as acrodin caps on teeth. Example: *Cheirolepis* (only known genus).

Chondrichthyes (chondrichthyan): Fish with skeleton composed largely of globular calcified cartilage, lacking endochondral bone, having the skin covered by very small placoid scales, such as sharks, rays, and chimaerids (holocephalans).

Chondrostei (chodrostean): Group of primitive ray-finned fishes (actinopterygians) that has developed a largely cartilaginous skeleton, although they still retain dermal bone in the skull, scales, and parts of the pectoral fin girdle. Once all primitive ray-finned fishes (for example, palaeoniscoids) were included in this group, but now it is used strictly to include the paddlefishes and the sturgeons.

Clade (cladistics): A hypothetical cluster of related organisms sharing derived characteristics (synapomorphies). The study of grouping organisms by shared derived features is called *phylogenetic systematics* or *cladistics.*

Claspers: Male reproductive structures used for internal fertilization that are attached to the pelvic fins. In fishes, they are found in chondrichthyans and some placoderm groups.

Climatiiformes (climatiiform): Order of acanthodian fishes having external dermal bones around the shoulder girdle, often expressed as elaborately ornamented as pinnal and lorical plates around the pectoral fin insertion. Examples: *Climatius, Parexus, Euthacanthus.*

Coccosteomorph (coccosteid): Group or grade of arthrodian placoderms having well-developed jaws with predatory cusps and broad head shields and trunk shields with long median dorsal plates, often with a posterior spine developed. Example: *Coccosteus.*

Cochliodontiformes (cochliodonts): Order of Paleozoic chondrichthyans within the Holocephalomorpha, having crushing tooth plates with only two or three teeth in each jaw. They are closely related to chimaerids as seen by their presence of prepelvic tenaculae preceding the claspers of males. Examples: *Cochliodus, Erismacanthus.*

Conodont: Group of Paleozoic-Triassic protochordates characterized by a wormlike body with bony rods supporting the tail and having a set of phosphatic jawlike elements in the head region, probably used for food capture or filtering. Their microscopic phosphatic elements are commonly found in marine sediments and are widely used for dating rock sequences.

Cosmine: A layer of shiny external bony tissue found on the dermal bones of some extinct sarcopterygian fishes. It has an outer enameloid layer over a dentin layer that houses a system of flask-shaped cavities open to the surface by pores and interconnected to each other by canals; believed to be part of a complex vascular system for the skin.

Crossopterygyii (crossopterygian): An old term no longer used by fish systematists, to denote a subclass of lobe-finned predatory osteichthyan fishes characterized by their kinetic skulls, the braincase being divided into ethmosphenoid and oticco-occipital divisions. Recent discoveries have shown that Crossopterygii is not a monophyletic grouping.

Crown group: Derived clade of organisms sharing synapomorphies, includes and is bracketed by living taxa as well as extinct forms.

Dentin: Hard, dense tissue found in the teeth of tetrapods, as well as in the teeth, spines, and dermal bones of many primitive fishes. There are many kinds of fossil dentin, based on the arrangement of nutritive canals and orientation of fibers within the dentin layer.

Dermal bone: Bone formed in the dermis of the skin; often has a surface ornament. Includes all the outside skull bones of a fish and many platelike bones in skulls of tetrapods. Scales are just small dermal bones that cover the fish's body.

Dermosphenotic: Bone in the skull of certain actinopterygian fishes that tflanks the skull roof and meets the cheek unit behind the eye. It carries the main lateral line canal down to the cheek.

Dinichthyidae (dinichthyids): Family of large arthrodiran placoderms that had pachyosteomorph trunk shields and large pointed cusps on the upper and lower jawbones. Some forms had lost the spinal plate from the armor. Includes the largest ever placoderms, such as *Dunkleosteus* and *Gorgonichthys*, believed to be up to 8 m long.

Dipnoi (dipnoan, lungfishes): These "twice-breathing" fishes have both gills and lungs and the ability to breathe air. Early lungfishes also had a mosaic of bones forming the skull roof not comparable to the skull roof patterns of other osteichthyans. The dentition on most forms is of hard crushing tooth plates or a fine shagreen of denticles covering the palate, lower jawbones, and on bones resting on the ventral gill arches. Examples: *Dipterus, Neoceratodus.*

Dipteridae (dipterid): Family of Devonian lungfishes with well-developed tooth plates having many rows of teeth, two dorsal fins present, and with cosmine retained in part of the dermal skeleton. Example: *Dipterus.*

Distal: Pertaining to part of an organism that is the farthest away, usually used as meaning far from the head (for example, the hand is the distal end of the arm).

Dorsal: Top (the dorsal view of a car is that looking down on its roof). The dorsal surface of a bone on the surface seen from above when in life position.

Durophagous: Feeding by crushing prey, particularly hard shelled-items. Such creatures usually have hardened tooth plates rather than sharp pointed teeth.

Edestoidei (edestid): Group of fossil sharks that thrived in the Late Paleozoic (especially in the Permian) characterized by their well-developed symphysial tooth whorls formed of laterally compressed bladelike teeth. Little is known of their skulls or body shape. Examples: *Helicoprion, Edestus, Agassizodus.*

Endochondral bone: Bone formed from a cartilage core, replaces the cartilage frame, for example, the limb bones.

Eusthenopteridae (= Tristicopteridae): Family of advanced osteolepiform fishes that have lost the cosmine layer from their bones and scales, have deep narrow heads, and most have lost the extratemporal bone from the skull. The scales are rounded with a boss (protrusion) on the basal surface. Examples: *Eusthenopteron, Marsdenichthys, Eusthenodon.*

Frontals: Paired bones in the midline of the skull roof in osteichthyan fishes and tetrapods. The frontals enclose the pineal opening, or foramen, in most fishes and tetrapods.

Galeaspida (galeaspid): Class of armored jawless fishes that lived in China (south and north China terranes) during the Silurian to Late Devonian. They are characterized by a single-piece head shield with large median opening in front of the eyes and pineal opening.

Genus (pl. genera): Taxonomic term, a group of species having shared features uniting them as a group.

Gill arches: Series of bones or cartilages that supports the gills of fishes, they articulate dorsally with the braincase and meet ventrally in a median basibranchal series. The first arch is the hyoid arch (hyomandibular, epihyal, ceratohyal, hypohyal, etc.). Often referred to as the visceral skeleton.

Gondwana: Ancient supercontinent comprising Australia, Antarctica, South America, Africa, India, and parts of southeast Asia. Existed from Ordovician-Triassic times, with remnants remaining after the breakup of Pangaea; the Australia-Antarctica connection lasted until about 90 million years ago.

Gnathostomata (gnathostomes): Refers to all jawed vertebrates.

Head shield: Bones covering the head in placoderm fishes, including the skull roof and cheek units.

Heterochrony: A mechanism for evolutionary transformation arising from accrued changes in morphology. These changes are derived from altering the timing of development of an organism. Changes due to earlier maturation, retaining juvenile features into the adult of a derived species are called *neotenic* or *paedomorphic*, those by prolongation of growth can result are called *peramorphic.*

Heterostracan: A group of jawless extinct armored fishes (Ordovician-Devonian) having multiple plates forming the bony shield around the body. Examples: *Pteraspis, Drepanaspis.*

Holocephali: Group of cartilaginous fishes (chondricthyans) having upper jaws fused to braincase, opercular cover, and modified durophagous dentition. Includes chimaeras and ratfishes.

Holocephalomorpha: Group of cartilaginous fishes (chondrichthyans) including modern holocephalans and some extinct primitive forms allied to them, such as cochliodont, petalodont, and iniopterygian fishes.

Holoptychiidae (holoptychiid): Family of extinct porolepiform (dipnomorphan) fishes that have rounded scales, have a nariodal bone in the snout, and have lost cosmine from their skeletons. Examples: *Holoptychius, Glyptolepis.*

Holostei (holostean): Term once used for a grade of ray-finned fishes more advanced than the palaeoniscoids but less derived than teleosteans. Used today to denote a group including bowfins and gars.

Hybodontodoidei (hybodonts): Extinct sharks of a particular monophyletic lineage having stout dorsal fin spines and teeth with multicuspid crowns, also with enlarged cranial denticles in the skin. Carboniferous to end of Cretaceous.

Hyomandibular: Large dorsal bone of the hyoid arch that braces jaw articulation in primitive fishes, lying against the palatoquadrate. In advanced sarcopterygians, it has two articulations to the braincase and in tetrapods becomes modified to turn upward and becomes the stapes of the inner ear.

Inioptergii (iniopterygians): Group of Late Paleozoic chondrichthyan fishes from North America placed within the Holocephalomorpha, having dorsal fins located high up the body wall with robust spines. Their tooth plates are sometimes fused to the braincase.

Intracranial joint: Joint between two divisions of the braincase in most sarcopterygian fishes, can be secondarily lost in some lineages.

Jugal: Generally large bone in the cheek of osteichthyans, part of the infraorbital series under the eye, and it carries the infraorbital sensory-line canal.

Lacrimal bone (lachrymal): Often small anteriorly located bone in the cheek of osteichthyans, part of the infraorbital series under the eye, and it carries the infraorbital sensory-line canal.

Lamniformes (lamnid): Group of sharks having no nictitating eyelid, includes mako, great white, and porbeagle sharks.

Lateral: Side. Externally facing surface of an object or organism.

Lateral-line: Main sensory-line canal that runs along the body of fishes and carries onto the skull roof.

Lepidotrichia: Fin rays, bony rods supporting the fins that are formed in the dermis in osteichthyan fishes.

Lophosteiformes (lophosteids): Poorly known group of Late Silurian / Early Devonian basal osteichthyans, believed to have been closely allied to actinopterygians, they lacked acrodin caps on teeth and peg-and-socket articulations on scales.

Maxilla (maxillary): Upper jawbone of most osteichthyan fishes, formed in the dermis, and bearing teeth.

Monophyletic: Group of organisms that can be demonstrated as a closely related clade due to shared derived features (synapomorphies).

Naris (nares): External nostrils.

Nasal bone: Dermal bone in the snout of fishes carrying supraorbital sensory line canal, often notched for the narial openings.

Nasophypohysial opening: Single opening on the top of the head in lampreys and osteostracans that opens ventrally to the nostrils and hypophyseal duct.

Nectonic: Free-swimming.

Neopterygyii (neopterygians): Group of derived actinopterygian fishes having fin-rays equal in number to their supports in the dorsal and anal fins, consolidated upper pharangeal dentition, and a reduced or lost clavicle.

Neoselachii: Clade of derived sharks with multilayered enameloid in their teeth crowns and root canals that penetrate the base.

Neoteleosteii: Clade of derived actinopterygians having among other features special constrictor muscles that work the pharyngeal dental plates.

Neurocranium: Braincase (endocranium), a block of cartilage or series of bony ossifications that encloses and protects the brain, either perichondrally or endochondrally ossified.

Occipital: Pertaining to the neck region or rear section of the head or braincase.

Onychodontiformes (onychodontids): A clade of extinct sarcopterygian fishes that lived in the Devonian Period, sharing well-developed laterally compressed tooth whorls on the lower jaws. Also called Struniiformes. Examples: *Onychodus, Strunius.*

Operculogular series: Series of flat platelike dermal bones covering the gill arches of osteichthyan fishes, includes the opercular, subopercular, branchiostegal rays, and gular plates.

Orbit: Hole in the skull for the eye. Orbital notch is the margin of a bone bordering the eye.

Orbital fenestra.: Centrally located hole in the head shield of some placoderms (antiarchs, acanthothoracids) for the eyes and nares.

Ornamentation: Refers to the elaborate patterns made up of tubercles, nodes, or reticulating ridges seen on the surface of dermal bones, especially well-developed on some placoderms and jawless fishes.

Osteolepiformes: Paraphyletic grouping of Devonian-Permian tetrapodomorphan osteichthyan fishes sharing a single pair of external nares and well-developed choana in the palate. Within this group are the subgroups of elpistostegalians, including tetrapods. Includes monophyletic families like Tristichopteridae, Megalichthyidae. Family Osteolepididae may not be monophyletic.

Otolith: Ear stones. In fishes, composed of vaterite and occurring in osteichthyans. In chondrichthyans and placoderms, statoliths may be present inside the inner ear, but these are not formed the same way as true otoliths.

Pachyosteomorph: Grade group of arthrodiran placoderms with the trunk shield open posteriorly for the pectoral fin.

Paedomorphosis: The retention of subadult characters in descendent adult forms, eventually resulting in new species being formed. A kind of heterochrony.

Palaeacanthaspid. See Acanthothoracid.

Palaeoniscoidei (palaeoniscid): Term used mainly as grade grouping (paraphyletic clade) of basal actinopterygian fishes, Devonian-Cretaceous forms.

Paleobiology: The study of fossils as once-living organisms, involving reconstruction of their biological processes. Paleozoology applies to the study of extinct animals.

Paleogeography: The study of past continental positions and past environments using information derived largely from fossil distributions combined with paleomagnetic data and lithlogical indicators of past temperature and humidity.

Paleontology (UK/Australia, palaeontology): The study of fossils.

Panderichthyids (=Elpistostegalia): Group of highly derived tetrapodomorphan fishes and basal tetrapods having large skulls with dorsally situated orbits. Some have lost the opercular bones Examples: *Tiktaalik* and tetrapods.

Pangaea: Ancient supercontinent formed by a merging of all continental landmasses in the Triassic Period; the split up of the continent had begun by the Late Jurassic.

Parallel evolution (convergence, homoplasy): Evolution of similar characters in two or more lineages (i.e., not homologous structures). Examples include horns evolving on ceratopsian dinosaurs and on bison or flippers on plesiosaurs and on whales.

Paraphyletic group: A set of organisms that are not monophyletic or sharing the derived characters but otherwise appear to be related. Includes grade groups that have basal members not related closely to the crown groups. Placoderms could well be a paraphyletic group.

Parasphenoid: Median, often-toothed bone of the palate in placoderms and osteichthyans.

Parietals: Paired skull bones that enclose the pineal region in osteichthyan fishes and tetrapods. In older literature, parietals of actinopterygains were often referred to as the frontals.

Parietal shield: A set of dermal bones covering the dorsal surface of the posterior half of the braincase in some sarcopterygians, also called oticco-occipital shield.

Pectoral: Pertaining to the chest. Used to describe the regions around the trunk of an animal where the front paired limbs or fins originate (e.g., pectoral fin, pectoral girdle).

Peramorphosis: Development of traits beyond that of the ancestral adult in new species, by extending growth stages or growth rates. A kind of heterochrony.

Percomorpha: Group of advanced actinopterygian teleostean fishes, within the Acanthopterygii.

Perichondral bone: Bone ossified from the perichondrium (a layer of tissue that surrounds the cartilage of bone as it develops), often thinly laminated noncellular bone that surrounds soft tissues.

Petalichthyida: Major group of placoderms having an elongated flat skull, often with two pairs of nuchals, trunk shield lacks posterior laterals. Example: *Lunaspis.*

Petrodentine: Hypermineralized dense tissue that forms the center of cusps in many lungfish tooth plates.

Pharynx (pharyngeal): In fishes, refers to the buccal cavity of the mouth and rear extension of this cavity into the gill arch area.

Pholidophoridiformes: A basal group of extinct teleostean (actinopterygian) fishes. Example: *Pholidophorus.*

Phyllolepida (phyllolepids): A group of dorsoventrally flattened arthrodiran placoderms having a very large central nuchal plate with radiating sensory-lines on the head shield. Examples: *Phyllolepis, Cowralepis.*

Phylogeny (phylogenetic): Pertains to the evolution of a group of organisms, the study of their relationships (phylogenetic systematics is often called cladistics).

Pineal opening: Small opening, sometimes paired, on the top of the head between the eyes of many vertebrates, sometimes covered by a pineal plate or series of plates. The pineal and parapineal organs are light-sensitive structures.

Placodermi (placoderms): A group of Silurian-Devonian jawed vertebrates having interlocking and overlapping bony plates covering the head and trunk regions. Examples: *Dunkleosteus, Bothriolepis.*

Porolepiformes: A group of dipnomorphan osteichthyan fishes having small eyes, broad, flat heads, and cheeks with prespiracular plates present. Teeth have dendrodont style of complex infolding. Examples: *Holotychius, Porolepis.*

Posterior: The back end or rear part of a surface or object.

Premaxilla (premaxillary): Paired median toothed dermal bones of the upper jaw in osteichthyans and tetrapods.

Proximal: The part of the organism that is closest to the head (e.g., the proximal end of the arm is near the shoulder).

Ptyctodontida (ptyctodonts): Group of placoderm fishes having durophagus dentition, short trunk shields, and very large eyes. Males have dermal claspers, and females have recently been found with embryos, indicating they bore their young live.

Rhenanida: Group of mainly Early Devonian flattened placoderms often with tesserated sections of dermal armor. Example: *Gemuendina.*

Rhizodontida: Group of mostly gigantic predatory tetrapodomophan fishes that lived from the Middle Devonian-Carboniferous. They had stiff paddle-like pectoral fins with long unbranched lepidotrichia. Examples: *Rhizodus, Strepsodus.*

Sarcopterygii: Group of osteichthyan fishes having robust paired fins, pectoral fin has a single bone (humerus) articulating with the shoulder girdle. Includes coelacanths, lungfishes, and extinct groups like onychodontids, osteolepiforms, porolepiforms, and rhizodontids.

Shagreen: From a word for untanned leather, refers to the rough skin of fishes such as sharks and rays that is covered with small particles.

Sinolepidoids: Group of antiarch placoderms with large ventral fenestra in the trunk shield, head with long occipital region, and having segmented pectoral fins. Examples: *Sinolepis, Grenfellaspis.*

Spines: Refers to elongated and often ornamented bony structures preceding the fins of some fishes. Spines may be deeply inserted into the musculature of the fish or locked into shoulder girdle structures by being fused to other bones.

Spiracles: A small gill-slit opening in front of the main gill series in chondrichthyans. In basal osteichthyans, it might be an opening in the skull roof bones or between the skull and cheek plate leading to the spiracular chamber. Use for accessory air-breathing in some forms. Example: *Polypterus.*

Stem group: Members of a lineage at the base of the tree, mostly generalized forms lacking the defining specializations of the crown group.

Swim-bladder: Internal gaseous exchange organ within osteichthyan fishes that regulates depth in the water column. Sometimes acts as accessory air-breathing device and directly related to the origin of lungs.

Symplectic bone: Bone in hyoid arch of some fishes between the hyomandibular and ceratohyal bones.

Synapomorphy: Shared derived characteristic used to define a group of organisms as a monophyletic group. Synapomorphies define clades.

Taxon (pl. taxa): A classificatory unit, e.g., species, genus, or family. The study of classifying organisms is called taxonomy.

Teleosteii (teleosteans): Clade of derived actinopterygian fishes sharing uroneural bones in the tail, a mobile premaxilla, unpaired basibranchial tooth plates, and some other derived features. The largest living group of vertebrates, with close to 29,000 species described.

Tetrapod: Four-legged vertebrates, includes amphibians, reptiles, birds, and mammals. Some forms may secondarily lose limbs, e.g., snakes, caecelians.

Tetrapodomorpha: Derived group of sarcopterygian fishes (and including all tetrapods) that has single pair of external nares and a choana present in the palate. Examples: *Gogonasus, Eusthenopteron, Tiktaalik.*

Thelodonti (thelodonts): Clade of extinct Silurian-Devonian jawless fishes characterized by having the body and head covered with ornamented scales with often elaborate sculptured crowns. Many species known just from their scales. Examples: *Thelodus, Turinia.*

Trunk shield: Ring of interlocking overlapping bony plates around the front part of the body in placoderm fishes. Joins or articulates with the head shield.

Uroneural bones: Special bones found in the tail of teleostean fishes, they are elongated neural arch structures.

Ventral: Pertaining to the belly side or underside of an organism or bone.

Viviparity (viviparous): Live bearing. Giving birth to live young rather than laying eggs.

Xenacanths: Group of extinct Mid-Late Paleozoic sharks characterized by having dual crowned teeth with prominent bony knobs (torus) on the base. Many have elaborate neck spines and ornamented dorsal fin spines.

Yunnanolepidoids: Group of generalized antiarch placoderms with small orbital fenestrae, large head shields with long occipital regions, and lacking jointed pectoral fins. Only known from the Silurian / Early Devonian of China and Vietnam. Examples: *Yunnanolepis, Chuchinolepis.*

Bibliography

Chapter 1. Earth, Rocks, Evolution, and Fish

Janvier, P. 1996. *Early vertebrates.* Clarendon Press, Oxford.

Kumar, S., and Hedges, S.B. 1998. A molecular timescale for vertebrate evolution. *Nature* 392: 917–20.

Martin, A.P., Naylor, G.J.P., and Palumbi, S.R. 1992. Rates of mitochondrial DNA evolution in sharks are slow compared with mammals. *Nature* 357: 153–55.

Nelson, J. 2006. *Fishes of the world.* 4th ed. Wiley, Hoboken, New Jersey.

Romer, A.F.S., and Parsons, T.S. 1977. *The vertebrate body.* 5th ed. W.B. Saunders Co., Philadelphia.

Trinajstic, K., Marshall C., Long J.A., and Bifield, K. 2007. Exceptional preservation of nerve and muscle tissues in Late Devonian placoderm fish and their evolutionary implications. *Biology Letters* 3: 197–200.

Young, G.C. 1981. Biogeography of Devonian vertebrates. *Alcheringa* 5: 225–43.

Chapter 2. Glorified Swimming Worms

Briggs, D.E.G., Clarkson, E.N.K., and Aldridge, R.J. 1983. The conodont animal. *Lethaia* 16: 1–14.

Delsuc, F., Brinkmann, H., Chourrout, D., and Philippe, H. 2006. Tunicates and not cephalochordates are the closest living relative of vertebrates, *Nature* 439: 965–68.

Donoghue, P.C.J., and Aldridge, R.J. 2001. Origin of a mineralised skeleton. Pp. 85–105 in Ahlberg, P.E. (ed) *Major events in early vertebrate evolution: palaeontology, phylogeny, genetics, and developmental biology.* Systematics Association, London.

Gabbott, S.E., Aldridge, R.J., and Theron, J.N. 1995. A giant conodont with preserved muscle tissue from the Upper Ordovician of South Africa. *Nature* 374: 800–803.

Gans, C. 1989. Stages in the origin of vertebrates: analysis by means of scenarios. *Biological Reviews* 64: 221–65.

Gans, C., and Northcutt, R.G. 1983. Neural crest and the origins of vertebrates: a new head. *Science* 220: 268–74.

Gans, C., and Northcutt, R.G. 1985. Neural crest: the implication for comparative anatomy. *Fortschritte der Zoologie* 30: 507–14.

Hanken, J., and Hall, B.K. 1993. *The vertebrate skull.* vols 1–3. University of Chicago Press, Chicago.

Hanken, J., and Thorogood, P. 1993. Evolution and development of the vertebrate skull: the role of pattern formation. *Trends in Ecology and Evolution* 8: 9–15.

Holland, L.Z., and Holland, N.D. 2001. *Amphioxus* and the evolutionary origin of neural crest and the midbrain/hindbrain boundary. Pp. 15–32 in Ahlberg, P.E. (ed), *Major events in early vertebrate evolution: palaeontology, phylogeny, genetics, and developmental biology.* Systematics Association, London.

Janvier, P. 1981. The phylogeny of the craniata, with particular reference to the significance of fossil agnathans. *Journal of Vertebrate Paleontology* 1: 121–71.

Janvier, P. 1995. Conodonts join the club. *Nature* 374: 761–62.

Jeffries, R.P.S. 1979. The origin of chordates—a methodological essay. Pp. 443–77 in House, M.R. (ed), *The origin of the major invertebrate groups.* Academic Press, London.

Jollie, M. 1982. What are the Calcichordata? and the larger question of the origin of the Chordates. *Zoological Journal of the Linnean Society* 75: 167–88.

Kemp, A. 2002. Hyaline tissue of themally altered conodont elements and the enamel of vertebrates. *Alcheringa* 26: 23–36.

Long, J.A., and Burrett, C.F. 1989. Tubular phosphatic microproblematica from the Early Ordovician of China. *Lethaia* 22: 439–46.

Mallatt, J., and Chen, J. 2003. Fossil sister-group of craniates: predicted and found. *Journal of Morphology* 258: 1–31.

Northcutt, R.G., and Gans, C. 1983. The genesis of neural

crest and epidermal placodes: a reinterpretation of vertebrate origins. *Quarterly Review of Biology* 58: 1–28.

Ørvig, T. 1967. Phylogeny of tooth tissues: evolution of some calcified tissues in early vertebrates. Pp. 1:45–110, in Miles, A.E. (ed), *Structural and chemical organisation of teeth*. Academic Press, London.

Purnell, M.A. 1993. Feeding mechanisms in conodonts and the function of the earliest vertebrate hard tissues. *Geology* 21: 375–77.

Purnell, M. 1995. Microwear on conodont elements and macrophagy in the first vertebrates. *Nature* 374: 798–800.

Reif, W.-E. 1982. The evolution of dermal skeleton and dentition in vertebrates: the odontode regulation theory. *Evolutionary Biology* 15: 287–368.

Repetski, J.E. 1978. A fish from the Upper Cambrian of North America. *Science* 200: 529–31.

Sansom, I.J., Smith, M.M., and Smith, M.P. 1996. Scales of thelodont and shark-like fishes from the Ordovician of Colorado. *Nature* 379: 628–30.

Sansom, I.J., Smith, M.M., and Smith, M.P. 2001. The Ordovician radiation of vertebrates. Pp. 156–72 in Ahlberg, P.E. (ed), *Major events in early vertebrate evolution: palaeontology, phylogeny, genetics, and developmental biology*. Systematics Association, London.

Sansom, I.J., Smith, M.P., Armstrong, H.A. and Smith, M.M. 1992. Presence of the earliest vertebrate hard tissues in conodonts. *Science* 256: 1308–11.

Sansom, I.J., and Smith, M.P. 2005. Late Ordovician vertebrates from the Bighorn Mountain of Wyoming, USA, *Palaeontology* 48: 31–48.

Sansom, R.S., Gabbott, S. & Purnell, M.A. 2010. Non-random decay of chordate characters causes bias in fossil interpretation. *Nature* 463: 797–800.

Shu, D. A palaoentological perspective of vertebrate origins. *Chinese Science Bulletin* 48: 725–35.

Shu, D., Conway Morris, S., Han, J., Zhang, Z.-F., Yasui, K., Janvier, P., Chen, L., Zhang, X.-L., Liu, J.-N., and Liu, H.-Q. 1999. Head and backbone of the Early Cambrian vertebrate *Haikouichthys*. *Nature* 421: 527–29.

Shu, D.G., Conway Morris, S., and Zhang, X.L. 1995. A *Pikaia*-like chordate from the Lower Cambrian of China. *Nature* 374: 798–800.

Shu, D., Luo, H.-L., Conway Morris, S., Zhang, X., Hu, S.-X., Chen, L., Han, J., Zhu, M., Li, Y. and Chen, L.-Z. 1999. Lower Cambrian vertebrates from South China. *Nature* 402: 42–46.

Smith, M.M., and Hall, B. K. 1990. Development and evolutionary origins of vertebrate skeletogenic and odontogenic tissues. *Biological Reviews* 65: 277–373.

Smith, M.P. 1990. The Conodonta—palaeobiology and evolutionary history of a major Palaeozoic chordate group. *Geological magazine* 127: 365–69.

Smith, M.P., Sansom, I.J., and Cochrane, K.D. 2001. The Cambrian origin of vertebrates. Pp. 67–84 in Ahlberg, P.E. (ed), *Major events in early vertebrate evolution: palaeontology, phylogeny, genetics, and developmental biology*. Systematics Association, London.

Vickers-Rich, P. 2007. The Nama Fauna of Southern Africa. Pp. 69–87 in Fedonkin, M.A., Gehling, J.G., Grey, K., Narbonne, G.M., Vickers-Rich, P. (eds), *The rise of animals: evolution and diversification of the kingdom Animalia*. Johns Hopkins University Press, Baltimore.

Young, G.C. 1997. Ordovician microvertebrate remains from the Amadeus Basin, Central Australia. *Journal of Vertebrate Paleontology* 17: 1–25.

Young, G.C., Karatajute-Talimaa, V.N., and Smith, M.M. 1996. A possible Late Cambrian vertebrate from Australia. *Nature* 383: 810–12.

Chapter 3. Jawless Wonders

Afanassieva, O.B. 1991. *The osteostracans of the USSR (Agnatha)*. Akademia Nauka, Moscow.

Afanassieva, O.B. 1992. Some peculiarities of osteostracan ecology. Pp. 61–70 in Mark-Kurik, E. (ed), *Fossil fishes as living animals*. Academy of Sciences, Estonia.

Afanassieva, O., and Janvier, P. 1985. *Tannuaspis, Tuvaspis* and *Ilemoraspis*, endemic ostracoderm genera from the Silurian and Devonian of Tuva and Khakassia (USSR). *Geobios* 18: 493–506.

Arsenault, M., and Janvier, P. 1991. The anaspid-like craniates of the Escuminac Formation (Upper Devonian) from Miguasha (Québec, Canada), with remarks on anaspid-petromyzontid relationships. Pp. 19–40 in Chang, M.M., Liu, Y.H., and Zhang, G.R. (eds), *Early vertebrates and related problems of evolutionary biology*. Science Press, Beijing.

Belles-Isles, M., and Janvier, P. 1984. Nouveaux Osteostraci du Dévonien inférieur de Podolie (R.S.S. D'Ukraine). *Acta Palaeontologica Polonica* 4: 157–66.

Blieck, A. 1975. *Althaspis anatirostra* nov. sp., Ptéraspide du Devonien inférieur du Spitsberg. *Colloques Recherches Sommelier de Société Geologiques de France* 3: 74–77.

Blieck, A. 1980. Le genre *Rhinopteraspis* Jaekel (vertébrés, Héterostracés) du Dévonien inférieur: systématique, morphologie, répartition. *Bulletin du Muséum national d'histoire naturelle* 4, ser. 2, sec. C, 1:25–47.

Blieck, A. 1981. Le genre *Protopteraspis* Leriche (vertébrés, Héterostracés) du Devonien inférieur nord-Atlantique. *Palaeontographica* 173A: 141–59.

Blieck, A. 1982a. Les Hétérostracés (vertébrés, Agnathes)

del'horizon Vogti (Groupe de Red Bay, Dévonien inférieur du Spitsberg). *Cahiers de Paléontologie,* Paris, 1–51.

Blieck, A. 1982b. Les Hétérostracés (vertébrés, Agnathes) du Devonien inférieur du Nord de la France et Sud de la Belgique (Artois-Ardenne). *Annales de Société de Geologique de Belgique* 105: 9–23.

Blieck, A. 1982c. Les grandes lignes de la biogéographie de Hétérostracés du Silurien supérieur–Devonien inférieur dans le domaine nord-Atlantique. *Paleogeography, Palaeoclimatology, Palaeoecology* 38: 283–316.

Blieck, A. 1983. Biostratigraphie du Dévonien inférieur du Spitsberg: données complementaires sur les Hétérostracés (vertébrés, Agnathes) du Groupe de Red Bay. *Bulletin du Muséum national d'histoire naturelle,* 5th series, C, 1: 75–111.

Blieck, A. 1984. *Les Hétérostracés Ptéraspidiformes, Agnathes du Silurien-Dévonien du continent nord-Atlantique et des blocs avoisinants.* Cahiers de Paléontologie (Vertébrés), Paris

Blieck, A., and Goujet, D. 1973. *Zascinaspis laticephala* nov. sp. (Agnatha, Heterostraci) du Dévonien inférieur du Spitsberg. *Annales de Paléontologie (Vertébrés-Invertébrés)* 69: 43–56.

Blieck, A., and Heintz, N. 1979. The heterostracan faunas in the Red Bay Group (Lower Devonian) of Spitsbergen and their biostratigraphical significance: a review including new data. *Bulletin of the Geological Society of France* 7th series, 21: 169–81.

Blieck, A., and Heintz, N. 1983. The cyathaspids of the Red Bay Group (Lower Devonian) of Spitsgergen. The Downtonian and Devonian Vertebrates of Spitsbergen, XIII. *Polar Research* 1: 49–74.

Blieck, A., and Janvier, P. 1991. Silurian vertebrates. *Palaeontology, Special Papers* 44: 345–89.

Blom, H. 2008. A new anaspid fish from the Middle Silurian Cowie Harbour Fish Bed of Stonehaven, Scotland. *Journal of Vertebrate Paleontology* 28: 594–600.

Blom, H., Märss, T., and Miller, C.G. 2002. Silurian and lowermost Devonian birkeniid anaspids from the Northern Hemisphere. *Transactions of the Royal Society of Edinburgh: Earth Sciences* 92: 263–323.

Broad, D.S. 1973. Amphiaspidiformes (Heterostraci) from the Silurian of the Canadian Arctic Archipelago. *Bulletin of the Geological Society of Canada* 222: 35–51.

Broad, D.S., and Dineley. D.L. 1973. *Torpedaspis,* a new Upper Silurian and Lower Devonian genus of Arctic Archipelago. *Bulletin of the Geological Society of Canada.* 222: 35–51.

Damas, H. 1944. La branchie préspiraculaire des Cephalaspides. *Annales de la Société Royale Zoologique de Belgique* 85: 89–102.

Denison, R.H. 1947. The exoskeleton of *Tremataspis. American Journal of Science* 245: 337–67.

Denison, R.H. 1951a. The evolution and classification of the Osteostraci. *Fieldiania Geology* 11: 156–96.

Denison, R.H., 1951b. The exoskeleton of early Osteostraci. *Fieldiana Geology* 11: 198–218.

Denison, R.H. 1952. Early Devonian fishes from Utah: I. Osteostraci. *Fieldiana Geology* 11: 265–287.

Denison, R.H. 1955. The Early Devonian Vertebrates from the Knoydart Formation of Nova Scotia. *Fieldiana Zoology* 37: 449–64.

Denison, R.H. 1963. New Silurian Heterostraci from Southeastern Yukon. *Fieldiana Geology* 14: 105–41.

Denison, R.H. 1966. *Cardipeltis,* an Early Devonian agnathan of the order Heterostraci. *Fieldiana Geology* 16: 89–116.

Denison, R.H. 1967. Ordovician vertebrates from the Western United States. *Fieldiana Geology* 16: 131–92.

Denison, R.H. 1970. Revised classification of the Pteraspididae with description of new forms from Wyoming. *Fieldiana Geology* 20: 1–41.

Denison, R.H. 1971. On the tail of Heterostraci (Agnatha). *Forma et functio* 4: 87–99.

Denison, R.H. 1973. Growth and wear of the shield in Pteraspididae (Agnatha). *Palaeontographica* 143A: 1–10.

Dineley, D.L. 1964. New specimens of *Traquairaspis* from Canada. *Palaeontology* 7: 210–19.

Dineley, D.L. 1967. The Lower Devonian Knoydart faunas. *Zoological Journal of the Linnean Society* 47: 15–29.

Dineley, D.L. 1968. Osteostraci from Somerset Island. *Bulletin of the Geological Survey of Canada* 165: 49–63.

Dineley, D.L. 1976. New species of *Ctenaspis* (Ostracodermi) from the Devonian of Arctic Canada. Pp. 26–44 in Churcher, C.S. (ed), *Athlon, essays on palaeontology in honour of Loris Shano Russell.* Royal Ontario Museum, Toronto.

Dineley, D.L. 1988. The radiation and dispersal of Agnatha in Early Devonian time. *Canadian Society of Petroleum Geologists Memoir* 14 (3): 567–77.

Dineley, D.L. 1994. Cephalaspids from the Lower Devonian of Prince of Wales Island, Canada. *Palaeontology* 37: 61–70.

Dineley, D.L., and Loeffler, E.J. 1976. Ostracoderm faunas of the Delorme and associated Siluro-Devonian Formations, North Territories, Canada. *Palaeontology,* special papers 18: 1–214.

Elliott, D.K. 1984. A new subfamily of the Pteraspididae (Agnatha, Heterostraci) from the Upper Silurian and Lower Devonian of Arctic Canada. *Palaeontology* 27:169–97.

Elliott, D.K. 1987. A reassessment of *Astraspis desiderata,* the oldest North American vertebrate. *Science* 237: 190–92.

Elliott, D.K. 1994. New pteraspidid (Agnatha: Heterostraci) from the Lower Devonian Water Canyon Formation of Utah. *Journal of Paleontology* 68: 176–79.

Elliott, D.K., and Dineley, D.L. 1983. New species of *Protopteraspis* (Agnatha, Heterostraci) from the (?) Upper Silurian to Lower Devonian of Northwest Territories, Canada. *Journal of Paleontology* 57: 474–94.

Elliott, D.K., and Dineley, D.L. 1985. A new heterostracan from the Upper Silurian of Northwestern Territories, Canada. *Journal of Vertebrate Paleontology* 5: 103–10.

Elliott, D.K., and Dineley, D.L. 1991. Additional information on *Alainaspis* and *Boothaspis,* Cythathaspids (Agnatha: Heterostraci) from the Upper Silurian of Northwest Territories, Canada. *Journal of Palaeontology* 65: 308–13.

Elliott, D.K., and Loeffler, E.J. 1989. A new agnathan from the Lower Devonian of Arctic Canada, and a review of the tessellated heterostracans. *Palaeontology* 32: 883–91.

Forey, P.L., and Janvier, P. 1993. Agnathans and the origin of jawed vertebrates. *Nature* 361: 129–34.

Gagnier, P.-Y. 1989. The oldest vertebrate: a 470-million-year-old jawless fish, *Sacabambaspis janvieri,* from the Ordovician of Bolivia. *National Geographic Research* 5: 250–53.

Gagnier, P.-Y. 1995. Ordovician vertebrates and agnathan phylogeny. *Bulletin du Muséum national d'histoire naturelle* 17: 1–37.

Gagnier, P.-Y., and Blieck, A. 1992. On *Sacabambaspis janvieri* and the vertebrate diversity in Ordovian seas. Pp. 21–40 in Mark-Kurik, E. (ed), *Fossil fishes as living animals.* Academy of Sciences, Estonia.

Gagnier, P.-Y., Blieck, A., and Rodrigo, G.S. 1986. First Ordovician vertebrate from South America. *Geobios* 19: 629–34.

Gross, W. 1967. Uber Thelodontier-Schuppen. *Palaeontographica* 127A:1–67.

Gross, W. 1971. Unterdevonische Thelodontier- und Acanthodier-Schuppen aus Westaustralien. *Paläontologische Zeitschrift* 45: 97–106.

Halstead, L.B. 1973. The heterostracan fishes. *Biological Reviews* 48: 279–332.

Halstead, L.B. 1987. Agnathan extinctions in the Devonian. Pp. 150:7–11 in Buffetaut, E., Jaeger, J.J., and Mazin, J.M. (eds), *Les extinctions dans l'histoire des vertébrés. Mémoires de la Société géologique de France,* nouvelle série, Paris.

Heintz, A. 1939. Cephalaspida from the Downtonian of Norway. *Norske Videnskaps Akademiens Skrifter (Matematiske-naturvidenskapslige Klasse)* 1939: 1–119.

Heintz, A. 1968. New agnathans from Ringerike Sandstone, *Norske Videnskaps Akademiens Skrifter (Matematiske-naturvidenskapslige Klasse)* 1969: 1–28.

Heintz, N. 1960. The Downtonian and Devonian vertebrates of Spitsbergen: X. Two new species of the genus *Pteraspis* from the Wood Bay series in Spitsbergen. *Norsk Polarinstitut Skriffter* 117: 1–13.

Heintz, N. 1962. The Downtonian and Devonian vertebrates of Spitsbergen: XI. *Gigantaspis,* a new genus of family Pteraspididae from Spitsbergen. A preliminary report. *Norsk Polarisntitut Arbok* 1960: 22–27.

Heintz, N. 1968. The pteraspid *Lyktaspis* n. g. from the Devonia of Vestspitsbergen. *Nobel Symposium* 4: 73–80, Stockholm.

Janvier, P. 1975a. Les yeux des cyclostomes fossiles et le problème de l'origine des Myxinoides. *Acta Zoologica* 56: 1–9.

Janvier, P. 1975b. Anatomie et position systématique des galeaspides (Vertebrata, Cyclostomata), Céphalaspidomorphes du Dévoniens inférieur du Yunnan (Chine). *Bulletin du Muséum national d'histoire naturelle* 3rd series, 278: 1–16.

Janvier, P. 1985. Les Cephalspides du Spitzberg. *Cahiers de Paleontologie, Editions du C.N.R.S.,* Paris.

Janvier, P. 1988. Un nouveau céphalaspide (osteostraci) du Dévonien inférieur de Podolie (R.S.S. D'Ukraine). *Acta Palaeontologica Polonica* 33: 353–58.

Janvier, P. 1990. La structure del'exosquelette des Galeaspida (Verteberata*). Comptes Rendues Academie Sciences, Paris,* 2nd series, 310: 655–59.

Janvier, P. 1996. The dawn of vertebrates: characters versus common ascent in the rise of the current vertebrate phylogenies. *Palaoentology* 39: 259–87.

Janvier, P. 2001. Ostracoderms and the shaping of the ganthostome characters. Pp. 172–186 in Ahlberg, P.E. (ed), *Major events in early vertebrate evolution: palaeontology, phylogeny, genetics, and developmental biology.* Systematics Association, London.

Janvier, P. 2007. Homologies and evolutionary transitions in early vertebrate history. Pp. 57–121 in Anderson, J.S., and Dieter-Sues, H. (eds), *Major transitions in vertebrate evolution.* Indiana University Press, Bloomington.

Janvier, P., and Blieck, A. 1979. New data on the internal anatomy of the Heterostraci (Agnatha) with general remarks on the phylogeny of the Craniata. *Zoologica Scripta* 8: 287–96.

Janvier, P., and Lelièvre, H. 1994. A new tremataspid osteostracan, *Aestiaspis viitaensis* n.g., n. sp., from the Silurian of Saaremaa, Estonia. *Proceedings of the Estonian Academy of Science, Geology* 43 (3): 122–28.

Janvier, P., Thanh, T.-D., and Phuong, T.-H. 1993. A new Early Devonian galeaspid from Bac Thai Province, Vietnam. *Palaeontology* 36: 297–309.

Karatajute-Talimaa, V.N. 1978. *Telodonti Silura i Devona S.S.S.R. i Shpitsbergena.* Mosklas, Vilnius, Lithuania.

Kiaer, J. 1924. The Downtonian fauna of Norway: I. Anaspida with a geological introduction, *Videnskappsselskapets Skrifter (Matematiske-naturvidenskapslige Klasse)* 6: 1–139.

Kiaer, J. 1932. The Downtonian and Devonian vertebrates of Spitsbergen: IV. Suborder Cyathaspida. *Skriffter Svalbard Ishavet* 52: 1–26.

Kiaer, J., and Heintz, A. 1935. The Downtonian and Devonian vertebrates of Spitsbergen: V. Suborder Cyathaspida, I: tribe Poraspidei, family Poraspidae. *Skriffter Svalbard Ishavet* 40: 1–138.

Khonsari, R.H., Li, B., Vernier, P., Northcutt, R.G., and Janvier, P. 2009. Agnathan brain anatomy and cranite phylogeny. *Acta Zoologica* 90: S52–S68.

Liu, S.-F. 1983. Agnatha from Sichuan, China. *Vertebrata PalAsiatica* 21: 97–102.

Liu, S.-F. 1986. Fossil Eugaleaspida from Guanxi. *Vertebrata PalAsiatica* 24: 1–9.

Liu, Y.-H. 1965. New Devonian agnathans of Yunnan. *Vertebrata PalAsiatica* 9: 125–134.

Liu, Y.-H. 1973. On the new forms of Polybranchiaspiiformes and Petalichthyida from the Devonian of Southwest China. *Vertebrata PalAsiatica* 11: 132–43.

Liu, Y.-H. 1975. Lower Devonian agnathans of Yunnan and Sichuan. *Vertebrata PalAsiatica* 13: 202–16.

Liu, Y.-H., and Wang J.-Q. 1985. A galeaspid (Agnatha) *Antiquisagittaspis cornuta* gen. et sp. nov. from the Lower Devonian of Guanxi, China. *Vertebrata PalAsiatica* 23: 247–254.

Lund, R., and Janvier, P. 1986. A second lamprey from the Lower Carboniferous (Namurian) of Bear Gulch, Montana (USA). *Geobios* 19: 647–52.

Mark-Kurik, E. 1966. On some alteration of the exoskeleton in the Psammosteidae (Agnatha). Pp. 56–60 in *The organism and its environment in the past* [in Russian], Nauka, Moscow.

Mark-Kurik, E. 1992. Functional aspects of the armour in the early vertebrates. Pp. 107–16 in Mark-Kurik, E. (ed), *Fossil fishes as living animals*, Academy of Sciences, Estonia.

Märss,T., 1979. Lateral-line sensory system of the Ludlovian *thelodont Phlebolepis elegans* Pander [in Russian]. *Eesti NSV Teaduste Akadeemia Toimetised* 28: 108–11.

Maars, T., Turner, S., and Karatajute-Talimaa, V. 2007. Agnatha: II. Thelodonti. In Schultze, H.-P. (ed), *Handbook of paleoichthyology*. Verlag F. Pfeil, Munich.

Novitskaya, L.I. 1986. Fossil agnathans of USSR—Heterostracans: cyathaspids, amphiaspids, pteraspids [in Russian]. *Akademia Nauk SSSR, Trudy Paleontologicheskogo Instituta* 219:1–159, Akademia Nauka, Moscow.

Novitskaya, L. 1992. Heterostracans: their ecology, internal structure and ontogeny. Pp 51–60 in Mark-Kurik, E. (ed), *Fossil fishes as living animals*. Academy of Sciences, Estonia.

Obruchev, D.V., and Karatajute-Talimaa, V.N. 1967. Vertebrate faunas and correlation of the Ludlovian-Lower Devonian in Eastern Europe. *Zoological Journal of the Linnean Society* 47: 5–15.

Obruchev, D.V., and Mark-Kurik, E. 1965. Devonian Psammosteids (Agnatha, Psammosteidae) of the USSR [in Russian]. *Eesti NSV Teaduste Akadeemia Geoloogia Instituut.*

Ørvig, T. 1969. Thelodont scales from the Grey Hoek Formation of Andree Land, Spitsbergen. *Norsk geologische tiddskrifter* 49: 387–401.

Pan, J., and Chen L. 1993. Geraspididae, a new family of Polybranchiaspidida (Agnatha) from Silurian of Northern Anhui. *Vertebrata PalAsiatica* 31: 225–30.

Powrie, J. 1870. On the earliest known vestiges of vertebrate life; being a description of the fish remains of the Old Red Sandstone of Forfarshire. *Transactions of the Geological Society of Edinburgh* 1: 284–301.

Ritchie, A. 1964. New lights on the Norwegian Anaspida. *Norske Videnskaps Akademiens Skrifter (Matematiske-naturvidenskapslige Klasse)* 14:1–35.

Ritchie, A. 1967. *Ateleaspis tessellate* Traquair, a non-cornuate cephalaspid from the Upper Silurian of Scotland. *Zoological Journal of the Linnean Society, London* 47: 69–81.

Ritchie, A. 1968a. New evidence on *Jamoytius kerwoodi*, an important ostracoderm from the Silurian of Lanarkshire, Scotland. *Palaeontology* 11: 21–39.

Ritchie, A. 1968b. *Phlebolepis elegans* Pander, an Upper Silurian thelodont of Oesel, with remarks on the morphology of thelodonts. Pp. 4: 81–88 in Ørvig, T. (ed), *Current problems in lower vertebrate phylogeny, Nobel Symposium*, Almqvist and Wiksell, Stockholm.

Ritchie, A. 1980. The Late Silurian anaspid genus *Rhyncholepis* from Oesel, Estonia, and Ringerike, Norway. *American Museum Novitates* 2699: 1–18.

Ritchie. A. 1984. Conflicting interpretations of the Silurian agnathan *Jamoytius*. *Scottish Geology* 20: 249–56.

Ritchie, A. 1985. *Arandaspis prionotolepis*. The Southern four-eyed fish. Pp. 95–106 in Rich, P., and van Tets, G. (eds), *Kadimakura*. Pioneer Design Studios, Lilydale, Victoria, Australia.

Ritchie, A., and Gilbert-Tomlinson J. 1977. First Ordovician vertebrates from the southern hemisphere. *Alcheringa* 1: 351–68.

Sansom, I., Smith, M.M., and Smith, M.P. 2001. The Ordovician radiation of vertebrates. Pp. 156–71 in Ahlberg, P.E. (ed), *Major events in early vertebrate evolution: palaeontology, phylogeny, genetics, and developmental biology*. Systematics Association, London.

Sansom, R. 2007. A review of the problematic osteostracan genus *Auchenaspis* and its role in thyestidian evolution. *Palaeontology* 50: 1001–11.

Sansom, R. 2009. Phylogeny, classification and character polarity of the Osteostraci (Vertebrata). *Journal of Systematic Paleontology* 7: 95–115.

Smith, M.M., and Coates, M.I. 2001. The evolution of vertebrate dentitions: phylogenetic pattern and developmental models. Pp. 223–40 in Ahlberg, P.E. (ed), *Major events in early vertebrate evolution: palaeontology, phylogeny, genetics, and developmental biology.* Systematics Association, London.

Smith, M.P., and Sansom, I.J. 1995.The affinity of *Anatolepis* Bockelie and Fortey. *Geobios* 28(2): S61–S63.

Stensiö, E.A. 1927. The Downtonian and Devonian vertebrates of Spitsbergen: I. Family Cephalaspidae. *Skrifter om Svalbard og Ishavet* 12: 1–391.

Stensiö, E.A. 1932. *The cephalaspids of Great Britain.* British Museum (Natural History), London.

Stensiö, E.A. 1939. A new anaspid from the Upper Devonian of Scaumenac Bay, Canada, with remarks on other anaspids. *Kungliga Svenska Vetenskapakadamiens Handlingar* 3rd series, 18: 3–25.

Stensiö, E.A. 1964. Les cyclostomes fossiles ou Ostracodermes. Pp. 4:92–382 in Piveteau, J. (ed), *Traité de paléontologie.* Masson, Paris.

Stensiö, E.A. 1968. The Cyclostomes with special reference to the diphyletic origin of the Petromyzontida and Myxinoidea. *Nobel Symposium* 4: 13–71.

Tarlo, L.B. 1962. The classification and evolution of the Heterostraci. *Acta Palaeontolgia Polonica* 7 (1–2): 249–90.

Tarlo, L.B. 1964. Psammosteiformes (Agnatha)—a review with descriptions of new material from the Lower Devonian of Poland: I. General Part. *Acta Palaeontologia Polonica* 13: 1–135.

Tarlo, L.B. 1965. Psammosteiformes (Agnatha)—a review with descriptions of new material from the Lower Devonian of Poland: II. Systematic Part. *Acta Palaeontologia Polonica* 15: 1–168.

Turner, S. 1970. Fish help trace continental movements. *Spectrum* 79: 810.

Turner, S. 1973. Siluro-Devonian thelodonts from the Welsh Borderland. *Journal of the Geological Society of London* 129: 557–84.

Turner, S. 1976. Fossilium Catalogus: I. Animalia, Pars 122 Thelodonti (Agnatha). Dr W. Junk B.V., 's-Gravenhage 1–35.

Turner, S. 1982a. Thelodonts and correlation. Pp. 128–33 in Rich, P.V., and Thompson, E.M. (eds), *The fossil vertebrate record in Australasia.* Monash Offset Printing, Clayton, Australia.

Turner, S. 1982b. A new articulated thelodont (Agnatha) from the Early Devonian of Britain. *Palaeontology* 25: 879–89.

Turner, S. 1986. Vertebrate fauna of the Silverband Formation, Grampians, western Victoria. *Proceedings of the Royal Society of Victoria* 98: 53–62.

Turner, S. 1991. Monophyly and interrelationships of the Thelodonti. Pp. 87–120 in Chang, M.M., Liu, Y.H., and Zhang, G.R. (eds), *Early vertebrates and related problems of evolutionary biology.* Science Press, Beijing, China.

Turner, S. 1992. Thelodont lifestyles. Pp. 21–40 in Mark-Kurik, E. (ed.), *Fossil fishes as living animals.* Academy of Sciences, Estonia.

Turner, S., and Dring, R.S. 1981. Late Devonian thelodonts (Agnatha) from the Gneudna Formation, Carnarvon Basin, Western Australia. *Alcheringa* 5: 39–48.

Turner, S., and Janvier, P. 1979. Middle Devonian Thelodonti (Agnatha) from the Khush-Yeilagh Formation, North-east Iran. *Geobios* 12: 889–92.

Turner, S., Jones, P.J., and Draper, J.J. 1981. Early Devonian thelodonts (Agnatha) from the Toko Syncline, western Queensland and a review of other Australian discoveries. *Bureau of Mineral Resources Journal of Australian Geology and Geophysics* 61: 51–69.

Turner, S., and Tarling, D.H. 1982. Thelodont and other agnathan distributions as a test of Lower Palaeozoic continental reconstructions. *Palaeogeography, Palaeoclimatology, Palaeoecology* 39: 295–311.

Turner, S., and Young, G.C. 1992. Thelodont scales from the Middle-Late Devonian Aztec Siltstone, southern Victoria Land, Antarctica. *Antarctic Science* 4: 89–105.

Van der Brughen, W., and Janvier, P. 1993. Denticles in thelodonts. *Nature* 364: 107.

Wang, J.-Q., and Zhu, M. 1994. *Zhaotongaspis janvieri* gen. et sp. nov., a galeaspid from Early Devonian of Zhaotong, northeastern Yunnan. *Vertebrata PalAsiatica* 32: 230–43.

Wang, N.-Z. 1984. Thelodont, acanthodian and chondrichthyan fossils from the Lower Devonian of Southwest China. *Proceedings of the Linnean Society of New South Wales* 107: 419–41.

Wang, N.-Z. 1991. Two new Silurian galeaspids (jawless craniates) from Zhejiang Province, China, with a discussion of galeaspid-gnathostome relationships. Pp. 41–65 in Chang, M.M., Liu, Y.H., and Zhang, G.R. (eds), *Early vertebrates and related problems of evolutionary biology.* Science Press, Beijing.

Wang, N.-Z., and Dong, Z. 1989. Discovery of late Silurian

microfossils of Agnatha and fishes from Yunnan, China. *Acta Palaeontologica Sinica* 8: 192–206.

Wang, S.-T., Dong, Z., and Turner, S. 1986. Middle Devonian Turinidae (Thelodont, Agnatha) from Western Yunnan, China. *Alcheringa* 10: 315–25.

Wang, S.-T., and Lan C. 1984. New discovery of polybranchiaspids from Yiliang County, Northeastern Yunnan. *Bulletin of the Geological Institute, Chinese Academy of Sciences* 9: 113–23.

Wang, S.-T., Xia, S., Chen, L., and, Du, S. 1980. On the discovery of Silurian Agnatha and Pisces from Chaoxian County, Anhui Province, and its stratigraphical significance. *Bulletin of the Geological Institute, Chinese Academy of Sciences* 2: 101–12.

Wangsjö, G. 1952. The Downtonian and Devonian vertebrates of Spitzbergen: IX. Morphologic and systematic studies of the Spitsbergen cephalaspids. *Norsk Polarinstitut Skriffter* 97: 1–615.

White, E.I. 1935. The ostracoderm *Pteraspis* Kner and the relationship of the agnathous vertebrates. *Philosophical Transactions of the Royal Society of London* B 527: 381–457.

White, E.I. 1938. New pteraspids from South Wales. *Quarterly Journal of the Geological Society of London* 94: 85–115.

White, E.I. 1946. The genus *Phialaspis* and the "*Psammosteus* Limestones." *Quarterly Journal of the Geological Society of London* 101: 207–42.

White, E.I. 1950a. The vertebrate faunas of the Lower Old Red Sandstones of the Welsh Borders. *Bulletin of the British Museum (Natural History), Geology* 1: 51–67.

White, E.I. 1950b. *Pteraspis leatherensis* White, a Dittonian zone fossil. *Bulletin of the British Museum (Natural History), Geology* 1: 69–89.

White, E.I. 1961. The Old Red Sandstone of Brown Clee Hill and the adjacent area: II. Palaeontology. *Bulletin of the British Museum (Natural History), Geology* 5: 243–310.

White, E.I. 1973. Form and growth in *Belgicaspis* (Heterostraci). *Palaeontographica* 143A: 11–24.

Wilson, M.V.H., and Caldwell, M.W. 1993. New Silurian and Devonian fork-tailed "thelodonts" are jawless vertebrates with stomachs and deep bodies. *Nature* 361: 442–44.

Wilson, M.V.W., Hanke, G.F., and Märss, T. 2007. Paired fins of jawless vertebrates and their homologies across the "agnathan"-gnathostome transition. Pp. 122–149 in Anderson, J.S., and Dieter-Sues, H. (eds), *Major transitions in vertebrate evolution.* Indiana University Press, Bloomington.

Young, G.C. 1991. The first armoured agnathan vertebrates from the Devonian of Australia. Pp. 67–85 in Chang, M.M., Liu, Y.H. and Zhang, G.R. (eds), *Early vertebrates and related problems of evolutionary biology.* Science Press, Beijing.

Zhu, M. 2000. Catalogue of Devonian vertebrates in China, with notes on bio-events. *Courier Forschungsinstitut Senckenberg* 223: 373–90.

Zhu, M., and Zhikun, G. 2006. Phylogenetic relationships of galeaspids (Agnatha). *Vertebrata PalAsiatica*, 44: 1–27.

Chapter 4. Armored Fishes and Fish with Arms

Ahlberg, P., Trinajstic, K., Johanson, Z. and Long, J. 2009. Pelvic claspers confirm chondrichthyan-like internal fertilisation in arthrodires. *Nature* 458: 888–89.

Anderson, P.S.L. 2008. Shape variation between arthrodire morphotypes indicates possible feeding niches. *Journal of Vertebrate Paleontology* 28: 961–69.

Anderson, P.S.L., and Westneat, M.W. 2007. Feeding mechanics and bite force modelling of the skull of *Dunkleosteus terrelli*, an ancient apex predator. *Biology Letters* 3: 76–79.

Arsenault, M., Desbiens, S., Janvier, P., and Kerr, J. 2004. New data on the soft tissues and external morphology of *Bothriolepis canadensis* (Whiteaves, 1880), from the Upper Devonian of Miguasha, Quebec. Pp. 439–54 in Arratia, G., Wilson, M.V.H., and Cloutier, R. (eds), *Recent advances in the origin and early radiation of vertebrates.* Verlag F. Pfeil, Munich.

Carr, R.K. 1991. Reanalysis of *Heintzichthys gouldii* (Newberry), an aspinothoracid arthrodire (placodermi) from the Famennian of northern Ohio, with a review of brachythoracid systematics. *Zoological Journal of the Linnean Society* 103: 349–90.

Carr, R.K., and Hlavin, W.J. in press. Two new species of *Dunkleosteus* Lehman, 1956, from the Ohio Shales Formation (USA, Famennian) and the Kettle Point Formation (Canada, Upper Devonian) and a cladistic analysis of the Eubrachythoraci (Placodermi, Arthrodira). *Zoological Journal of the Linnean Society.*

Carr, R.K., Johanson, Z., and Ritchie, A. 2009. The phyllolepid placoderm *Cowralepis mclachlani:* insights into the evolution of feeding mechanism in jawed vertebrates. *Journal of Morphology* 270: 775–804.

Chaloner, W. G., Forey, P.L., Gardiner, B.G., Hill, A.J., and Young, V.T. 1980. Devonian fish and plants of the Bokkeveld Series of South Africa. *Annals of the South African Museum* 81: 127–57.

Denison, R. H. 1941. The soft anatomy of *Bothriolepis. Journal of Paleontology* 15: 553–61.

Denison, R.H. 1958. Early Devonian fishes from Utah: III. Arthrodira. *Fieldiana Geology* 11: 461–551.

Denison, R.H. 1975. Evolution and classification of placoderm fishes. *Breviora* 432: 553–615.

Denison, R.H. 1978. Placodermi. Pp. 1–128: in Schultze, H.-P. (ed), *Handbook of paleoichthyology.* Gustav Fischer Verlag, Stuttgart, Germany.

Denison, R.H. 1983. Further consideration of placoderm evolution. *Journal of Vertebrate Paleontology* 3: 69–83.

Denison, R.H. 1984. Further consideration of the phylogeny and classification of the order Arthrodira (Pisces: Placodermi). *Journal of Vertebrate Paleontology* 4: 396–412.

Dennis, K.D., and Miles, R.S. 1979a. A second eubrachythoracid arthrodire from Gogo, Western Australia. *Zoological Journal of the Linnean Society* 67: 1–29.

Dennis, K.D., and Miles, R.S. 1979b. Eubrachythoracid arthrodires with tubular rostral plates from Gogo, Western Australia. *Zoological Journal of the Linnean Society* 67: 297–328.

Dennis, K.D., and Miles, R.S. 1980. New durophagous arthrodires from Gogo, Western Australia. *Zoological Journal of the Linnean Society* 69: 43–85.

Dennis, K.D., and Miles, R.S. 1981. A pachyosteomorph arthrodire from Gogo, Western Australia. *Zoological Journal of the Linnean Society* 73: 213–58.

Dennis, K.D., and Miles, R.S. 1982. A eubrachythoracid arthrodire with a snub-nose from Gogo, Western Australia. *Zoological Journal of the Linnean Society* 75: 153–66.

Dennis-Bryan, K., 1987. A new species of eastmanosteid arthrodire (Pisces: Placodermi) from Gogo, Western Australia. *Zoological Journal of the Linnean Society* 90: 1–64.

Dennis-Bryan, K., and Miles, R.S. 1983. Further eubrachythoracid arthrodires from Gogo, Western Australia. *Zoological Journal of the Linnean Society* 67: 1–29.

Desmond, A.J. 1974. On the coccosteid arthrodire *Millerosteus minor. Zoological Journal of the Linnean Society* 54: 277–98.

Dineley, D.L., and Liu, Y.-H. 1984. A new actinolepid arthrodire from the Lower Devonian of arctic Canada. *Palaeontology* 27: 875–88.

Downs, J., and Donohue, P. 2009. Skeletal histology of *Bothriolepis canadensis* (Placodermi, Antiarchi) and evolution of the skeleton at the origin of jawed vertebrates. *Journal of Morphology* 270: 1364–80.

Dunkle, D.H., and Bungart, P.A. 1939. A new arthrodire from the Cleveland Shale Formation. *Scientific Publications of the Cleveland Museum of Natural History* 8(2): 13–28.

Dunkle, D.H., and Bungart, P.A. 1940. On one of the least known of the Cleveland Shale Arthrodira. *Scientific Publications of the Cleveland Museum of Natural History* 8(3): 29–47.

Dupret, V. 2008. First wuttagoonaspid (Placodermi, Arthrodira) from the Lower Devonian of Yunnan, South China. Origin, dispersal and palaeobiogeograhic sifgnificance. *Journal of Vertebrate Palaeontology* 28:12–20.

Dupret, V., and Zhu, M. 2008. The earliest phyllolepid (Placodermi, Arthrodira), gavinaspis convergens, from the late Lochkovian (Lower Devonian) of Yunnan, South China. *Geology magazine* 145: 257–78.

Dupret, V., Zhu, M., and Wang, J.-Q. 2009. The morphology of *Yujiangolepis liujingensis* (Placodermi, Arthrodira) from the Pragian of Guangxi (South China) and its phylogenetic significance. *Zoological Journal of the Linnean Society* 157: 70–82.

Eastman, C.R. 1907. Devonic fishes of the New York formations. *Memoirs of the New York State Museum* 10: 1–235.

Forey, P.L., and Gardiner, B.G. 1986. Observations on *Ctenurella* (Ptyctodontida) and the classification of placoderm fishes. *Zoological Journal of the Linnean Society* 86: 43–74.

Gardiner, B.G. 1984. The relationships of placoderms. *Journal of Vertebrate Palaeontology* 4: 379–95.

Gardiner, B.G., and Miles, R.S. 1975. Devonian fishes of the Gogo Formation, Western Australia. *Colloques internationale du C.N.R.S.* 218: 73–79.

Gardiner, B.G. and Miles, R.S. 1990. A new genus of eubrachythoracid arthrodire from Gogo, Western Australia. *Zoological Journal of the Linnean Society* 99: 159–204.

Gardiner, B.G. and Miles, R.S. 1994. Eubrachythoracid arthrodires from Gogo, Western Australia. *Zoological Journal of the Linnean Society* 112: 443–77.

Goujet, D. 1972. Nouvelles observations sur la joue d'*Arctolepis* (Eastman) et d'autres Dolichothoraci. *Annales de Paléontologie* 58: 3–11.

Goujet, D. 1973. *Sigaspis,* un nouvel arthrodire du Dévonien inferieur du Spitsberg. *Palaoentographica* 143A:73–88.

Goujet, D. 1975. *Dicksonosteus,* un nouvel arthrodire du Dévonien inférieur du Spitsberg. Remarques sur le squelette visceral des Dolichothoraci. *Colloques internationale du C.N.R.S.* 218: 81–99.

Goujet, D. 1984a. Les poissons placoderms du Spitsberg. Arthrodires Dolichothoraci de la Formation de Wood Bay (Dévonien inférieur). *Cahiers de Paléontologie, Section Vértébres.* Editions du centre national de la recherche scientifique, Paris.

Goujet, D. 1984b. Placoderm interrelationships: a new interpretation, with a short review of placoderm classifications. *Proceedings of the Linnean Society of New South Wales* 107: 211–43.

Goujet, D., Janvier, P., and Suarez-Riglos, M., 1985. Un nouveau Rhénanide (vertebrata, Placodermi) de la Formation de Belén (Dévonien moyen), Bolivie. *Annales de Paléontologie.* 71: 35–53.

Goujet, D., and Young, G.C. 2004. Placoderm anatomy and phylogeny: new insights. Pp. 109–26 in Arratia, G., Wilson, M., and Cloutier, R. (eds), *Recent advances in the origin and early radiation of vertebrates*. Verlag F. Pfeil, Munich.

Gross, W., 1931. *Asterolepis ornata* Eichwald, und das Antiarchi-problem. *Palaeontographica* 75: 1–62.

Gross, W. 1932. Die Arthrodira Wiuldungens. *Geologische und Palaeontologische Abhandlungen* 19: 5–61.

Gross, W., 1937. Die Wirbeltiere des rheinischen Devonbs: II. *Abhandlungen der Preussischen Geologischen Landesanstadlt* 176: 5–83.

Gross, W. 1958. Uber die älteste Arthrodiran-Gattung. *Notizblatt des Hessisches Landesamt Bodenforsch, Weisbaden.*

Gross, W. 1959. Arthrodiran aus dem Obersilur der Prager Mulde. *Palaeontographica* 113A: 1–35.

Gross, W. 1961. *Lunaspis broilli* und *L. heroldi* aus dem Hunsruckscheifer (Unterdevon; Rheinland). *Notizblatt des Hessisches Landesamt Bodenforsch, Weisbaden* 89: 17–43.

Gross, W. 1962. Neuuntersuchung der Dolichothoraci aus dem Unterdevon von Overath bei Köln. *Paläontologische Zeitschrift* 45–63.

Gross, W. 1963. *Gemuendina stuertzi* Traquair. Neuuntersuchung. *Notizblatt des Hessischen Landesamtes für Bodenforschung, Wiesbaden* 91: 36–73.

Gross, W. 1965. Uber die Placodermen-gattungung *Asterolepis* und *Tiarsapis* aus dem Devon Belgiens und einen fraglichen *Tiaraspis*—Rest aus dem Devon Spitzbergens. *Institut Royal des Sciences naturelles Belgique, Bulletin* 41: 1–19.

Heintz, A. 1932. The structure of *Dinichthys*, a contribution to our knowledge of the Arthrodira. Pp. 115–224 in *The Bashford Dean memorial volume*. The American Museum of Natural History, New York.

Heintz, A. 1934. Revision of the Estonian Arthrodira: I. Family Homosteidae Jaekel. *Archiv für die Naturkunde Estlands* 10 (1): 180–290.

Heintz, A. 1968. The spinal plate in *Homosteus* and *Dunkleosteus*. *Nobel Symposium* 4: 145–51.

Hemmings, S.K. 1978. The Old Red sandstone antiarchs of Scotland. *Pterichthyodes* and *Microbrachius* [monograph]. Palaeontographical Society, London.

Hills, E.S. 1936. On certain endocranial structures in *Coccosteus*. *Geological magazine* 73: 213–26.

Ivanov, A.O. 1988. A new genus of arthrodires from the Upper Devonian of Timan. *Paleontological Journal* 117–20.

Janvier, P. 1978. The Upper Devonian of the Middle East, with special reference to the Antiarchi of the Antalya "Old Red Sandstone." Pp. 2:331–40 in Izdar, E., and Nakoman, E. (eds), *Sixth colloquium on geology of the Aegean region.* Piri Reis International Contributions Series, Piri Reis University, Istanbul, Turkey.

Janvier, P. 1979. Les vertébrés Dévoniens de l'Iran central: III. Antiarches. *Geobios* 12: 605–8.

Janvier, P. 1995. The branchial articulation and pectoral fin in antiarchs (Placodermi). *Bulletin du Muséum national d'histoire naturelle* 17: 143–61.

Janvier, P., and Marcoux, J. 1977. Les grès rouges de l'Armutgözlek Tepe: leur faune de poissons (Antiarches, Arthrodires et Crossoptérygiens) d'age Dévonien supérieur (Nappes d'Anatalya, Tuarides occidentales, Turquie). *Géologie Méditerranéene* 4: 183–88.

Janvier, P., and Pan, J. 1982. *Hyrcanaspis bliecki* n.g. n. sp., a new primitive euantiarch (Antiarcha, Placodermi) from the Eifelian of northeastern Iran, with a discussion on antiarch phylogeny. *Neues Jahrbuch für Geologie und Palaeontologie Ablandlungen* 164: 364–92.

Janvier, P., and Ritchie, A. 1977. Le genre *Groenlandaspis* Heintz (Pisces, Placodermi, Arthrodira) dans le Dévonien d'Asie. *Colloques Researches Academie des Sciences de Paris* series D, 284: 1385–88.

Johanson, Z. 1997. New *Remigolepis* (Placodermi, Antiarchi) from Canowindra, New South Wales, Australia. *Geological magazine* 134: 813–46.

Johanson, Z. 1998. The Upper Devonian *Bothriolepis* (Antiarchi, Placodermi) from near Canowindra, New South Wales, Australia. *Records of the Australian Museum* 50: 315–48.

Johanson, Z. 2002. Vascularization of the osteostracan and antiarch (Placodermi) pectoral fin: similarities, and implications for placoderm relationships. *Lethaia* 35: 169–86.

Johanson, Z., Carr, R., and Ritchie, A. (in press). Fusion, gene misexpression and homeotic transformations in vertebral development of the gnathostome stem group (Placodermi). *International Journal of Developmental Biology.*

Lehman, J.P. 1956. Les arthrodires du Dévonien supérieur du Tafilalt (sud marocain). *Notes et Mémoires Services Géologique de Maroc.* 129: 1–70.

Lehman, J.P. 1977. Nouveaux arthrodires du Tafilalt et de ses environs. *Annales de Paléontologie* 63: 105–32.

Lelièvre, H. 1984a. *Atlantidosteus hollardi* n.gen. n.sp., nouveau Brachythoraci (Vértébrés, Placodermes) du Dévonien inférieur du Maroc presarharien. *Bulletin du Muséum national d'histoire naturelle* 4: 197–208.

Lelièvre, H. 1984b. *Antineosteus lehmani* n.gen. n.sp., nouveau Brachythoraci du Dévonien inférieur du Maroc presarharien. *Annales de Paléontologie* 70: 115–58.

Lelièvre, H. 1988. Nouveau matérial d'*Antineosteus lehmani* Lelièvre 1984 (Placoderme, Brachythoraci) et d'Acanthodiens du Dévonien inférieur (Emsien) d'Algerie. *Bulletin du Muséum national d'histoire naturelle* 4: 287–302.

Lelièvre, H. 1991. New information on the structure and the systematic position of *Tafilalichthys lavocati* (Placoderme, Arthrodire) from the Late Devonian of Tafilalt, Morocco. Pp. 121–30 in Chang, M.M., Liu, Y.H., and Zhang, G.R. (eds), *Early vertebrates and related problems of evolutionary biology.* Science Press, Beijing.

Lelièvre, H., Janvier, P., and Goujet, D. 1981. Les vertébrés Dévoniens de l'Iran central: IV. Arthrodires et ptyctodontes. *Geobios* 14: 677–709.

Lelièvre, H., Feist, R., Goujet, D. and Blieck, A. 1987. Les vertébrés Dévoniens de la Montagne Noire (sud de la France) et leur apport à la phylogenie des Pachyosteomorphes (Placodermes, Arthrodires). *Palaeovertebrata* 17: 1–26.

Liu, H.-T. 1955. *Kiangyouosteus,* a new arthrodiran fish from Szechuan, China. *Acta Palaeontologica Sinica* 3: 261–74.

Liu. S.-F. 1973. New materials of *Bothriolepis shaokuanensis* and the age of the fish bearing beds. *Vertebrata PalAsiatica* 11: 36–42.

Liu, S.-F. 1974. The significance of the discovery of yunnanolepidoid fauna from Guanxi. *Vertebrata PalAsiatica* 12: 144–148.

Liu, S.-F. 1981. Occurrence of *Lunaspis* in China. *Chinese Science Bulletin* 26: 829.

Liu, S.-F. 1982a. Preliminary note of the Arthrodira from Guanxi, China. *Vertebrata PalAsiatica* 20: 106–14.

Liu, S.-F. 1982b. An arthrodire endocranium. *Vertebrata PalAsiatica* 20: 271–75.

Liu, T.-S., and P'an, K. 1958. Devonian fishes from the Wutung Series near Nanking, China. *Palaeontologica Sinica* 141: 1–41.

Liu, Y.-H. 1962. A new species of *Bothriolepis* from Yunnan. *Vertebrata PalAsiatica* 6: 80–85.

Liu, Y.-H. 1963. On the Antiarchi from Chutsing, Yunnan. *Vertebrata PalAsiatica* 7: 39–45.

Liu, Y.-H. 1979. On the arctolepid arthrodires from the Lower Devonian of Yunnan. *Vertebrata PalAsiatica* 17: 23–34.

Liu, Y.-H. 1991. On a new petalichthyid, *Eurycaraspis incilis* gen. et sp. nov., (placodermi, Pisces) from the Middle Devonian of Zhanyi, Yunnan. Pp. 139–77 in Chang, M.M., Liu, Y.H., and Zhang, G.R. (eds), *Early vertebrates and related problems of evolutionary biology.* Science Press, Beijing.

Liu, Y.-H., and Wang, J.-Q. 1981. On three new arthrodires from the Middle Devonian of Yunnan. *Vertebrata PalAsiatica* 19: 295–304.

Long, J.A. 1983. New bothriolepid fishes from the Late Devonian of Victoria, Australia. *Palaeontology* 26: 295–320.

Long, J.A. 1984a. A plethora of placoderms: the first vertebrates with jaws? Pp. 185–210 in Archer, M., and Clayton, G. (eds), *Vertebrate zoogeography and evolution in Australasia,* Hesperian Press, Carlisle, Australia.

Long, J.A. 1984b. New phyllolepids from Victoria and the relationships of the group. *Proceedings of the Linnean Society of New South Wales* 107: 263–304.

Long, J.A. 1984c. New placoderm fishes from the Early Devonian Buchan Group, eastern Victoria. *Proceedings of the Royal Society of Victoria* 96: 173–86.

Long, J.A. 1988a. New information on the Late Devonian arthrodire *Tubonasus* from Gogo, Western Australia. *Memoirs of the Association of Australasian Palaeontologists* 7: 81–85.

Long, J.A. 1988b. A new camuropiscid arthrodire (Pisces: Placodermi) from Gogo, Western Australia. *Zoological Journal of the Linnean Society* 94: 233–58.

Long, J.A. 1990. Two new arthrodires (placoderm fishes) from the Upper Devonian Gogo Formation, Western Australia. *Memoirs of the Queensland Museum* 28: 51–63.

Long, J.A. 1994a. A second incisoscutid arthrodire from Gogo, Western Australia. *Alcheringa* 18: 59–69.

Long, J.A. 1995a. A new groenlandaspidid arthrodire (Pisces; Placodermi) from the Middle Devonian Aztec Siltstone, southern Victoria Land, Antarctica. *Records of the Western Australian Museum* 17: 35–41.

Long, J.A. 1995b. A new plourdosteid arthrodire from the Late Devonian Gogo Formation, Western Australia: systematics and phylogenetic implications. *Palaeontology* 38: 1–24.

Long, J.A., and Werdelin, L. 1986. A new species of *Bothriolepis* (Placodermi, Antiarcha) from Tatong, Victoria, with descriptions of others from the state. *Alcheringa* 10: 355–99.

Long, J.A., and Young, G.C. 1988. Acanthothoracid remains from the Early Devonian of New South Wales, including a complete sclerotic capsule and pelvic girdle. *Memoirs of the Association of Australasian Palaeontologists* 7: 65–80.

Long, J.A., Trinajstic, K., and Johanson, Z. 2009. Devonian arthrodire embryos and the origin of internal fertilisation in vertebrates. *Nature* 457: 1124–26.

Long, J.A., Trinajstic, K., Young, G.C., and Senden, T. 2008. Live birth in the Devonian Period. *Nature* 453: 650–52.

Luksevics, E.V. 1991. New *Remigolepis* (Pisces, Antiarchi) from the Famennian deposits of the central Devonian field (Russia, Tula region) [in Russian]. *Daba un Muzejs* 3: 51–56.

Malinovskaya, S. 1973. *Stegolepis* (Antiarchi, Placodermi) a new Middle Devonian genus from central Kazakhstan [in Russian]. *Palaeontological Journal* 7: 189–99.

Malinovskaya, S. 1977. Taxonomical status of antiarchs from central Kazakhstan. Pp. 29–35 in *Essays on phylogeny and systematics of fossil fishes and agnathans.* Akademia Nauka, Moscow.

Malinovskaya, S. 1992. New Middle Devonian antiarchs (Placodermi) of central Kazakhstan. Pp. 177–84 in Mark-Kurik, E. (ed.) *Fossil fishes as living animals,* Academy of Sciences, Estonia.

Mark-Kurik, E. 1973a. *Actinolepis* (Arthrodira) from the Middle Devonian of Estonia. *Palaeontographica* 143A: 89–108.

Mark-Kurik, E. 1973b. *Kimaspis,* a new palaeacanthaspid from the early Devonian of central Asia. *Eesti NSV teaduste akadeemia toimetised* 22: 322–30.

Mark-Kurik, E. 1977. The structure of the shoulder girdle in early ptyctodontids. Pp. 61–70 in Menner, V.V. (ed), *Ocherki po filogenii i sistematike iskopaemykh ryb i beschelyustnykh,* Akademia Nauka, Moscow.

Mark-Kurik, E. 1992. The inferognathal in the Middle Devonian arthrodire *Homostius. Lethaia* 25: 173–78.

Miles, R.S. 1966a. The placoderm fish *Rhachiosteus pterygiatus* Gross and its relationships. *Transactions of the Royal Society of Edinburgh* 66: 377–92.

Miles, R.S. 1966b. *Protitanichthys* and some other coccosteomorph arthrodires from the Devonian of North America. *Kungliga Svenska Vetenskapakadamiens Hanlingar* 10: 1–49.

Miles, R.S. 1967a. Observations on the ptyctodontid fish *Rhamphodopsis* Watson. *Zoological Journal of the Linnean Society* 47: 99–120.

Miles, R.S. 1967b. The cervical joint and some aspects of the origin of the Placodermi. *Colloques internationale du C.N.R.S.* 163: 49–71.

Miles, R.S. 1968.The Old Red Sandstone antiarchs of Scotland. Family Bothriolepididae [monograph]. The Palaeontographical Society, London.

Miles, R.S. 1969. Features of placoderm diversification and the evolution of the arthrodire feeding mechanism. *Transactions of the Royal Society of Edinburgh* 68: 123–70.

Miles, R.S. 1971. The Holonematidae (placoderm fishes): a review based on new specimens of Holonema from the Upper Devonian of Western Australia. *Philosophical Transactions of the Royal Society of London* 263B: 101–234.

Miles, R.S. 1973. An actinolepid arthrodire from the Lower Devonian Peel Sound Formation, Prince of Wales Island. *Palaeontographica* 143A: 109–18.

Miles, R.S., and Dennis, K. 1979. A primitive eubrachythoracid arthrodire from Gogo, Western Australia. *Zoological Journal of the Linnean Society* 66: 31–62.

Miles, R.S., and Westoll, T.S. 1968. The placoderm fish *Coccosteus cuspidatus* Miller ex Agassiz from the Middle Old Red Sandstone of Scotland: I. Descriptive morphology. *Transactions of the Royal Society of Edinburgh* 67: 373–476.

Miles, R.S., and Young, G.C. 1977. Placoderm interrelationships reconsidered in the light of new ptyctodontids from Gogo, Western Australia. *Linnean Society Symposium Series* 4: 123–98.

Ørvig, T. 1957. Notes on some Palaeozoic lower vertebrates from Spitsbergen and North America. *Norsk geologische tiddskrifter* 37: 285–353.

Ørvig, T. 1960. New finds of acanthodians, arthrodires, crossopterygians, ganoids and dipnoans in the Upper Middle Devonian Calcareous Flags (Oberer Plattenkalk) of the Bergisch-Paffrath Trough (Part 1). *Paläontologische Zeitschrift* 34: 295–335.

Ørvig, T. 1962. Y a-t-il une relation directe entre les arthrodires ptyctodontides et les holocephales? *Colloques internationale du C.N.R.S.* 104: 49–61.

Ørvig, T. 1969. Vertebrates of the Wood Bay group and the position of the Emsian-Eifelian boundary. *Lethaia* 2: 273–328.

Ørvig, T. 1975. Description with special reference to the dermal skeleton, of a new radotinid arthrodire from the Gedinnian of Arctic Canada. *Colloques internationale du C.N.R.S.* 218: 41–71.

Ørvig, T. 1980a. Histologic studies of ostracoderms, placoderms and fossil elasmobranchs: III. Structure and growth of gnathalia in certain arthrodires. *Zoologica Scripta* 9: 141–59.

Ørvig, T. 1980b. Histologic studies of ostracoderms, placoderms and fossil elasmobranchs: IV. Ptyctodontid tooth plates and their bearing on holocephalan ancestry: the condition in chimaerids. *Zoologica Scripta* 14: 55–79.

Pageau, Y. 1969. Nouvelle faune ichthyologique de Dévonian moyen dans les Grès de Gaspé (Quebec): II. Morphologie et systematique. *Le Naturaliste Canadien* 96: 399–478, 805–89.

Pan, J. 1981. Devonian antiarch biostratigraphy of China. *Geological magazine* 118: 69–75.

Pan, J., Hou, F., Cao, J., Gu, Q., Liu, S.-Y., Wang, J., Gao, L., and Liu, C. 1987. *Continental Devonian of Ningxia and its biotas.* Geological Publishing House, Beijing.

Pan J., Wang, S.-T., Liu, S.-Y., Gu, G.-C., and Jia, H. 1980. Discovery of Devonian *Bothriolepis* and *Remigolepis* in Ninxia. *Acta Geologica Sinica* 54: 176–86.

Panteleyev, N. 1992. New remigolepids and high armoured antiarchs of Kirgizia. Pp. 185–92 in Mark-Kurik, E. (ed), *Fossil fishes as living animals.* Academy of Sciences, Estonia.

Ritchie, A. 1973. *Wuttagoonaspis* gen. nov., an unusual arthrodire from the Devonian of western New South Wales, Australia. *Palaeontographica* 143A: 58–72.

Ritchie, A. 1975. *Groenlandaspis* in Antarctica, Australia and Europe. *Nature* 254: 569–73.

Ritchie, A. 1984. A new placoderm, *Placolepis* gen. nov. (Phyllolepidae) from the Late Devonian of New South Wales, Australia. *Proceedings of the Linnean Society of New South Wales* 107: 321–53.

Ritchie, A. 2007. *Cowralepis,* a new genus of phyllolepid fish (Pisces, Placodermi) from the Late Middle Devonian of New South Wales, Australia. *Proceedings of the Linnean Society of New South Wales* 126: 215–59.

Ritchie, A., Wang, S.-T., Young, G.C., and Zhang, G.-R. 1992. The Sinolepidae, a family of antiarchs (placoderm fishes) from the Devonian of South China and eastern Australia. *Records of the Australian Museum* 44: 319–70.

Schultze, H.-P. 1973. Large Upper Devonian arthrodires from Iran. *Fieldiana Geology* 23: 53–78.

Schultze, H.-P. 1984. The head-shield of *Tiaraspis subtilis* (Gross) (Pisces, Arthrodira). *Proceedings of the Linnean Society of New South Wales* 107: 355–65.

Smith, M.M., and Johanson, Z. 2003. Separate evolutionary origins of teeth from evidence in fossil jawed vertebrates. *Science* 299: 1235–36.

Stensiö, E.A. 1925. On the head of macropetalichthyids with certain remarks on the head of other arthrodires. *Field Museum Natural History Publication Series* 4: 87–197.

Stensiö, E.A. 1931. On the Upper Devonian vertebrates of East Greenland. *Meddelesler om Grønland.* 86: 1–212.

Stensiö, E.A. 1936. On the Placodermi of the Upper Devonian of East Greenland. Supplement to Part 1. *Meddelesler om Grønland* 97 (1): 1–52.

Stensiö, E.A. 1939. On the Placodermi of the Upper Devonian of East Greenland. Second supplement to Part 1. *Meddelesler om Grønland* 97 (3): 1–33.

Stensiö, E.A. 1942. On the snout in arthrodires. *Kungliga Svenska Vetenskapakadamiens Hanlingar* 20(3): 1–32.

Stensiö, E.A. 1944. Contributions to the knowledge of the vertebrate fauna of the Silurian and Devonian of Western Podolia: II. Notes on two arthrodires from the Downtonian of Podolia. *Arkiv für Zoologi* 35: 1–83.

Stensiö, E. A. 1948. On the Placodermi of the Upper Devonian of East Greenland: II. Antiarchi: subfamily Bothriolepinae. With an attempt at a revision of the previously described species of that family. *Meddelelser om Grønland* 139, *Palaeozoologica Groenlandica,* 2 : 1–622.

Stensiö, E.A. 1959. On the pectoral fin and shoulder girdle of the arthrodires. *Kungliga Svenska Vetenskapakadamiens Hanlingar* 8: 1–229.

Stensiö, E.A. 1963. Anatomical studies on the arthrodiran head: I. Preface, geological and geographical distribution, the organization of the arthrodires, the anatomy of the head in the Dolichothoraci, Coccosteomorphi and Pachyosteomorphi. Taxonomic appendix. *Kungliga Svenska Vetenskapakadamiens Hanlingar* 4 (9) 2: 1–419.

Stensiö, E.A. 1969a. Anatomie des arthrodires dans leur cadre systématique. *Annales de Paléontologie* 57: 151–86.

Stensiö, E.A. 1969b. Elasmobranchiomorphi Placodermata Arthrodires. Pp. 71–692 in Piveteau, J.P. (ed), *Traite de paléontologie.* Masson, Paris.

Trinajstic, K. 1995. The role of heterochrony in the evolution of the eubrachythoracid arthrodires with special reference to *Compagospiscis croucheri* and *Incisoscutum ritchiei* from the Late Devonian Gogo Formation, Western Australia. *Geobios, Memoire Special* 19: 125–28.

Trinajstic, K. 1999. New anatomical information on *Holonema* (Placodermi) based on material from the Frasnian Gogo Formation and the Givetian-Frasnian Gneudna Formation, Western Australia. *Geodiversitas* 21: 69–84.

Trinajstic, K.M., and Hazelton, M. 2007. Ontogeny, phenotypic variation and phylogenetic implications of arthrodires from Gogo Formation, Western Australia. *Journal of Vertebrate Paleontology* 27: 571–83.

Trinajstic, K.M., and McNamara, K.J.M. 1999. Heterochrony in the Late Devonian arthrodiran fishes *Compagopiscis* and *Incisoscutum. Records of the Western Australian Museum Supplement* 57: 93–106.

Trinajstic, K., and Long, J.A. 2009. A new genus and species of ptyctodont (Placodermi) from the Late Devonian Gneudna Formation, Western Australia, and an analysis of ptyctodont phylogeny. *Geological magazine* 146: 743–86.

Upeniece, I., and Upenieks, J. 1992. Young Upper Devonian antiarch (*Asterolepis*) individuals from the Lode quarry, Latvia. Pp. 167–76 in Mark-Kurik, E. (ed), *Fossil fishes as living animals.* Academy of Sciences, Estonia.

Vezina, D. 1986. Les plaques gnathales de *Plourdosteus canadensis* (Placodermi, Arthrodira) du Dévonien supérieur du Québec (Canada): remarques sue la croissance dentaire et al mécanique masticatrice. *Bulletin de la Muséum de l'histoire naturelle* 4th series, 8: 367–91.

Vezina, D. 1990. Les Plourdosteidae fam. nov. (Placodermi, Arthrodira) et leurs relations phyletiques au sein des Brachythoraci. *Canadian Journal of Earth Sciences* 27, 677–83.

Wang, J. 1982. New materials of Dinichthyidae. *Vertebrata PalAsiatica* 20: 181–86.

Wang, J. 1991a. The Antiarchi from the Early Silurian of Hunan. *Vertebrata PalAsiatica* 29: 240–44.

Wang, J. 1991b. New material of *Hunanolepis* from the Middle Devonian of Hunan, Pp. 213–47 in Chang, M.M., Liu, Y.H., and Zhang G.R. (eds), *Early vertebrates and related problems of evolutionary biology.* Science Press, Beijing.

Wang, J. 1991c. A fossil Arthrodira from Panxi, Yunnan. *Vertebrata PalAsiatica* 29: 264–75.

Wang, J., and Wang, N. 1983. A new genus of Coccosteidae. *Vertebrata PalAsiatica* 21: 1–8.

Wang, J., and Wang, N. 1984. New materials of Arthrodira from the Wuding Region, Yunnan. *Vertebrata PalAsiatica* 22: 1–7.

Wang, S.-T. 1987. A new antiarch from the Early Devonian of Guanxi. *Vertebrata PalAsiatica* 25: 81–90.

Wang, S.-T, and Cao, R., 1988. Discovery of Macropetalichthyidae from the Lower Devonian in Western Yunnan. *Vertebrata PalAsiatica* 26: 73–75.

Watson, D.M.S. 1934. The interpretation of arthrodires. *Proceedings of the Zoological Society of London* 3: 437–64.

Watson, D.M.S. 1938. On *Rhamphodopsis,* a ptyctodont from the middle Old Red Sandstone of Scotland. *Transactions of the Royal Society of Edinburgh* 59: 397–410.

Werdelin, L., and Long, J.A. 1986. Allometry in *Bothriolepis canadensis* Whiteaves (Placodermi, Antiarcha) and its significance to antiarch evolution. *Lethaia* 19: 161–69.

Westoll, T.S. 1945. The paired fins of placoderms. *Transactions of the Royal Society of Edinburgh* 61: 381–98.

Westoll, T.S. 1967. *Radotina* and other tesserate fishes. *Zoological Journal of the Linnean Society* 47: 83–98.

Westoll, T.S., and Miles, R.S. 1963. On an arctolepid fish from Gemünden. *Transactions of the Royal Society of Edinburgh* 65: 139–53.

White, E.I. 1952. Australian arthrodires. *Bulletin of the British Museum (Natural History), Geology* 1: 249–304.

White, E.I. 1978. The larger arthrodiran fishes from the area of the Burrinjuck Dam, N.S.W. *Transactions of the Zoological Society of London* 34: 149–262.

White, E.I., and Toombs, H.A. 1972. The buchanosteid arthrodires of Australia. *Bulletin of the British Museum (Natural History), Geology* 22: 379–419.

Woodward, A. S. 1941. The head shield of a new macropetalichthyid (*Notopetalichthys hillsi*; gen. et sp. nov.) from the Middle Devonian of Australia. *Annals and Magazine of Natural History* 8: 91–96.

Young, G.C. 1979. New information on the structure and relationships of *Buchanosteus* (Placodermi: Euarthrodira) from the Early Devonian of New South Wales. *Zoological Journal of the Linnean Society* 66: 309–52.

Young, G.C. 1980. A new Early Devonian placoderm from New South Wales, Australia, with a discussion of placoderm phylogeny. *Palaeontographica* 167A: 10–76.

Young, G.C. 1981. Biogeography of Devonian vertebrates. *Alcheringa* 5: 225–43.

Young, G.C. 1983. A new antiarchan fish (Placodermi) from the Late Devonian of southeastern Australia. *Bureau of Mineral Resources Journal of Australian Geology and Geophysics* 8: 71–81.

Young, G.C. 1984a. An asterolepidoid antiarch (placoderm fish) from the Early Devonian of the Georgina Basin, central Australia. *Alcheringa* 8: 65–80.

Young, G.C. 1984b. Comments on the phylogeny and biogeography of antiarchs (Devonian placoderm fishes), and the use of fossils in biogeography. *Proceedings of the Linnean Society of New South Wales* 107: 443–73.

Young, G.C. 1984c. Reconstruction of the jaws and braincase in the Devonian placoderm fish *Bothriolepis. Palaeontology* 27: 625–61.

Young, G.C. 1985. New discoveries of Devonian vertebrates from the Amadeus Basin, central Australia. *BMR Journal of Australian Geology and Geophysics* 9: 239–54.

Young, G.C. 1986. The relationships of placoderm fishes. *Zoological Journal of the Linnean Society* 88: 1–57.

Young, G.C. 1988a. Antiarchs (placoderm fishes) from the Devonian Aztec Siltstone, southern Victoria Land, Antarctica. *Palaeontographica* A202: 1–125.

Young, G.C. 1988b. New occurrences of phyllolepid placoderms from the Devonian of central Australia. *Bureau of Mineral Resources Journal of Australian Geology and Geophysics* 10: 363–76.

Young, G.C. 1990. New antiarchs (Devonian placoderm fishes) from Queensland, with comments on placoderm phylogeny and biogeography. *Memoirs of the Queensland Museum* 28: 35–50.

Young, G.C. 2003. A new species of *Atlantidosteus* Lelièvre, 1984 (placodermi, Arthrodira, Brachythoraci) from the Middle Devonian Broken River area (Queensland, Australia). *Geodiversitas* 25: 681–94.

Young, G.C. 2004a. Large brachythoracid arthrodires (placoderm fishes) from the early Devonian of Wee Jasper, New South Wales, Australia, with a discussion of basal brachythoracid characters. *Journal of Vertebrate Paleontology* 24: 1–17.

Young, G.C. 2004b. Homosteiid remains (placoderm fishes). From the Early Devonian of the Burrinjuck area, New South Wales. *Alcheringa* 28: 129–46.

Young, G.C. 2004c. A Devonian brachythoracid arthrodire skull (placoderm fish) from the Broken River area, Queensland. *Proceedings of the Linnean Society of New South Wales* 125: 43–56.

Young, G.C. 2005. Early Devonian arthrodire remains (Placodermi? Holonematidae) from the Burrinjuck area, New South Wales, Australia. *Geodiversitas* 27: 201–19.

Young, G.C. 2008a. The relationships of antiarchs (Devonian

placoderm fishes): evidence supporting placoderm monophyly. *Journal of Vertebrate Paleontology* 28: 626–36.

Young, G.C. 2008b. Number and arrangement of extraocular muscles in primitive gnathostomes: evidence from extinct placoderm fishes. *Biology Letters* 4: 410–14.

Young, G.C., and Goujet, D. 2003. Devonian fish remains from the Dulcie Sandstone and Cravens Peak beds, Georgina basin, central Australia. *Records of the Western Australian Museum Supplement* 65: 1–85.

Young, G.C., and Zhang, G. 1992. Structure and function of the pectoral joint and operculum in antiarchs, Devonian placoderm fishes. *Palaeontology* 35: 443–64.

Young, G.C., and Zhang, G. 1996. New information on the morphology of yunnanolepid antiarchs (placoderm fishes) from the Early Devonian of South China. *Journal of Vertebrate Paleontology* 16: 623–41.

Young, V.T. 1983. Taxonomy of the arthrodire *Phylctaenius* from the Lower or Middle Devonian of Campbellton, New Brunswick, Canada. *Bulletin of the British Museum (Natural History), Geology* 37: 1–35.

Zhang, G. 1978. The antiarchs from the Early Devonian of Yunnan. *Vertebrata PalAsiatica* 16: 147–86.

Zhang, G. 1984. New form of Antiarchi with primitive brachial process from Early Devonian of Yunnan. *Vertebrata PalAsiatica* 22: 81–91.

Zhang, G., and Liu, S. 1978. Fossil *Bothriolepis* from Yujiang Formation of Kwangsi. *Vertebrata PalAsiatica* 16: 4–6.

Zhang, G., and Liu, Y.-G. 1991. A new Antiarchi from the Upper Devonian of Jianxi, China. Pp. 67–85 in Chang, M.M., Liu, Y.H. and Zhang G.R. (eds), *Early vertebrates and related problems of evolutionary biology*. Science Press, Beijing.

Zhang, M. (Chang Meeman). 1980. Preliminary note on a Lower Devonian antiarch from Yunnan, China. *Vertebrata PalAsiatica* 18: 179–90.

Zhu, M. 1991. New information on *Diandongpetalichthys* (Placodermi: Petalichthyida). Pp. 179–94 in Chang, M.M., Liu, Y.H. and Zhang G.R. (eds), *Early vertebrates and related problems of evolutionary biology*. Science Press, Beijing.

Chapter 5. Sharks and Their Cartilaginous Kin

Anderson, M.E., Long, J.A., Gess, R.W. and Hiller, N.1999. An unusal new fossil shark (Pisces: Chondrichthyes) from the Late Devonian of South Africa. *Records of the Western Australian Museum Supplement* 57: 151–56.

Bendix-Almgreen, S.E. 1966. New investigations on *Helicoprion* from the Phosphoria Formation of South-east Idaho, U.S.A. *Biologia Danske Videnskabernes Selskabs Skrifter* 14: 1–54.

Bendix-Almgreen, S.E. 1975. The paired fins and shoulder girdle in *Cladoselache*, their morphology and phyletic significance. *Colloques internationale du C.N.R.S.* 218: 111–123.

Burrow, C.J., Hovestadt, D.C., Hovestadt-Euler, M., Turner, S., and Young, G.C. 2008. New information on the Devonian shark *Mcmurdodus*, based on material from western Queensland, Australia. *Acta Palaeontologica Polonica* 58: 155–63.

Carvalho, M.R. de, Maisey, J.G., Grande, L., 2004. Freshwater stingrays of the Green River Formation of Wyoming (early Eocene), with the description of a new genus and species and an analysis of its phylogenetic relationships (Chondrichthyes, Myliobatiformes). *Bulletin of the American Museum of Natural History* 284: 1–136.

Coates, M.I., and Sequiera, S.E.K. 2001. A new stethacanthid chondrichthyan from the Lower Carboniferous of Bearsden, Scotland. *Journal of Vertebrate Paleontology* 21: 438–59.

De Pomeroy, A.M. 1994. Mid-Devonian chondrichthyan scales from the Broken River, North Queensland, Australia. *Memoirs of the Queensland Museum* 37: 87–114.

Dick, J.R.F. 1978. On the Carboniferous shark *Tristychius arcuatus* Agassiz from Scotland. *Transactions of the Royal Society of Edinburgh* 70: 63–109.

Dick, J.F.R., and Maisey, J.G. 1980. The Scottish Lower Carboniferous shark *Onychoselache traquairi*. *Palaeontology* 23: 363–74.

Downes, J.P., and Daeschler, E. B. 2001. Variation within a large sample of *Ageleodus pectinatus* teeth (Chondrichthyes) from the Late Devonian of Pennsylvania, U.S.A. *Journal of Vertebrate Paleontology* 21: 811–14.

Duffin, C. J., and Ward, D.J.1983. Neoselachian sharks teeth from the Lower Carboniferous of Britain and the Lower Permian of the U.S.A. *Palaeontology* 26: 93–110.

Eastman, C.R. 1899. Descriptions of new species of *Diplodus* teeth from the Devonian of northeastern Illinois. *Journal of Geology* 7: 489–93.

Ehret, D.J., Hubbell, G., and McFadden, B. 2009. Exceptional preservation of the White Shark *Carcharodon* (Lamniformes, Lamnidae) from the Early Pliocene of Peru. *Journal of Vertebrate Paleontology* 29: 1–13.

Friman, L. 1983. *Ohiolepis*-Schuppen aus dem unteren Mitteldevon der Eifel (Rheinisches Schiefergebirge). *Neues Jahrbuch für Geologie und Paläontologie Monatshefte* H.4: 228–36.

Ginter, M. 1990. Late Famennian sharks from the Holy Cross Mountains, central Poland. *Acta Geologica Polonica* 40: 69–81.

Ginter, M. 2001.Chondrichthyan biofacies in the Late Famennian of Utah and Nevada. *Journal of Vertebrate Paleontology* 21: 714–29.

Ginter, M., and Ivanov, A. 1992. Devonian phoebodont shark teeth. *Acta Palaeontologica Polonica* 37: 55–75.

Goto, M. 1987. *Chlamydoselachus angineus*—a living cladodont shark. *Report of Japanese Group for Elasmobranch Studies* 23: 11–13.

Grogan, E.D., and Lund, R. 2000. *Debeerius ellefseni* (fam. nov., gen. nov. sp. nov.), an autodiastylic chondrichthyan from the Mississippian Bear Gulch Limestone of Montana (USA), the relationships of the Chondrichthyes, and comments on gnathostome evolution. *Journal of Morphology* 243: 219–45.

Grogan, E.D., and Lund, R. 2008. A basal elasmobranch, *Thrinacoselache gracia* n. gen. and sp. (Thrinacodontidae, new family) from the Bear Gulch Limestone, Serpukhovian of Montana, USA. *Journal of Vertebrate Paleontology* 28: 970–88.

Gross, W. 1938. Das Kopfskelett von *Cladodus wildungensis* Jaekel: II. Der Kieferborgen. Anhang: *Protacrodus vetustus* Jaekel. *Senckenbergiana* 20: 123–45.

Gross, W. 1973. Kleinschuppen, Flossenstacheln und Zahne aus europischen und nordamerikanischen Bonebeds des Devons. *Palaeontographica* 142A: 51–155.

Hairapetian, V., Ginter, M., and Yazdi, M. 2008. Early Frasnian sharks from Iran. *Acta Geologica Polonica* 58: 173–79.

Hampe, O. 1988. Uber die Bezahnung des Orthacanthus (Chondrichthyes: Xenacanthida; Oberkarbon Unterperm). *Paläontologische Zeitschrift* 62: 285–96.

Hampe, O. 1989. Revision der *Triodus* Arten (Chondrichthyes: Xenacanthida) aus dem saarpfalzischen Rotliegenden (Oberkarbon Perm, SW Deutschland) aufgrund ihrer Bezahnung. *Paläontogische Zeitschrift* 63: 79–101.

Hampe, O. 1991. Histological investigations on fossil teeth of the shark order Xenacanthida (Chondrichthyes: Elasmobranchii) as revealed by flourescence microscopy. *Leica Scientific and Technical Information* 10 (1): 17–27.

Hampe, O. 1993a. Growth anomalies in xenacanthid teeth. P. 29 in Turner, S. (ed), *Abstracts of the IGCP 328 Gross Symposium*. Université des Sciences et Technologies de Lille, France.

Hampe, O. 1993b. Variation of xenacanthid teeth in the Permo-carboniferous deposits of the Saar-Nahe Basin (SW Germany). Pp. 37–51 in Heidetke, U. (ed), *New research on Permo-Carboniferous faunas*. Pollichia-Buch, Bad Durkheim, Germany.

Hampe, O. 1994. Neue erkenntnisse zur permokarbonischen Xenacanthiden-Fauna (Chondrichthyes, Elasmobranchii) und deren Verbreitung im südwestdeutschen Saar-Nahe-Becken. *Neues Jahrbuch für Geologie und Palaeontologie Ablandlungen* 192: 53–87.

Hampe, O., and Ivanov, A. 2007. Bransonelliformes: a new order of the Xenacanthimorpha (Chondrichthyes, Elasmobranchii). *Fossil Record* 10: 190–94.

Hampe, O., and Long, J.A., 1999. The histology of Middle Devonian chondrichthyan teeth from southern Victoria Land, Antarctica. *Records of the Western Australian Museum Supplement* 57: 23–36.

Hanke, G.F., and Wilson, M.V.H. 2004. New teleostome fishes and acanthodian systematics. Pp. 289–316 in Arratia, G., Wilson, M.V.H., and Cloutier, R. (eds), *Recent advances in the origin and early radiation of vertebrates*. Verlag F. Pfeil, Munich.

Hansen, M.C. 1978. A presumed lower dentition and spine of a Permian petalodontiform chondrichthyan, *Megactenopetalus kaibabanus*. *Journal of Paleontology* 52: 55–60.

Hansen, M. 1988. Microscopic chondrichthyan remains from Pennsylvanian marine rocks of Ohio and adjacent regions. *Ichthyolith Issues* 1: 6–7.

Harris, J. E. 1951. *Diademodus hydei*, a new fossil shark from the Cleveland Shale. *Proceedings of the Zoological Society of London* 120: 683–97.

Janvier, P. 1976. Description de restes d'Elasmobranches (Pisces) du Dévonien moyen de Bolivie. *Palaeovertebrata* 7: 126–32.

Janvier, P. 1987. Les vertébrés siluriens et dévoniens de Bolivie: remarques particulières sur le Chondrichthyens. In IV Congresso Latinamericano de Palaeontologia, Santa-Cruz 1: 159–78.

Janvier, P., and Dingerkus, G. 1991. Le synarcual de *Pucapampella* Janvier et Suarez-Riglos: une prevue de l'existence d'Holocephales (Vertebrata, Chondrichthyes) dès le Dévonien moyen. *Compte-Rendus de l'Académie des Sciences, Paris* 312: 549–52.

Johnson, G.D. 1980. Xenacanthodii (Chondrichthyes) from the Tecovas Formation (Late Triassic) of west Texas. *Journal of Paleontology* 54: 923–32.

Johnson, G.D. 1984. A new species of Xenacanthodii (Chondrichthyes, Elasmobranchii) from the Late Pennsylvanian of Nebraska. *Carnegie Museum of Natural History, Special Publications* 9: 178–86.

Karatajute-Talimaa, V.N. 1973. *Elegestolepis grossi* gen. et sp. nov., ein neuer typ der Placoidschuppe aus dem oberen Silur der Tuwa. *Palaeontographica* 143A: 35–50.

Karatajute-Talimaa, V. 1992. The early stages of the dermal skeleton formation of chondrichthyans. Pp. 223–43 in Mark-Kurik, E. (ed), *Fossil fishes as living animals*. Academy of Sciences, Estonia.

Karatajute-Talimaa, V.N., Novitskaya, L.I., Rozman, K.S., and Sodov, Z. 1990. *Mongolepis*—a new Lower Silurian genus of elasmobranchs from Mongolia. *Palaeontological Journal* 1: 37–48.

Klug, S. 2009. A new palaeospinacid shark (Chondrichthyes, Neoselachii) from the Upper Jurassic of Southern Germany. *Journal of Vertebrate Paleontology* 29: 326–35.

Leu, M.R. 1989. A Late Permian freshwater shark from eastern Australia. *Palaeontology* 32: 265–86.

Liu, G.-B., and Wang, Q., 1994. New material of *Sinohelicoprion* from Changxing, Zhejiang Province. *Vertebrata PalAsiatica* 32: 245–47.

Long, J.A. 1990. Late Devonian chondrichthyans and other microvertebrate remains from northern Thailand. *Journal of Vertebrate Paleontology* 10: 59–71.

Long, J.A., and Young, G.C. 1995. Sharks from the Middle-Late Devonian Aztec Siltstone, southern Victoria Land, Antarctica. *Records of the Western Australian Museum* 17: 287–308.

Lund, R. 1974a. *Squatinactis caudispinatus,* a new elasmobranch from the Upper Mississippian of Montana. *Annals of the Carnegie Museum of Natural History* 45: 43–55.

Lund, R. 1974b. *Stethacanthus altonensis* (Elasmobranchii) from the Bear Gulch Limestone of Montana. *Annals of the Carnegie Museum of Natural History* 45: 161–78.

Lund, R. 1977a. New information on the evolution of the bradyodont chondrichthyans. *Fieldiana Geology* 33: 521–39.

Lund, R. 1977b. A new petalodont (Chondrichthyes, Bradyodonti) from the Upper Mississippian of Montana. *Annals of the Carnegie Museum of Natural History* 46: 129–55.

Lund, R. 1977c. *Echinochimera meltoni,* new genus and species (Chimaeriformes) from the Mississippian of Montana. *Annals of the Carnegie Museum of Natural History* 46: 195–221.

Lund, R. 1980.Viviparity and interuterine feeding in a new holocephalan fish from the Lower Carboniferous of Montana. *Science* 209: 697–99.

Lund, R. 1982. *Harpagofututor volsellorhinus* new genus and species (Chondrichthyes, Chondrenchelyiformes) from the Namurian Bear Gulch Limestone, Chondrenchelys problematica Traquair (Visean), and their sexual dimorphism. *Journal of Paleontology* 56: 938–58.

Lund, R. 1983. On a dentition of *Polyrhizodus* (Chondrichthyes, Petalodontiformes) from the Namurian Bear Gulch Limestone of Montana. *Journal of Vertebrate Paleontology* 3: 1–6.

Lund, R. 1985a. Stethacanthid elasmobranch remains from the Bear Gulch Limestone (Namurian E2b) of Montana. *American Museum Novitates* 2828: 1–24.

Lund, R. 1985b. The morphology of *Falcatus falcatus* (St. John and Worthen), a Mississippian stethacanthid chondrichthyan from the Bear Gulch Limestone of Monatana. *Journal of Vertebrate Paleontology* 5: 1–19.

Lund, R. 1986a. New Mississippian holocephalan (Chondrichthyes) and the evolution of the Holocephali. Pp. 195–205 in Rusell, D.E., Santoro, J.-P., and Sigogneau-Russell, D. (eds), *Teeth revisited: Proceedings of the VIIth International Symposium on dental morphology,* Paris.

Lund, R. 1986b. On *Damocles serratus* nov. gen. et sp. (Elasmobranchii: Cladodontida) from the Upper Mississippian Bear Gulch Limestone of Montana. *Journal of Vertebrate Paleontology* 6: 12–19.

Lund, R. 1986c. The diversity and relationships of the Holocephali. Pp. 97–106 in Uyeno, T., Arai, R., Taniuchi, T., and Matsuura, K. (eds), *Indo-Pacific fish biology: Proceedings of the Second International Conference on Indo-Pacific Fishes.* Ichthyological Society of Japan.

Lund, R. 1989. New petalodonts (Chondrichthyes) from the Upper Mississippian Bear Gulch Limestone (Namurian E2b) of Montana. *Journal of Vertebrate Paleontology* 9: 350–68.

Lund, R., 1990. Chondrichthyan life history styles as revealed by the 320 million years old Mississippian of Montana. *Environmental Biology of Fishes* 27: 1–19.

Mader, H. 1986. Schuppen und Zahne von Acanthodiern und Elasmobranchiern aus dem Unter-Devon Spaniens (Pisces). *Gottinger Arbeiten zur Geologie und Paläontologie* 28: 1–59.

Mader, H., and Schultze, H.-P. 1987. Elasmobranchier-reste aus dem Unterkarbon des Rheinisches Schieferbirges und des Harzes (W-Deutchsland). *Neues Jahrbuch für Paläontologie und Geologie, Abhandlang* 175: 317–46.

Maisey, J.G. 1975. The interrelationships of the phalacanthous selachians. *Neues Jahrbuch für Geologie und Palaeontologie Ablandlungen* 9: 553–67.

Maisey, J.G. 1977a. Structural notes on a cladoselachian dorsal spine. *Neues Jahrbuch für Geologie und Palaontologie, Monaltschafte* 47–55.

Maisey, J.G. 1977b. The fossil selachian fishes *Palaeospinax* Egerton 1872, and *Nemacanthus* Agassiz 1837. *Zoological Journal of the Linnean Society* 60: 259–73.

Maisey, J.G. 1978. Preservation and prefossilisation of fossil finspines. *Neues Jahrbuch für Geologie und Palaontologie, Monaltschafte* 1978: 595–99.

Maisey, J.G. 1980. An evaluation of jaw suspension in sharks. *American Museum Novitates* 2706: 1–17.

Maisey, J.G. 1981. Studies on the Palaeozoic selachian genus *Ctenacanthus* Agassiz: I. Historical review and revised diagnosis of *Ctenacanthus,* with a list of referred taxa. *American Museum Novitates* 2718: 1–22.

Maisey, J.G. 1982. Studies on the Palaeozoic selachian genus *Ctenacanthus* Agassiz: II. *Bythiacanthus* St. John and Worthen, *Amelacanthus*, new genus, *Eunemacanthus* St. John and Worthen, Sphenacanthus Agassiz, and Wodnika Münster. *American Museum Novitates* 2722: 1–24.

Maisey, J.G. 1983. Some Pennsylvanian chondrichthyan spines from Nebraska. *Transactions of the Nebraska Academy of Sciences* 11: 81–84.

Maisey, J.G. 1984a. Studies on the Palaeozoic selachian genus *Ctenacanthus* Agassiz: III. Nominal species referred to *Ctenacanthus*. *American Museum Novitates* 2774: 1–20.

Maisey, J.G. 1984b. Chondrichthyan phylogeny: a new look at the evidence. *Journal of Vertebrate Paleontology* 4: 359–71.

Maisey, J.G. 1989a. *Hamiltonichthys mapesi* g. and sp. nov. (chondrichthyes: Elasmobranchii) from the Upper Pennsylvanian of Kansas. *American Museum Novitates* 2931: 1–42.

Maisey, J.G. 1989b. Visceral skeleton and musculature of a Late Devonian shark. *Journal of Vertebrate Paleontology* 9: 174–90.

Maisey, J.G. 2001a. A primitive chondrichthyan braincase from the Middle Devonian of Bolivia. Pp. 263–88 in Ahlberg, P.E. (ed), *Major events in early vertebrate evolution: paleontology, phylogeny, genetics, and development*. Taylor and Francis, New York.

Maisey, J.G. 2001b. CT-scan reveals new cranial features in Devonian chondrichthyan "*Cladodus*" *wildungensis*. *Journal of Vertebrate Paleontology* 21(4): 807–10.

Maisey, J.G. 2004a. Endocranial morphology in fossil and recent chondrichthyans. Pp. 139–70 in Arratia, G., Wilson, M., and Cloutier, M. (eds), *Recent advances in the origin and radiation of early vertebrates*. Verlag F. Pfeil, Munich.

Maisey, J.G. 2004b. Morphology of the braincase in the broadnose sevengill shark *Notorynchus* (Elasmobranchii, Hexanchiformes), based on CT scanning. *American Museum Novitates* 3351: 1–52.

Maisey, J.G. 2005. Braincase of the upper Devonian shark *Cladodoides wildungensis* (Chondrichthyes, Elasmobranchii), with observations on the braincase in early chondrichthyans. *Bulletin of the American Museum of Natural History* 288: 1–103.

Maisey, J.G. 2008. The fossil selachian fishes *Palaeospinax* Egerton, 1872, and *Nemacanthus* Agassiz, 1837. *Zoological Journal of the Linnean Society* 60: 259–73.

Maisey, J.G., and Anderson, M.E. 2001. A primitive chondrichthyan braincase from the Early Devonian of South Africa. *Journal of Vertebrate Paleontology* 21: 702–13.

Maisey, J.G., and de Carvalho, M.R. 1997. A new look at old sharks. *Nature* 385: 779–80.

Maisey, J., Miller, R., and Turner, S. 2009. The braincase of the chondrichthyan *Doliodus* from the Lower Devonian Campbellton Formation of New Brunswick, Canada. *Acta Zoologica* 90 (Supple.): S109–S22.

Maisey, J.G., Naylor, G.J.P., and Ward. D.J. 2004. Mesozoic elasmobranchs, neoselachian phylogeny and the rise of modern elasmobranch diversity. Pp. 17–56 in Arratia, M., and Tintori, A. (eds), *Mesozoic Fishes 3–systematics, paleoenvironments and biodiversity*. Verlag F. Pfeil, Munich.

Miller, R.F., Cloutier, R. and Turner, S. 2003. The oldest articulated chondrichthyan from the Early Devonian Period. *Nature* 425: 501–4.

Moy-Thomas, J.A. 1936. The structure and affinities of the fossil elasmobranch fishes from the Lower Carboniferous rocks of Glencartholm, Eskdale. *Proceedings of the Zoological Society of London* 1936: 761–88.

Moy-Thomas, J.A. 1939. The early evolution and relationships of the elasmobranchs. *Biological Reviews* 14: 1 –26.

Neilsen, E. 1932. Permo-Carboniferous fishes from East Greenland. *Meddelesler om Grønland* 86: 1–63.

Neilsen, E. 1952. On new or little known Edestidae from the Permian and Triassic of East Greenland. *Meddelesler om Grønland* 144: 1–55.

Newberry, J.S., and Worthen, A.H. 1870. Descriptions of fossil vertebrates. *Geological Survey of Illinois* 4: 347–74.

Obruchev, D. 1953. Studies on the edestids and the works of A.P. Karpinski. U.S.S.R. *Academy of Sciences Paleontological Journal* 45: 1–86.

Oelofsen, B. 1981. The fossil record of the Class Chondrichthyes in southern Africa. *Palaeontologica Africana* 24: 11–13.

Ørvig, T. 1967. Histologic studies of ostracoderms and fossil elasmobranchs: II. On the dermal skeleton of two late Palaeozoic elasmobranchs. *Arkivi Zoologica Kungliga Svenska Vetenskapakamiens* 19: 1–39.

Ossian, C. 1976. Redescription of *Megactenopetalus kaibabanus* David 1944 (Chondrichthyes, Petalodontidae) with comments on its geographic and stratigraphic distribution. *Journal of Paleontology* 50: 392–97.

Patterson, C. 1965. The phylogeny of the chimaeroids. *Philosophical Transactions of the Royal Society of London* 249 B: 101–219.

Patterson, C. 1968. *Menaspis* and the bradyodonts. *Nobel Symposium* 4: 171–205.

Pfeil, F.H. 1983. Zahnmorphologische Untersuchungen an rezenten und fossilen Haien der Ordnungen Chlamydoselachiformes und Echinorhiniformes. *Palaeoichthyologica* 1: 13–15.

Pradel, A., Langer, M., Maisey, J.G., Geffard-Kuriyama, D.,

Cloetens, P., Janvier, P., and Tafforeau, P. 2009. Skull and brain of a 300-million-year-old chimaerid fish revealed by synchrotron tomography. *Proceedings of the National Aacdemy of Sciences* 106: 5224–228.

Reif, W-E. 1978. Types of morphogenesis of the dermal skeleton in fossil sharks. *Paläontologische Zeitschrift* 52: 110–28.

Reif, W.-E. 1982. The evolution of the dermal skeleton and dentition in vertebrates. The odontode regulation theory. *Evolutionary Biology* 15: 287–368.

Reif, W.-E. 1985. Squamation and ecology of sharks. *Courier Forschungsinstitut Senckenberg* 78: 1–255.

Romer, A.S. 1964. The braincase of the Palaeozoic elasmobranch *Tamiobatis*. *Bulletin of the Museum of Comparative Zoology* 131: 89–106.

Schaeffer, B. 1967. Comments on elasmobranch evolution. Pp. 3–35 in Gilbert, P.W., Mathewson, R.F., and Rall, D.P. (eds), *Sharks, skates, and rays*. Johns Hopkins Press, Baltimore.

Schaeffer, B. 1981. The xenacanth shark neurocranium, with comments on elasmobranch monophyly. *Bulletin of the American Museum of Natural History* 169: 1–66.

Schaeffer, B., and Williams, M. 1977. Relationships of fossil and living elasmobranchs. American Zoologist 17: 101–9.

Schneider, J. 1988. Grundlagen der morphologie, taxonomie und biostratigraphie isolieter xenacanthodier-zähne (Elasmobranchii). *Heiberger Forschungschefte* 1988: 71–80.

Stensiö, E.A. 1937. Notes on the endocranium of a Devonian *Cladodus*. *Bulletin of the Geological Institute of Uppsala* 27: 128–44.

Stritzke, R. 1986. Xenacanthid shark teeth in Middle Devonian Limestones of the Rhenish Schiefergebirge, West Germany. *Journal of Paleontology* 60: 1134–35.

St. John, O.H., and Worthen, A.H. 1875. Descriptions of fossil fishes. *Geological Survey of Illinois* 6: 245–488.

Teichert, C. 1940. *Helicoprion* in the Permian of Western Australia. *Journal of Paleontology* 14: 140–49.

Traquair, R.H. 1884. Description of a fossil shark (*Ctenacanthus costellatus*) from the Lower Carboniferous rocks of Eskdale, Dumfriesshire. *Geological magazine* 1(3): 7–8.

Turner, S. 1982. Middle Palaeozoic elasmobranchs remains from Australia. *Journal of Vertebrate Paleontology* 2: 117–31.

Turner, S. 1983. Taxonomic note on "*Harpago*." *Journal of Vertebrate Paleontology* 3: 38.

Turner, S. 1985. Remarks on the early history of chondrichthyans, thelodonts, and some "higher elasmobranchs." *Geological Survey of New Zealand Record* 9: 93–95.

Turner, S. 1990. Lower Carboniferous shark remains from the Rockhampton district, Queensland. *Memoirs of the Queensland Museum* 28: 65–73.

Turner, S. 1993. Palaeozoic microvertebrates from eastern Gondwana. Pp. 174–207 in Long, J. (ed), *Palaeozoic vertebrate biostratigraphy and biogeography*. Belhaven Press, London.

Turner, S., and Hansen, M.C. in press. *Ageleodus pectinatus* and other Lower Carboniferous shark remains from the Narrien Range, central Queensland. *Journal of Vertebrate Paleontology*.

Turner, S., and Young, G.C., 1987. Shark teeth from the Early-Middle Devonian Cravens Peak Beds. *Alcheringa* 11: 233–44.

Williams, M.E. 1985. The "Cladodont level" sharks of the Pennsylvanian Black Shales of central North America. *Palaeontographica* 190A: 83–158.

Williams, M. 1992. Jaws, the early years. Feeding behavior in Cleveland Shale sharks. *Explorer*, The Cleveland Museum of Natural History, Ohio, summer: 4–8.

Woodward, A.S. 1891. *Catalogue of the fossil fishes in the British Museum (Natural History), Cromwell Rd, SW: II. Elasmobranchii*. Trustees, British Museum, of Natural History, London.

Young, G.C. 1982. Devonian sharks from south-eastern Australia and Antarctica. *Palaeontology* 25: 817–43.

Zangerl, R. 1973. Interrelationships of early chondrichthyans. *Journal of the Linnean Society of London* 1: S1–S14.

Zangerl, R. 1979. New chondrichthyans from the Mazon Creek fauna (Pennsylvanian) of Illinois. Pp. 449–500 in Nitecki, M. (ed), *Mazon Creek faunas*. Academic Press, New York.

Zangerl, R. 1981. Paleozoic Chondrichthyes. In Schultze, H.-P. (ed), *Handbook of paleoichthyology*, part 3A. Gustav Fischer Verlag, Stuttgart, Germany.

Zangerl, R., and Case, G.R. 1973. Iniopterygia, a new order of chondrichthyan fishes from the Pennsylvanian of North America. *Fieldiana Geology* 6: 1–67.

Zangerl, R., and Case, G.R. 1976. *Cobelodus aculeatus* (Cope), an anacanthous shark from Pennsylvanian Black Shales of North America. *Palaeontographica* 154 (A): 107–57.

Zangerl, R., and Williams, M. 1975. New evidence on the nature of the jaw suspension in Palaeozoic anacanthous sharks. *Palaeontology* 18: 333–41.

Zidek, J. 1973. Oklahoma paleoichthyology: II. Elasmobranchii (*Cladodus*, minute elements of cladoselachian derivation, *Dittodus, Petrodus*). *Oklahoma Geology Notes* 33: 87–103.

Zidek, J. 1976. Oklahoma paleoichthyology: IV. Chondrichthyes. *Oklahoma Geology Notes* 36: 175–92.

Zidek, J. 1990. Xenacanth genera: how many and how to tell them apart? Pp. 30–32 in abtracts volume of *Symposium on "New results on Permocarboniferous fauna."* Pfalzmuseum für Naturkunde, Bad Durkheim, Germany.

Chapter 6. Spiny-Jawed Fishes

Bernacsek, G.M., and Dineley, D.L. 1977. New acanthodians from the Delorme Formation (Lower Devonian) of Northwest Territories, Canada. *Palaeontographica* 158A: 1–25.

Burrow, C.J. 2004. Acanthodian fishes with dentigerous jawbones: the Ischnacanthiformes and *Acanthodopsis. Fossils and Strata* 50: 8–22.

Burrow, C.J., and Young, G.C. 1999. An articulated teleostome fish from the Late Silurian (Ludlow) of Victoria, Australia. *Records of the Western Australian Museum Supplement* 57: 1–14.

Burrow, CJ., and Young, G.C. 2004. Diplacanthid acanthodians from Aztec Siltstone (late Middle Devonian) of southern Victoria Land, Antarctica. *Fossils and Strata* 50: 23–43.

Dean, B. 1907. Notes on acanthodian sharks. *American Journal of Anatomy* 7: 209–22.

Denison, R.H. 1976. Note on the dentigerous jaw bones of Acanthodii. *Neues Jahrbuch für Geologie und Paläontologie Monatshefte* 395–99.

Denison, R.H. 1979. Acanthodii. In Schultze, H.-P. (ed). *Handbook of paleoichthyology*, part 5. Gustav Fischer Verlag, Stuttgart, Germany.

Forey, P.L., and Young, V.T. 1985. Acanthodian and coelacanth fish from the Dinantian of Foulden, Berwickshire, Scotland. *Transactions of the Royal Society of Edinburgh, Earth Sciences* 76: 53–59.

Fritsch, A. 1893. *Fauna der Gaskohle und Kalksteine der Permformation Böhemens. Band III, Heft 2: Selachii (Traquaira, Protacanthodes, Acanthodes). Actinopterygii (Megalichthys, Trissolepis)*. Selbstverlag, Prague.

Gagnier, P.-Y., and Wilson, M.V.H. 1996. Early Devonian acanthodians from Northern Canada. *Palaeontology* 39: 241–58.

Gross, W. 1940. Acanthodier und Placodermen aus Heterostius-Schichten Estlands und Lettlands. *Annales Societatis rebus naturae investigandis in Universitate Tartuensi* 46: 1–79.

Gross, W. 1947. Die Agnathan und Acanthodier des obersilurischen Beyrichienkalks. *Palaeontographica* 96A: 91–161.

Gross, W. 1971. Downtonische und Dittonische Acanthodier-Reste des Ostseegebietes. *Palaeontographica* 136A: 1–82.

Hancock, A., and Atthey, T. 1868. Notes on the remains of some reptiles and fishes from the shales of the Northumberland coal-field. *Annals of the Magazine of Natural History* 1(4): 266–78, 346–78.

Hanke, G. 2001. A revised interpretation of the anatomy and relationships of *Lupopsyrus pygmaeus* (Acanthodii, Climatiiformes?). *Journal of Vertebrate Paleontology* 21: 58A.

Hanke, G. 2002. *Paucicanthus vanelsti* gen. et sp. nov., an Early Devonian (Lochkovian) acanthodian that lacks paired fin-spines. *Canadian Journal of Earth Sciences* 39: 1071–83.

Hanke, G.F., Davis, S.P., and Wilson, M.V.H. 2001. New species of the acanthodian genus *Tetanopsyrus* from northern Canada and comments on related taxa. *Journal of Vertebrate Paleontology* 21: 740–53.

Hanke, G.F., and Wilson, M.V.H. 2004. New teleostome fishes and acanthodian systematics. Pp. 289–316 in Arratia, G., Wilson, M.V.H., and Cloutier, R. (eds), *Recent advances in the origin and early radiation of vertebrates.* Verlag F. Pfeil, Munich.

Hanke, G.F., Wilson, M.W., and Lindow, K.L.A. 2001. New species of Silurian acanthodians from the Mackenzie Mountains, Canada. *Canadian Journal of Earth Sciences* 38: 1517–29.

Heyler, D. 1969. *Vertébrés de l'Autunien de France.* Cahiers de Paléontologie, C.N.R.S., Paris.

Janvier, P. 1974. Preliminary report on Late Devonian fishes from central Iran. *Geological Survey of Iran* 31: 5–48.

Janvier, P., and Melo, J.H.G. de. 1988. Acanthodian fish remains from the Upper Silurian or Lower Devonian of the Amazon Basin, Brazil. *Palaeontology* 31: 771–77.

Janvier, P., and Melo, J.H.G. de. 1992. New acanthodian and chondrichthyan remains from the Lower and Middle Devonian of Brazil. *Neues Jahrbuch für Geologie und Palaeontologie Abhandlungen* 164: 364–92.

Jarvik, E. 1977. The systematic position of acanthodian fishes. Pp. 199–225 in Andrews, S.M., Miles, R.S., and Walker, A.D. (eds), *Problems in vertebrate evolution.* Academic Press, London.

Liu, S.-F. 1973. Some new acanthodian fossil materials from the Devonian of South China. *Vertebrata PalAsiatica* 11: 144–47.

Long, J.A. 1983. A new diplacanthoid acanthodian from the Late Devonian of Victoria. *Memoirs of the Association of Australasian Palaeontologists* 1: 51- 65.

Long, J.A. 1986a. A new Late Devonian acanthodian fish from Mt. Howitt, Victoria, Australia, with remarks on acanthodian biogeography. *Proceedings of the Royal Society of Victoria* 98: 1–17.

Long, J.A. 1986b. New ischnacanthid acanthodians from the Early Devonian of Australia, with a discussion of acanthodian interrelationships. *Zoological Journal of the Linnean Society* 87: 321–39.

Long, J.A. 1990. Fishes. Pp. 255–78 in McNamara, K.J. (ed), *Evolutionary trends.* Belhaven Press, London.

Long, J.A., Burrow, C.J., and Ritchie, A. 2004. A new Late Devonian acanthodian fish from the Hunter Formation near Grenfell, New South Wales. *Alcheringa* 28: 147–56.

Miles, R.S. 1964. A reinterpretation of the visceral skeleton of *Acanthodes. Nature* 204: 457–59.

Miles, R.S. 1965. Some features in the cranial morphology of acanthodians and the relationship of the Acanthodii. *Acta Zoologica Stockholm* 46: 233–55.

Miles, R.S. 1966. The acanthodian fishes of the Devonian Plattenkalk of the Paffrath Trough in the Rhineland with an appendix containing a classification of the Acanthodii and a revision of the genus *Homalacanthus. Arkiv für Zoologi* (Stockholm) 2nd series 18: 147–94.

Miles, R.S. 1970. Remarks on the vertebral column and caudal fin of acanthodian fishes. *Lethaia* 3: 343–62.

Miles, R.S. 1973a. Relationships of acanthodians. Pp. 63–104 in Greenwood, P.H., Miles, R.S., and Patterson, C. (eds), *Interrelationships of fishes*. Academic Press: London.

Miles, R.S. 1973b. Articulated acanthodian fishes from the Old Red Sandstone of England, with a review of the structure and evolution of the acanthodian shoulder girdle. *Bulletin of the British Museum of Natural History (Geology)* 24: 113–213.

Ørvig, T. 1967. Some new acanthodian material from the Lower Devonian of Europe. *Zoological Journal of the Linnean Society* 47: 131–53.

Ørvig, T. 1973. Acanthodian dentition and its bearing on the relationships of the group. *Palaeontographica* 143A: 119–50.

Schultze H.-P., 1982. Ein primitiver Acanthodier (Pisces) aus dem Unterdevon Lettlands. *Paläontologische Zeitschrift* 56: 95–105.

Schultze, H.-P. 1990. A new acanthodian from the Pennsylvanian of Utah, USA, and the distribution of otoliths in vertebrates. *Journal of Vertebrate Paleontology* 10: 49–58.

Valiukevicius, J. J. 1979. Acanthodian scales from the Eifelian of Spitsbergen. *Palaeontology Journal* 4: 482–92.

Valiukevicius, J. J. 1985. *Acanthodians from the Narva Regional stage of the main Devonian field* [in Russian with English summary]. Mosklas, Vilnius.

Valiukevicius, J.J. 1992. First articulated *Poracanthodes* from the Lower Devonian of Severnaya Zemlya. Pp. 193–214 in Mark-Kurik. E. (ed), *Fossil fishes as living animals*. Academy of Sciences, Estonia.

V'yushkova, L. 1992. Fish assemblages and facies in the Telegitian Suprahorizon of Salair. Pp. 281–88 in Mark-Kurik, E. (ed), *Fossil fishes as living animals*. Academy of Sciences, Estonia.

Wang, N.-Z., and Dong, Z.-Z. 1989. Discovery of Late Silurian microfossils of Agnatha and fishes from Yunnan, China. *Acta Palaeontologica Sinica* 28: 196–206.

Watson, D.M.S. 1937. The acanthodian fishes. *Philosophical Transactions of the Royal Society of London B* 228: 49–146.

Woodward, A.S. 1906. On a Carboniferous fish fauna from the Mansfield district. *Memoirs of the National Museum of Victoria* 1: 1–32.

Young, G.C. 1989. New occurrences of culmacanthid acanthodians (Pisces, Devonian) from Antarctica and southeastern Australia. *Proceedings of the Linnean Society of New South Wales* 111: 11–24.

Young, V.T. 1986. Early Devonian fish material from the Horlick Formation, Ohio Range, Antarctica. *Alcheringa* 10: 35–44.

Zajic, J. 1985. New finds of acanthodians (Acanthodii) from the Kounov Member (Stephanian B, central Bohemia). *Vestnik Ustredniho ustavu geologickeho* 60: 277–84.

Zajic, J. 1986. Stratigraphic position of finds of the acanthodians (Acanthodii) in Czechoslovakia. *Acta Universitatis Carolinae, Geologica Spinar* 2: 145–53.

Zidek, J. 1976. Kansas Hamilton Quarry (Upper Pennsylvanian) *Acanthodes*, with remarks on the previously reported North American occurrences of the genus. *The University of Kansas, Paleontological Contributions* Paper 83: 1–41.

Zidek, J. 1980. *Acanthodes lundi*, new species (Acanthodii) and associated coprolites from uppermost Mississippian Heath Formation of central Montana. *Annals of the Carnegie Museum* 49: 49–78.

Zidek J. 1981. *Machaeracanthus* Newberry (Acanthodii: Ischnacanthiformes)—morphology and systematic position. *Neues Jahrbuch für Geologie und Paläontologie Monatshefte* H12: 742–48.

Zidek, J. 1985. Growth in *Acanthodes* (Acanthodii: Pisces), data and implications. *Paläontologische Zeitschrift* 59: 147–66.

Chapter 7. An Epiphany of Evolution

Basden, A. M., and Young, G. C., 2001. A primitive actinopterygian neurocranium from the Early Devonian of southeastern Australia. *Journal of Vertebrate Paleontology* 21: 754–66.

Botella, H., Blom, H., Dorka, M., Ahlberg, P.E., and Janvier, P. 2007. Jaws and teeth of the earliest bony fishes. *Nature* 448: 583–86.

Burrow, C.J. 1995. A new lophosteiform (Osteichthyes) from the Lower Devonian of Australia. *Geobios Memoire special* 19: 327–33.

Clack, J. 2007. Devonian climate change, breathing, and the origin of the tetrapod stem group. *Integrative and Comparative Biology*. 47: 510–523.

Friedman, M., 2007. *Styloichthys* as the oldest coelacanth: implications for early osteichthyan interrelationships. *Journal of Systematic Palaeontology* 5: 289–343.

Gross, W. 1969. *Lophosteus superbus* Pander, ein Teleostome aus dem SilurOesels. *Lethaia* 2: 15–47.

Gross, W. 1971. *Lophosteus superbus* Pander: Zähne. Zahnknocken und besondere Schuppenformen. *Lethaia* 4: 131–52.

Schaeffer, B. 1968. The origin and basic radiation of the Osteichthyes. *Nobel Symposium* 4: 207–22.

Schultze, H.-P. 1968. Palaeoniscoidea-Schuppen aus dem Australiens und Kanadas und aus dem mitteldevon Spitzbergens. *Bulletin of the British Museum of Natural History (Geology)* 16(7): 343–68.

Wang, N.Z., and Dong, Z.Z. 1989. Discovery of Late Silurian microfossils of Agnatha and fishes from Yunnan, China. *Acta Paleontologica Sinica* 28: 192–206.

Zhu, M., and Schultze, H.P. 2001. Interrelationships of basal osteichthyans. Pp. 289–314 in Ahlberg, P.E. (ed), *Major events in early vertebrate evolution: paleontology, phylogeny, genetics and development.* Taylor and Francis, London.

Zhu, M., and Yu, X. 2009. Stem sarcopterygians have primitive polybasal fin articulation. *Biology Letters* 5: 372–75.

Zhu, M., Yu, X., and Janvier, P. 1999. A primitive fossil fish sheds light on the origin of bony fishes. *Nature* 397: 607–10.

Zhu, M., Zhoa, L.J., Lu, J., Qiao, T., and Qu, Q. 2009. The oldest articulated osteichthyan reveals mosaic gnathostome characters. *Nature* 458: 469–74.

Chapter 8. Primitive Ray-Finned Fishes

Aldinger, H. 1937. Permische Ganoidfishe aus Ostgronland. *Meddelesler om Grønland* 102: 1–392.

Beltan, L. 1978. Découverte d'une ichthyofaune dans le Carbonifére supérieur d'Uruguay rapports avec les faunes ichthyologiques contemporaires des autres régions du Gondwana. *Annales de la Societé Géologique du Nord* 97: 351–55.

Campbell, K.S., and Phuoc, L.D. 1983. A Late Permian actinopterygian fish from Australia. *Palaeontology* 26: 33–70.

Casier, E. 1952. Un paléoniscide du Faménnian inférieur de la Fagne: *Stereolepsis marginis,* gen. n.sp. *Bulletin de l'Institute Royale Sciences nationale de Belgique* 28: 1–10.

Casier, E. 1954. Note additionnelle relative à *"Stereolepis"* (= Osorioichthyes nov. nom) et à l'origine de l'interoperculaire. *Bulletin de l'Institute Royale Sciences nationale de Belgique* 30: 1–12.

Choo, B., Long, J.A., and Trinajstic, K. 2009. A new genus and species of basal actinopterygian fish from the Upper Devonian Gogo Formation of Western Australia. *Acta Zoologica* 90 (Supple. 1): 194–210.

Coates, M.I. 1993. New actinopterygian fish from the Namurian Manse Burn Formation of Bearsden, Scotland. *Palaeontology* 36: 123–46.

Coates, M.I. 1994. Actinopterygian and acanthodian fishes from the Visean of east Kirkton, west Lothian, Scotland. *Transactions of the Royal Society of Edinburgh* 84: 317–27.

Coates, M.I. 1995. Actinopterygians from the Namurian of Bearsden, Scotland, with comments on early actinopterygerian neurocrania. *Zoological Journal of the Linnean Society* 122: 27–60.

Dunkle, D.H. 1946. A new palaeoniscoid fish from the Lower Permian of Texas. *Journal of the Washington Academy of Sciences* 36: 402–9.

Dunkle, D., and Schaeffer, B. 1973. *Tegeolepis clarki* (Newberry), a palaeonisciform from the Upper Devonian Ohio Shale. *Palaeontographica* 143A: 151–8.

Esin, D. 1990. Species of Devonian palaeoniscoid fishes of the world. *Ichthyolith Issues* 3:14.

Friedman, M., and Brazeau, M. 2010. A reappraisal of the origin and basal radiation of the Osteichthyes. *Journal of Vertebrate Paleontology* 30: 36–56.

Gardiner, B.G. 1960. A revision of certain actinopterygian and coelacanth fishes, chiefly from the Lower Lias. *Bulletin of the British Museum (Natural History) Geology* 4: 239–384.

Gardiner, B.G. 1962. *Namaichthys schroederi* Gurich and other Palaeozoic fishes from South Africa. *Palaeontology* 5: 9–21.

Gardiner, B.G. 1963. Certain palaeoniscoid fishes and the evolution of the snout in actinopterygians. *Bulletin of the British Museum of Natural History (Geology)* 8: 258–325.

Gardiner, B.G. 1967. Further notes on palaeoniscoid fishes with a classification of the Chondrostei. *Bulletin of the British Museum of Natural History (Geology)* 14: 143–206.

Gardiner, B.G. 1969. New palaeoniscoid fish from the Witteberg Series of South Africa. *Zoological Journal of the Linnean Society* 48: 423–25.

Gardiner, B.G. 1973. Interrelationships of teleostomes. Pp. 195–35 in Greenwood, P.H., Miles, R.S., and Patterson, C. (eds), *Interrelationships of fishes.* Academic Press, London.

Gardiner, B.G. 1984. Relationships of the palaeoniscoid fishes, a review based on new specimens of *Mimia* and *Moythomasia* from the Upper Devonian of Western Australia. *Bulletin of the British Museum of Natural History (Geology)* 37: 173–428.

Gardiner, B.G. 1986. Actinopterygian fish from the Dinantian of Foulden, Berwickshire, Scotland. *Transactions of the Royal Society of Edinburgh* 76: 61–66.

Gardiner, B.G., and Bartram, A.W.H. 1977. The homologies of ventral cranial fissures in osteichthyans. Pp. 227–45 in Andrews, S.M., Miles, R.S., and Walker, A.D. (eds), *Problems in early vertebrate evolution.* Academic Press: London.

Gottfried, M.D. 1992. Functional morphology of the feeding

mechanism in a primitive palaeoniscoid-grade actinopterygian fish. Pp. 151–58 in Mark-Kurik, E. (ed) *Fossil fishes as living animals,* Academy of Sciences, Estonia.

Grande, L., and Bemis, W.E. 1998. A comprehensive phylogenetic study of amiid fishes (Amiidae) based on comparative skeletal anatomy. *Journal of Vertebrate Paleontology, Special Memoir* 4: 1–690.

Gross, W. 1953. Devonische Palaeonisciden-Reste in Mittel und Osteuropa. *Paläontologische Zeitschrifter* 27: 85–112.

Gross, W. 1968. Fraglich Actinopterygier- Schuppen aus dem Silur Gotlands. *Lethaia* 1: 184–218.

Jessen, H. 1968. *Moythomasia nitida* Gross und M. cf striata Gross, devonische Palaeonisciden aus dem Oberen Plattenkalk der Bergisch-Gladbach-Paffrather Mulde (Rheinisches Schiefergebirge). *Palaeontographica* 128A: 87–114.

Jessen, H. 1972a. Schltergürtel und Pectoralflosse bei Actinopterygiern. *Fossils and Strata* 1: 1–101.

Jessen, H. 1972b. Die Bauchschuppen von *Moythomasia nitida* Gross (Pisces, Actinopterygii). *Paläontologische Zeitschrift* 46: 121–32.

Kasantseva-Selezneva, A.A. 1974. Morpho-functional characteristics of the respiratory apparatus in the Palaeonisci. *Paleontological Journal* 4: 508–16.

Kasantseva-Selezneva, A.A. 1976a. Palaeonioscoid evolution. *Acta Biologica Yugoslavica—Ichthyologia* 8: 49–57.

Kasantseva-Selezneva, A.A. 1976b. Fulcral and keel scales in palaeoniscoids. *Paleontological Journal* 1: 124–26.

Kasantseva-Selezneva, A.A. 1977. System and phylogeny of the order Palaeonisciformes. Pp. 98–116 in Menner, V.V. (ed), *Ocherki po filogenii i sistematike iskopayemych ryb i beschelyustnych.* Akademia Nauka, Moscow.

Kasantseva-Selezneva, A.A. 1978. Difference in the jaw-opening mechanism in the higher and lower actinopterygians. *Journal of Ichthyology* 18: 78–85.

Kasantseva-Selezneva, A.A. 1979. A new palaeoniscoid fish from the Permian of the Kuznetsk Basin. *Paleontological Journal* 2: 147–50.

Kasantseva-Selezneva, A.A. 1981. Late Palaeozoic palaeoniscoids of East Kazakhstan, systematics and phylogeny. *Trudy Paleontologicheskogo Instituta Akademii Nauka* 180: 1–139.

Kasantseva-Selezneva, A.A. 1982. Phylogeny of the lower actinopterygians. *Journal of Ichthyology* 22: 1–16.

Lauder, G.V. 1980. Evolution of the feeding mechanism in the primitive actinopterygian fishes: a functional anatomical analysis of *Polypterus, Lepisosteus* and *Amia. Journal of Morphology* 163: 283–317.

Lauder, G.V., and Liem, K. V. 1983. The evolution and interrelationships of the actinopterygian fishes. *Bulletin of the Museum of Comparative Zoology* 150, 195–197.

Lehman, J.-P. 1947. Description de quelques examplaires de *Cheirolepis canadensis* (Whiteaves). *Kungliga Svenska Vetenskapakadamiens Hanlingar* 24: 5–40.

Lehman, J.-P. 1966. Actinopterygii, Dipnoi, Crossopterygii, Brachiopterygii. Pp. 4:1–387, 4:398–420 in Piveteau, J. (ed), *Traité de paléontologie.* Masson, Paris.

Long, J.A. 1988a. New palaeoniscoid fishes from the Late Devonian and Early Carboniferous of Victoria. *Memoirs of the Association of Australasian Palaeontologists* 7: 1–64.

Long, J.A., Choo, B. and Young, G.C. 2008. A new basal actinopterygian fish from the Middle Devonian Aztec Siltstone of Antarctica. *Antarctic Science* 20: 393–412.

Lowney, K.A. 1980. A revision of the Family Haplolepidae (Actinopterygii, Palaeonisciformes) from Linton, Ohio (Westphalian D, Pennsylvanian). *Journal of Paleontology* 54: 942–53.

Lund, R., and Melton, W.G., Jr. 1982. A new actinopterygian fish from the Mississippian Bear Gulch Limestone of Montana. *Palaeontology* 25: 485–98.

Martin, K. 1873. Ein Beitrag zur Kenntniss fossiler Euganoiden. *Zeitschrift Deutsche Geologi Ges* 25: 699.

Merrilees, M.J., and Crossman, E.J. 1973. Surface pits in the family Esocidae. *Journal of Morphology* 141: 307–20.

Moy-Thomas, J.A. 1937. The palaeoniscoids from the cement stones of Tarras Waterfoot, Eskdale, Dumfriesshire. *Annals of the Magazine of Natural History* 10: 345–56.

Moy-Thomas, J.A. 1942. Carboniferous palaeoniscoids from East Greenland. *Annals of the Magazine of Natural History* 11: 737–59.

Moy-Thomas, J.A., and Dyne, M.B. 1938. Actinopterygian fishes from the Lower Carboniferous of Glencartholm, Eskdale, Dumfriesshire. *Transactions of the Royal Society of Edinburgh* 59: 437–80.

Nielsen, E. 1942. Studies on the Triassic fishes from East Greenland. I. *Glaucolepis* and *Boreosomus. Meddelesler om Grønland* 138: 1–403.

Nielsen, E. 1949. Studies on the Triassic fishes from East Greenland. II. *Palaeozoologica Groenlandica* 3: 1–309.

Nybelin, O. 1976. On the so-called postspiracular bones in crossopterygians, brachiopterygians and actinopterygians. *Acta Regiae Societatis Scientiarum et Litterarum Gothoburgensis: Zoologica* 10: 5–31.

Patterson, C. 1973. Interrelationships of holosteans. Pp. 235–305 in Greenwood, P.H., Miles, R.S., and Patterson, C. (eds), *Interrelationships of fishes.* Academic Press, London.

Patterson, C. 1982. Morphology and interrelationships of primitive actinopterygian fishes. *American Zoologist* 22: 241–59.

Pearson, D.M. 1982. Primitive bony fishes with especial reference to *Cheirolepis* and palaeonisciform actinopterygians. *Zoological Journal of the Linnean Society* 74: 35–67.

Pearson, D.M., and Westoll, T.S. 1979. The Devonian actinopterygian *Cheirolepis* Agassiz. *Transactions of the Royal Society of Edinburgh* 70: 337–99.

Poplin, C. 1974. Étude de quelques paléoniscidés pennsylvaniens du Kansas. *Cahiers de Paléontologie (section vertébrés)*, Editions du C.N.R.S., Paris.

Poplin, C. 1975. Remarques sur le system arteriel epibranchial chez les actinopterygians primitifs fossils. *Colloques internationale du C.N.R.S.* 218: 265–71.

Poplin. C. 1984. *Lawrenciella schaefferi*, n.gen. n. sp. (Pisces: Actinopterygii) and the use of endocranial characters in the classification of the Palaeonisciformes. *Journal of Vertebrate Paleontology* 4: 413–421.

Poplin, C., and Heyler, D. 1993. The marginal teeth of three primitive fossil actinoipterygians. Pp. 113–24 in Heidekte, U. (ed.), *New research on Permo-Carboniferous faunas.* Pollichia-Buch, Bad Durkheim, Germany.

Rayner, D. 1951. On the cranial structure of an early palaeoniscid *Kentuckia* gen. nov. *Transactions of the Royal Society of Edinburgh* 62: 53–83.

Reed, J.W. 1992. The actinopterygian *Cheirolepis* from the Devonian of Red Hill, Nevada, and its implications for acanthodian-actinopterygian relationships. Pp. 243–50 in Mark-Kurik, E. (ed.), *Fossil fishes as living animals.* Academy of Sciences, Estonia.

Schaeffer, B. 1973. Interrelationships of chondrosteans. Pp. 207–26 in Greenwood, P.H., Miles, R.S., and Patterson, C. (eds), *Interrelationships of fishes.* Academic Press, London.

Schultze, H.-P., 1977. Ausgangform und Entwicklung der rhombischen Schuppen der Osteichthyes (Pisces). *Paläontologische Zeitschrifter* 51: 152–68.

Schultze, H.-P., 1992. Early Devonian actinopterygians (Osteichthyes, Pisces) from Siberia. Pp. 233–42 in Mark-Kurik, E. (ed.), *Fossil fishes as living animals.* Academy of Sciences, Estonia.

Swartz, B.A., 2009. Devonian actinopterygian phylogeny and evolution based on a redescription of *Stegostrachelus finlayi. Zoological Journal of the Linnean Society* 156: 750–84.

Traquair, R., 1877. On new and little known fishes from the Edinburgh district: I. *Proceedings of the Royal Society of Edinburgh* 9: 427–45.

Traquair, R. 1877–1914. The ganoid fishes of the British Carboniferous formations [monograph]. *Palaeontographical Society (London)* 31: 1–159.

Trinajstic, K. 1999a. Scales of palaeoniscoid fishes (Osteichthyes: Actinopterygii) from the Late Devonian of Western Australia. *Records of the Western Australian Museum Supplement* 57: 93–106.

Trinajstic, K. 1999b. The Late Devonian palaeoniscoid *Moythomasia durgaringa* Gardiner and Bartram 1977. *Alcheringa* 23: 9–19.

Turner, S., and Long, J.A. 1987. Lower Carboniferous palaeoniscoids (Pisces: Actinopterygii) from Queensland. *Memoirs of the Queensland Museum* 25(1): 193–200.

Veran, M. 1988. Les éléments accessoires de l'arc hyoïdien des poissons téléostomes (Acanthodiens et Osteichthyens fossiles et actuels). *Mémoires du Muséum national d'histoire naturelle* (C) 54: 13–98.

Watson, D.M.S. 1928. On some points in the structure of the palaeoniscid and allied fish. *Proceedings of the Zoological Society of London,* 49–70.

Westoll, T.S., 1944. The Haplolepidae, a new family of Late Carboniferous bony fishes. A study in taxonomy and evolution. *Bulletin du Muséum national d'histoire naturelle* 83: 1–122.

White, E.I. 1927. The fish fauna of the Cementstones of Foulden, Berwickshire. *Transactions of the Royal Society of Edinburgh* 55: 255–87.

White, E.I. 1939. A new type of palaoniscid fish, with remarks on the evolution of the actinopterygian pectoral fins. *Proceedings of the Zoological Society of London* 109: 41–61.

Woodward, A.S. 1906. On a Carboniferous fish fauna from the Mansfield district. *Memoirs of the National Museum of Victoria* 1: 1–32.

Woodward, A.S., and White, E.I. 1926. The fossil fishes of the Old Red Sandstone of the Shetland Islands. *Transactions of the Royal Society of Edinburgh* 54: 567–71.

Chapter 9. Teleosteans, the Champions

Arratia, G. 1985. Late Jurassic teleosts (Actinopterygii, Pisces) from northern Chile and Cuba. *Palaeontographica* 189A: 29–61.

Arratia, G. 1997. *Basal teleosts ands teleostean phylogeny.* Verlag F. Pfeil, Munich.

Bean, L.B. 2006. The leptolepid fish *Cavenderichthys talbragarensis* (Woodward, 1895) from the Talbragar fish bed (Late Jurassic) near Gulgong, New South Wales. *Records of the Western Australian Museum* 23: 43–76.

Bellwood, D.R. 1996. The Eocene fishes of Monte Bolca: the earliest coral reef fish assemblage. *Coral Reefs* 15: 11–19.

Chang, M.M., and Maisey, J.M. 2003. Redescription of *Ellima branneri* and *Diplomystus shengliensis,* and relationships of some basal clupeomorphs. *American Museum Novitates* 3404 : 1–35.

Friedman, M. 2008. The evolutionary origin of flatfish asymmetry. *Nature* 454 : 209–12.

Friedman, M., Shimada, K., Martin, L.D., Everhart, M.J., Liston, J., Maltese, A., and Treibold, M. 2010. 100-million-year dynasty of giant planktivorous bony fishes in the Mesozoic seas. *Science* 327: 990–93.

Hilton, E.J. 2003. Comparative osteology and phylogenetic systematics of the fossil and living bony tongued fishes (Actinopterygii, Teleostei, Osteoglossomorpha). *Zoological Journal of the Linnean Society* 137: 1–100.

Hurley, I.A., Lockridge Mueller, R., Dunn, K.A., Schmidt, E.J., Friedman, M., Ho, R.K., Prince, V.E., Yang, Z.,Thomas, M.,G., and Coates, M.I. 2007. A new time-scale for ray-finned fish evolution. *Proceedings of the Royal Society B* 274: 489–98.

Inoua, J.G., Masaki, M., Tsukamoto, K., and Nishida, M. 2004. Mitogenomic evidence for the monophyly of elopomorph fishes (Teleostei) and the evolutionary origin of the leptocephalus larva. *Molecular Phylogenetics and Evolution* 32: 274–86.

Johnson, G.D., and Patterson, C. 1996. Relationships of lower euteleostean fishes. Pp. 251–332 in Stiassni, M.L.J., Parenti, L.R., and Johnson, G.D. (eds), *Interrelationships of fishes.* Academic Press, San Diego.

Li, G.Q., and Wilson, M.H.V. 1996. Phylogeny of the Osteoglossomorpha. Pp 163–74 in Stiassney, M.L.J., Parenti, L.R., and Johnstone, G.D. (eds), *Interrelationships of fishes.* Academic Press, San Diego.

Li, G.Q., Wilson, M.H.V., and Grande, L. 1997. Review of Eohiodon (Teleostei: osteoglossomopha) from western North America, with aphylogenetic reassessment of Hiodontidae. *Journal of Paleontology* 71: 1109–24.

Maisey, J.G. 1991. *Santana fossils, an illustrated atlas.* T.F.H. Publications, Neptune City, New Jersey.

McCook, L.J, Ayling, T., Cappo, M., Choat, J. H., et al. 2010. Adaptive management of the Great Barrier Reef: a globally significant demonstration of the benefits of marine reserves. *Proceedings of the National Academy of Sciences,* doi/10.1037/pnas.0909335107.

Mehta, R.S., and Wainwright, P.C. 2007. Raptorial jaws in the throat help moray eels to swallow large prey. *Nature* 449: 79–82.

Nolf, D. 1985. Otolithi piscium. Pp. 1–145 in Schultze, H.-P. (ed), *Handbook of paleoichthyology*, vol 10. Verlag F. Pfeil, Munich.

Patterson, C., 1975. The braincase of pholidophorid and leptolepid fishes, with a review of the actinopterygian braincase. *Philosophical Transactions of the Royal Society B* 269: 275–579.

Rosen, D.E, and Patterson, C. 1977. Review of ichthyodectiform and other Mesozoic teleost fishes and the theory and practice of classifying fossils. *Bulletin of the Museum of Natural History* 158: 81–172.

Chapter 10. The Ghost Fish and Other Primeval Predators

Agassiz, J.L.R. 1843–1844. *Recherches sur les poissons fossiles.* Neuchâtel, Switzerland.

Andrews, S.M. 1973. Interrelationships of crossopterygians. Pp.137–77 in Greenwood, P.H., Miles, R.S., and Patterson, C. (eds), *Interrelationships of fishes.* Academic Press, London.

Andrews, S.M. 1977. The axial skeleton of the coelacanth, *Latimeria.* Pp. 271–88 in Andrews, S.M., Miles, R.S., and Walker, A.D. (eds), *Problems in vertebrate evolution.* Academic Press, London.

Andrews, S.M., Long, J.A., Ahlberg, P.E., Campbell, K.S.W., and Barwick, R.E. 2006. *Onychodus jandemarrai,* new species, from the Late Devonian Gogo Formation of Western Australia. *Transactions of the Royal Society of Edinburgh, Earth Sciences* 96: 197–307.

Basden, A.M., and Young, G.C., 2001. A primitive actinopterygian neurocranium from the Early Devonian of Southeastern Australia. *Journal of Vertebrate Paleontology* 21: 754–66.

Basden, A.M., Young, G.C., Coates, M.I., and Ritchie, A., 2000. The most primitive osteichthyan braincase? *Nature* 403 185–88.

Forey, P.L. 1980. *Latimeria:* a paradoxical fish. *Proceedings of the Royal Society of London* series B 208: 369–84.

Forey, P.L. 1981. The coelacanth *Rhabdoderma* in the Carboniferous of the British Isles. *Palaeontology* 24: 203–29.

Friedman, M. 2007. *Styloichthys* as the oldest coelacanth: implications for early osteichthyan interrelationships. *Journal of Systematic Paleontology* 5: 289–343.

Jessen, H. 1973. Weitere Fishrestes aus dem Oberen Plattenkalk der Bergisch-Gladbach-Paffrather Mulde (Oberdevon, Rheinisches Schiefergebirge). *Palaeontographica* 143A: 159–87.

Jessen, H. 1980. Lower Devonian Porolepiformes from the Canadian Arctic with special reference to *Powichthys thorsteinssoni* Jessen. *Palaeontographica* 167A: 180–214.

Johanson, Z., Long, J.A., Janvier, P., and Talent, J. 2006. Oldest coelacanth from the Early Devonian of Australia. *Biology Letters* 3: 443–46.

Johanson, Z., Long, J.A., Janvier, P., Talent, J., and Warren, J.W. 2007. New onychodontiform (Osteichthyes; Sarcopterygii) from the Lower Devonian of Australia. *Journal of Paleontology* 81: 1034–46.

Long, J.A. 1991. Arthrodire predation by *Onychodus* (Pisces,

Crossopterygii) from the Upper Devonian Gogo Formation, Western Australia. *Records of the Western Australian Museum* 15: 369–71.

Lu, J., and Zhu, M. 2010. An onychodont fish (Osteichthyes, Sarcopterygii) from the Early Devonian of China and the evolution of the Onychodontiformes. *Proceedings of the Royal Society of London B* 277: 293–99.

Lund, R., and Lund, W. 1985. The coelacanths from the Bear Gulch Limestone (Namurian) of Montana and the evolution of the coelacanthiformes. *Bulletin of the Carnegie Museum of Natural History* 25: 1–74.

Schultze, H.-P., and Cumbaa, S.L. 2001. *Dialipina* and the characters of basal actinopterygians. Pp. 316–32 in Ahlberg, P.E. (ed), *Major events in early vertebrate evolution: paleontology, phylogeny, genetics and development*. Taylor and Francis, London.

Stensiö, E.A. 1937. On the Devonian coelacanthids of Germany with special reference to the dermal skeleton. *Kungliga Svenska Vetenskapakadamiens Hanlingar* (3)16(4): 1–56.

Zhu, M., and Fan, J. 1995. *Youngolepis* from the Xishanchun Formation (Early Lochovian) of Qujing, China. *Geobios memoire special* 19: 293–99.

Zhu, M., and Yu, X. 2002. A primitive fish close to the common ancestor of tetrapods and lungfish. *Nature* 418: 767–70.

Zhu, M. Yu, X., and Ahlberg, P.E. 2001. A primitive sarcopterygian fish with an eyestalk. *Nature* 410: 81–84.

Zhu, M., Yu, X., Wang, W., Zhao, W., and Jia, L. 2006. A primitive fish provides key characteristics bearing on deep osteichthyan phylogeny. *Nature* 441: 77–80.

Zhu, M., Zhoa, L.J., Lu, J., Qiao, T., and Qu, Q. 2009. The oldest articulated osteichthyan reveals mosaic gnathostome characters. *Nature* 458: 469–74.

Chapter 11. Strangers in the Bite

Ahlberg, P.E. 1989. Paired fin skeletons and relationships of the fossil group Porolepiformes (Osteichthyes: Sarcopterygii). *Zoological Journal of the Linnean Society* 96: 119–66.

Ahlberg, P.E. 1991. A re-examination of sarcopterygian interrelationships, with special reference to the Porolepiformes. *Zoological Journal of the Linnean Society* 103: 241–87.

Ahlberg, P.E. 1992a. A new holoptychioid porolepiform fish from the upper Frasnian of Elgin, Scotland. *Palaeontology* 35: 813–28.

Ahlberg, P.E. 1992b. The palaeoecology and evolutionary history of porolepiform sarcopterygians. Pp. 71–90 in Mark-Kurik, E. (ed), *Fossil fishes as living animals*. Academy of Sciences, Estonia.

Ahlberg, P.E., Johanson, Z., and Daeschler, E.B. 2001. The Late Devonian lungfish *Soederberghia* (Sarcopterygii, Dipnoi) from Australia and North America, and its biogeographical implications. *Journal of Vertebrate Paleontology* 21: 1–12.

Barwick, R.E., and Campbell, K.S.W. 1996. A Late Devonian dipnoan, *Pillararhynchus,* from Gogo, Western Australia, and its relationships. *Palaeontographica* 239A: 1–42.

Bemis, W.1984. Paedomorphosis and the evolution of the Dipnoi. *Palaeobiology* 10: 293–307.

Bernacsek, G.M. 1977. A lungfish cranium from the Middle Devonian of the Yukon Territory, Canada. *Palaeontographica* 157 B: 175–200.

Campbell, K.S.W. 1965. An almost complete skull roof and palate of the dipnoan *Dipnorhynchus sussmilchi* (Etheridge). *Palaeontology* 8: 634–37.

Campbell, K.S.W., and Barwick, R.E. 1982a. A new species of the lungfish *Dipnorhynchus* from New South Wales. *Palaeontology* 25: 509–27.

Campbell, K.S.W., and Barwick, R.E. 1982b. The neurocranium of the primitive dipnoan *Dipnorhynchus sussmilchi* (Etheridge). *Journal of Vertebrate Paleontology* 2: 286–327.

Campbell, K.S.W., and Barwick, R.E. 1983. Early evolution of dipnoan dentitions and a new species *Speonesydrion*. *Memoirs of the Association of Australasian Palaeontologists* 1: 17–49.

Campbell, K.S.W., and Barwick, R.E. 1984a. The choana, maxillae, premaxillae and anterior bones of early dipnoans. *Proceedings of the Linnean Society of New South Wales* 107: 147–70.

Campbell, K.S.W., and Barwick, R.E. 1984b. Speonesydrion, an Early Devonian dipnoan with primitive toothplates. *PalaeoIchthyologica* 2: 1–48.

Campbell, K.S.W., and Barwick, R.E. 1985. An advanced massive dipnorhynchid lungfish from the Early Devonian of New South Wales. *Records of the Australian Museum* 37: 301–16.

Campbell, K.S.W., and Barwick, R.E. 1987. Palaeozoic lungfishes—a review. *Journal of Morphology* 1: S93–S131.

Campbell, K.S.W., and Barwick, R.E. 1988a. Geological and palaeontological information and phylogenetic hypotheses. *Geological magazine* 125: 207–27.

Campbell, K.S.W., and Barwick, R.E. 1988b. *Uranolophus:* a reappraisal of a primitive dipnoan. *Memoirs of the Association of Australasian Palaeontologists* 7: 87–144.

Campbell, K.S.W., and Barwick, R.E. 1990. Palaeozoic dipnoan phylogeny: functional complexes and evolution without parsimony. *Paleobiology* 16: 143–69.

Campbell, K.S.W., and Barwick, R.E. 1991. Teeth and tooth

plates in primitive lungfish and a new species of *Holodipterus*. Pp. 429–40 in Chang, M.-M., Liu, Y. H., and Zhang, G.R. (eds), *Early vertebrates and related problems of evolutionary biology*. Science Press, Beijing.

Campbell, K.S.W., and Barwick, R.E. 1995. The primitive dipnoan dental plate. *Journal of Vertebrate Paleontology* 15: 13–27.

Campbell, K.S.W., and Barwick, R.E. 1998. A new tooth-plated dipnoan from the Upper Devonian Gogo Formation and its relationships. *Memoirs of the Queensland Museum* 42: 403–37.

Campbell, K.S.W., and Barwick, R.E. 1999. Dipnoan fishes from the Late Devonian Gogo Formation of Western Australia. *Records of the Western Australian Museum Supplement* 57: 107–38.

Campbell, K.S.W., Barwick, R.E., and Pridmore, P.E. 1995. On the nomenclature of the roofing and skull bones in primitive dipnoans. *Journal of Vertebrate Paleontology* 15: 13–27.

Campbell, K.S.W., and Bell, M.W. 1982. *Soederberghia* (Dipnoi) from the Late Devonian of New South Wales. *Alcheringa* 6: 143–52.

Campbell, K.S.W., and Smith, M.M. 1987. The Devonian dipnoan *Holodipterus:* dental variation and remodelling growth mechanisms. *Records of the Australian Museum* 38: 131–67.

Chang, M.-M. 1982. The braincase of *Youngolepis,* a Lower Devonian crossopterygian from Yunnan, south-western China. Papers in the Department of Geology, University of Stockholm, 1–113.

Chang, M.-M. 1992. Head exoskeletom and shoulder girdle of *Youngolepis*. Pp. 355–78 in Chang, M.M., Liu, Y.H. and Zhang, G.R. (eds), Early vertebrates and related problems of evolutionary biology. Science Press, Beijing.

Chang, M.-M., and Yu, X.-B. 1981. A new crossopterygian, *Youngolepis precursor,* gen. et sp. nov., from the Lower Devonian of E. Yunnan, China. *Scientia Sinica* 24: 89–97.

Chang, M.-M., and Yu, X.-B. 1984. Structure and phylogenetic significance of *Diabolichthys speratus* gen. et sp. nov., a new dipnoan-like form from the Lower Devonian of eastern Yunnan, China. *Proceedings of the Linnean Society of New South Wales* 107: 171–84.

Cheng, H. 1989. On the tubuli in Devonian lungfishes. *Alcheringa* 13: 153–66.

Churcher, C.S. 1995. Giant Cretaceous lungfish *Neoceratodus tuberculatus* from a deltaic environment in the Quseir (= Baris) Formation of Kharga Oasis, western desert of Egypt. *Journal of Vertebrate Paleontology* 15: 845–49.

Clement, A.M. 2008. A new genus of lungfish from the Givetian (Middle Devonian) of central Australia. *Acta Palaeontologica Polonica* 54: 615–26.

Clement, A., and Long, J.A. 2010a. Air-breathing adaptation in a marine Devonian lungfish. *Biology Letters* doi: 10.1098/rsbl.2009.1033.

Clement, A.M., and Long, J.A. 2010b. *Xeradipterus hatcheri,* a new holodontid lungfish from the Late Devonian (Frasnian) Gogo Formation, Western Australia, and other holodontid material. *Journal of Vertebrate Paleontology* 30:681–95.

Cloutier, R. 1996. Dipnoi (Akinetia: Sarcopterygii). Pp. 198–226 in Schultze, H.P., and Cloutier, R. (eds), *Devonian fishes and plants of Miguasha, Quebec, Canada*. Verlag F. Pfeil, Munich.

Den Blaauwen, J.L., Barwick, R.E., and Campbell, K.S.W. 2005. Structure and function of the tooth plates of the Devonian lungfish *Dipterus valenciennesi* from Caithness and the Orkney Islands. *Records of the Western Australian Museum* 23: 91–113.

Denison, R.H. 1968. Early Devonian lungfishes from Wyoming, Utah and Idaho. *Fieldiana Geology* 17: 353–413.

Denison, R.H. 1974. The structure and evolution of teeth in lungfishes. *Fieldiana Geology* 33: 31–58.

Friedman, M. 2007. The interrelationships of Devonian lungfishes (Sarcopterygii: Dipnoi) as inferred from neurocranial evidence and new data from the genus *Soederberghia* Lehman, 1959. *Zoological Journal of the Linnean Society* 151: 115–71.

Friedman, M. 2008. Cranial structure in the Devonian lungfish *Soederberghia* groenlandica and its implications for the interrelationships of "rhynchodipterids." *Earth and Environmental Transactions of the Royal Society of Edinburgh* 98: 179–98.

Gorizdro-Kulczyzka, Z. 1950. Les dipneustes dévoniens du Massif de Ste.-Croix. *Acta Geologica Polonica* 1: 53–105.

Gross, W. 1956. Uber Crossopterygier und Dipnoer aus dem Baltischen Oberdevon im Zusammenhang einer vergliechenden Untersuchung des porenkanalsystems palaozoischer Agnathen und Fische. *Kungliga Svenska Vetenskapakadamiens Handlingar* 5, 6: 5–140.

Gross, W. 1964. Uber die Randzahne des Mundes, die Ethmoidalregion des Schadels und die Unterkeifersymphyse von *Dipterus oervigi* n.sp. *Paläontogische Zeitschrift* 38: 7–25.

Jaekel, O. 1927. Der Kopf der Wirbeltiere. *Ergebnisse der Anatomie und Entwicklungsgeschichte* 27: 815–974.

Jarvik, E. 1972. Middle and Upper Devonian Porolepiformes from East Greenland with special reference to *Glyptolepis groenlandica* n.sp. *Meddelesler om Grønland* 187 (2): 1–295.

Lehman, J.-P. 1959. Les dipneustes du Devonien supérieur du Groenland. *Meddelesler om Grønland* 160 (4): 1–58.

Lehman, J.-P. 1966. Dipneustes. Pp. 245–300, vol. 4, part 3, in Piveteau, J. (ed.) *Traité de paleontologie*. Masson, Paris.

Lehman, J.-P. and Westoll, T.S. 1952. A primitive dipnoan fish from the Lower Devonian of Germany. *Proceedings of the Royal Society of London B* 140: 403–21.

Long, J.A. 1990a. Fishes. Pp. 255–78 in McNamara, K.J. (ed), *Evolutionary trends.* Belhaven Press, London.

Long, J.A. 1990b. Heterochrony and the origin of tetrapods. *Lethaia* 23: 157–63.

Long, J.A. 1992a. *Gogodipterus paddyensis* (Miles), gen. nov., a new chirodipterid lungfish from the Late Devonian Gogo Formation, Western Australia. *The Beagle, Records of the Northwest Territory Museum* 9: 11–20.

Long, J.A. 1992b. Cranial anatomy of two new Late Devonian lungfishes (Pisces: Dipnoi) from Mt. Howitt, Victoria. *Records of the Australian Museum* 44: 299–318.

Long, J.A. 1993. Cranial ribs in Devonian lungfish and the origin of dipnoan air-breathing. *Memoirs of the Association of Australasian Palaeontologists* 15: 199–209.

Long, J.A. 2006. *Swimming in stone—the amazing Gogo fossils of the Kimberley.* Fremantle Arts Centre Press, Fremantle, Western Australia.

Long, J.A., Barwick, R.E., and Campbell, K.S.W. 1997. Osteology and functional morphology of the osteolepiform fish, *Gogonasus andrewsae* Long, 1985, from the Upper Devonian Gogo Formation, Western Australia. *Records of the Western Australian Museum Supplement* 53: 1–93.

Long, J.A., and Campbell, K.S.W. 1985. A new lungfish from the Early Carboniferous of Victoria. *Proceedings of the Royal Society of Victoria* 97: 87–93.

Long, J.A., and Clement, A. 2009. The postcranial anatomy of two Middle Devonian lungfishes (Osteichthyes, Dipnoi) from Mt. Howitt, Victoria, Australia. *Memoirs of Museum Victoria* 66: 189–202.

Miles, R.S. 1977. Dipnoan (lungfish) skulls and the relationships of the group: a study based on new species from the Devonian of Australia. *Zoological Journal of the Linnean Society of London* 61: 1–328.

McKinney, M.L., and McNamara, K.J. 1991. *Heterochrony—the evolution of ontogeny.* Plenum Press, New York.

Moy-Thomas, J., and Miles, R.S. 1971. *Palaeozoic fishes.* 2nd ed. Chapman and Hall, London.

Ørvig, T., 1961. New finds of acanthodians, arthrodires, crossopterygians, ganoids and dipnoans in the upper Middle Devonian calcareous flags (Oberer Plattenkalk) of the Bergisch Gladbach Paffrath Trough. 2. *Paläontologische Zeitschrift* 35: 10–27.

Pander, C.R. 1858. Uber die Ctenodipterinen des devonischen Systems. St. Petersberg.

Pridmore, P.A., and Barwick, R.E. 1993. Post-cranial morphologies of the Late Devonian dipnoans *Griphognathus* and *Chirodipterus* and locomotor implications. *Memoirs of the Association of Australasian Palaeontologists* 15: 161–82.

Pridmore, P.A., Campbell, K.S.W., and Barwick, R.E. 1994. Morphology and phylogenetic position of the holodipteran dipnoans of the Upper Devonian Gogo Formation of northwestern Australia. *Philosophical Transactions of the Royal Society London B.* 344: 105–64.

Save-Soderbergh, G. 1937. On *Rhynchodipterus elginensis* n.g., n.sp., representing a new group of dipnoan-like Choanata from the Upper Devonian of East Greenland and Scotland. *Arkiv für Zoologi* 29: 1–8.

Save-Soderbergh, G. 1952. On the skull of *Chirodipterus wildungensis,* an Upper Devonian dipnoan from Wildungen. *Kungliga Svenska Vetenskapakadamiens Hanlingar* 3(4): 1–29.

Schultze, H.-P. 1969. *Griphognathus* Gross, ein langschnauziger Dipnoer aus dem Overdevon von Bergisch-Gladbach (Rheinisches Schiefergebirge) und von Lettland. *Geologica et Palaeontologica* 3: 21–79.

Schultze, H.-P. 1975. Das Axialskelett der Dipnoer aus dem Oberdevon von Bergisch-Gladbach (westdeutschland). *Colloques internationale du C.N.R.S.* 218: 149–57.

Schultze, H.-P. 1992. A new long-headed dipnoan (Osteichthyes) from the Middle Devonian of Iowa, USA. *Journal of Vertebrate Paleontology* 12: 42–58.

Schultze, H.-P. 2008. A Porolepiform Rhipidistian from the Lower Devonian of the Canadian Arctic. *Fossil Record* 3: 99–109.

Schultze, H.-P., and Campbell, K.S.W. 1987. Characterisation of the Dipnoi, a monophyletic group. *Journal of Morphology* 1(Supple.): 25–37.

Schultze, H.-P., and Marshall, C.R. 1993. Contrasting the use of functional complexes and isolated characters in lungfish evolution. *Memoirs of the Association of Australasian Palaeontologists* 15: 211–24.

Smith, M.M. 1977. The microstructure of the dentition and dermal ornament of three dipnoans from the Devonian of Western Australia: a contribution towards dipnoan interrelationships, and morphogenesis, growth and adaptation of skeletal tissues. *Philosophical Transactions of the Royal Society London* 281 B: 29–72.

Smith M.M. 1979. SEM of the enamel layer in oral teeth of fossil and extant crossopterygian and dipnoan fishes. *Scaning Electron Microscopy* 2: 483–90.

Smith, M.M. 1984. Petrodentine in extant and fossil dipnoan dentitions: microstructure, histogenesis and growth. *Proceedings of the Linnean Society of New South Wales* 107: 367–407.

Smith, M.M. 1992. Microstructure of enamel in the tusk teeth of *Youngolepis* compared with enamel in crossopterygians teeth and with a youngolepid-like tooth from the Lower Devonian of Vietnam. Pp. 341–53 in Chang, M.M., Liu, Y.H. and Zhang, G.R. (eds), *Early vertebrates and related problems of evolutionary biology.* Science Press, Beijing.

Smith, M.M., and Campbell, K.S.W. 1987. Comparative morphology, histology and growth of dental plates of the Devonian dipnoan *Chirodipterus. Philosophical Transactions of the Royal Society of London* 317: 329–63.

Song, C.-Q., and Chang, M.-M. 1992. Discovery of *Chirodipterus* (Dipnoi) from Lower Upper Devonian of Hunan, South China. Pp. 465–76 in Chang, M.M., Liu, Y.H., and Zhang, G.R. (eds), *Early vertebrates and related problems of evolutionary biology,* Science Press, Beijing.

Thomson, K.S., and Campbell, K.S.W. 1971. The structure and relationship of the primitive Devonian lungfish *Dipnorhynchus sussmilchi* (Ethridge). *Bulletin of the Peabody Museum of Natural History* 38: 1–109.

Wang, S.-T., Drapala, V., Barwick, R.E., and Campbell, K.S.W. 1993. The dipnoan species, *Sorbitorhynchus deleaskitus,*from the Lower Devonian of Guangxi, China. *Philosophical Transactions of the Royal Society of London,* 340 B: 1–24.

White, E., and Moy-Thomas, J.A. 1940. Notes on the nomenclature of fossil fishes: II. Homonyms. D.-L. *Annals of the Magazine of Natural History London* 6: 98–103.

Young, G.C., Barwick, R.E., and Campbell, K.S.W. 1989. Pelvic girdles of lungfishes (Dipnoi). Pp. 59–83 in Le Maitre, R.W. (ed), *Pathways in geology, essays in honour of Edwin Sherbon Hills.* Blackwell Scientific Publications, Carlton, Victoria.

Zhang, M., and Yu, X. 1981, A new crossopterygian, *Youngolepis praecursor* gen. et sp. nov. from Lower Devonian of East Yunnan, Yunnan. *Scientia Sinica* 24: 89–97.

Chapter 12. Big Teeth, Strong Fins

Ahlberg, P.E., and Johanson, Z. 1997. Second tristichopterid (Sarcopterygii, Osteolepiformes) from the Upper Devonian of Canowindra, New South Wales, Australia, and phylogeny of the Tristichopteridae. *Journal of Vertebrate Paleontology* 17: 653–73.

Ahlberg, P.E., and Johanson, Z. 1998. Osteolepiforms and the ancestry of tetrapods. *Nature* 395: 792–94.

Andrews, S.M. 1985. Rhizodont crossopterygian fish from the Dinantian of Foulden, Berwickshire, Scotland, with a re-evaluation of this group. *Transactions of the Royal Society of Edinburgh (Earth Sciences)* 76: 67–95.

Andrews, S.M., and Westoll, T.S. 1970a. The postcranial skeleton of *Eusthenopteron foordi* Whiteaves. *Transactions of the Royal Society of Edinburgh* 68: 207–329.

Andrews, S.M., and Westoll, T.S. 1970b. The postcranial skeleton of rhipidistian fishes excluding *Eusthenopteron. Transactions of the Royal Society of Edinburgh* 68: 391–489.

Holland T., and Long, J.A. 2009. On the phylogenetic position of *Gogonasus andrewsae* Long 1985, within the Tetrapodamorpha. *Acta Zoologica* 90 (Supple 1.): 285–96.

Holland T., Long, J.A., and Snitting, D. 2010. New information on the enigmatic tetrapodomorph fish *Marsdenichthys longioccipitus* (Long, 1985). *Journal of Vertebrate Paleontology* 30: 68–77.

Holland, T., Warren, A., Johanson, Z., Long, J.A, Parker, K., and Garvey, G. 2007. A new species of *Barameda* (Rhizodontida) and heterochrony in the rhizodontid pectoral fin. *Journal of Vertebrate Paleontology* 27: 295–315.

Janvier, P. 1980. Osteolepid remains from the Devonian of the Middle East, with particular reference to the endoskeletal shoulder girdle. Pp. 223–54 in Panchen, A.L. (ed), *The terrestrial environments and the origin of land vertebrates.* Systematics Assocation, Academic Press, London.

Janvier, P., Termier, G., and Termier, H. 1979. The osteolepiform rhipidistian fish *Megalichthys* in the Lower Carboniferous of Morocco, with remarks on the paleobiogeography of the Upper Devonian and Permo-Carboniferous osteolepidids. *Neues Jahrbuch für Geologie und Palaeontologie Monatshefte* 7–14.

Jarvik, E. 1944. On the dermal bones, sensory canals and pit-lines of the skull in *Eusthenopteron foordi* Whiteaves, with some remarks on *E. save-soderberghi* Jarvik. *Kungliga Svenska Vetenskapakadamiens Hanlingar* 3(21): 1–48.

Jarvik, E. 1948. On the morphology and taxonomy of the Middle Devonian osteolepid fishes of Scotland. *Kungliga Svenska Vetenskapakadamiens Hanlingar* 3(25): 1–301.

Jarvik, E. 1950a. On some osteolepiform crossopterygians from the Upper Old Red Sandstone of Scotland. *Kungliga Svenska Vetenskapakadamiens Hanlingar* 2: 1–35.

Jarvik, E. 1950b. Middle Devonian vertebrates from Canning Land and Wegeners Halvö (East Greenland): II. Crossopterygii. *Meddelesler om Grønland* 96 (4): 1–132.

Jarvik, E. 1952. On the fish-like tail in the ichthyostegid stegocephalians with descriptions of a new stegocephalian and a new crossopterygian from the Upper Devonian of East Greenland. *Meddelesler om Grønland* 114: 1–90.

Jarvik, E., 1954. On the visceral skeleton in *Eusthenopteron* with a discussion of the parasphenoid and palatoquadrate in fishes. *Kungliga Svenska Vetenskapakadamiens Hanlingar* (4) 5: 1–104.

Jarvik, E. 1960. *Théories de l'évolution des vertébrés reconsidérée la lumière des récentes découvertes sur les vertébrés inférieurs.* Masson and Cie, Paris.

Jarvik, E. 1963. The composition of the intermandibular division of the head in fish and tetrapods and the diphyletic origin of the tetrapod tongue. *Kungliga Svenska Vetenskapakadamiens Hanlingar* 4(9): 1–74.

Jarvik, E., 1964. Specializations in early vertebrates. *Annales de la Société Royale Zoologique de Belgique* 94: (1):11–95.

Jarvik, E., 1966. Remarks on the structure of the snout in *Megalichthys* and certain other rhipidistid crossopterygians. *Arkiv für Zoologi* 19: 41–98.

Jarvik, E. 1980. *Basic structure and evolution of vertebrates,* vols. 1 and 2. Academic Press, London.

Jarvik, E. 1985. Devonian osteolepiform fishes from East Greenland. *Meddelesler om Grønland* 13: 1–52.

Johanson, Z., and Ahlberg, P.E. 1997. A new tristichopterid (Osteolepiformes: sarcopterygii) from the Mandagery Sandstone (Late Devonian) near Canowindra, New South Wales. *Transactions of the Royal Society of Edinburgh: Earth Sciences* 88: 39–68.

Johanson, Z., and Ahlberg, P.E. 1998. A complete primitive rhizodontid from Australia. *Nature* 349: 569–73.

Johanson, Z., and Ahlberg, P.E. 2001. Devonian rhizodontids (Sarcopterygii; Tetrapodomorpha) from East Gondwana. *Transactions of the Royal Society of Edinburgh: Earth Sciences* 92: 43–74.

Long, J.A. 1985a. The structure and relationships of a new osteolepiform fish from the Late Devonian of Victoria, Australia. *Alcheringa* 9: 1–22.

Long, J.A. 1985b. A new osteolepidid fish from the Upper Devonian Gogo Formation, Western Australia. *Records of the Western Australian Museum* 12: 361–77.

Long, J.A. 1985c. New information on the head and shoulder girdle of *Canowindra grossi* Thomson, from the upper Devonian Mandagery sandstone, New South Wales. *Records of the Australian Museum* 37: 91–99.

Long, J.A. 1987. An unusual osteolepiform fish from the Late Devonian of Victoria, Australia. *Palaeontology* 30: 839–52.

Long, J.A. 1988. Late Devonian fishes from the Gogo Formation, Western Australia. *National Geographic Research* 4: 436–50.

Long, J.A. 1989. A new rhizodontiform fish from the Early Carboniferous of Victoria, Australia, with remarks on the phylogentic position of the group. *Journal of Vertebrate Paleontology* 9: 1–17.

Long, J.A., Campbell, K.S.W., and Barwick, R.E. 1996–97. Osteology and functional morphology of the osteolepiform fish *Gogonasus andrewsae* Long, 1985, from the Upper Devonian Gogo Formation, Western Australia. *Records of the Western Australian Museum Supplement* 53: 1–90.

Long, J.A., and Gordon, M. 2004. The greatest step in vertebrate history: a paleobiological review of the fish-tetrapod transition. *Physiological and Biochemical Zoology* 77: 700–19.

Long, J.A., Young, G.C., Holland, T., Senden, T.J., and Fitzgerald, E.M.C. 2006. An exceptional Devonian fish from Australia sheds light on tetrapod origins. *Nature* 444: 199–202.

Rosen, D.E., Forey, P.L., Gardiner, B.G., and Patterson, C. 1981. Lungfishes, tetrapods, palaeontology and plesiomorphy. *Bulletin of the American Museum of Natural History* 167: 159–276.

Schultze, H.-P. 1984. Juvenile specimens of *Eusthenopteron foordi* Whiteaves, 1881 (osteolepiform rhipidistian, Pisces), from the Upper Devonian of Miguashua, Quebec, Canada. *Journal of Vertebrate Paleontology* 4: 1–16.

Vorobyeva, E. 1962. Rhizodont crossopterygians from the Devonian main field of the USSR. *Trudy Palaontological Institute* 104: 1–108.

Vorobyeva, E. 1975. Formenvielfalt und Verwandtschaftsbeiziehungen der osteolepidida (crossopterygier, Pisces). *Paläontologische Zeitschrift* 49: 44–54.

Vorobyeva, E. 1977. Morphology and nature of evolution of crossopterygian fish. *Trudy Palaeontological Institute* 163: 1–239.

Vorobyeva, E. 1980. Observations on two rhipidistian fishes from the Upper Devonian of Lode, Latvia. *Zoological Journal of the Linnean Society* 70: 191–201.

Vorobyeva, E., and Obrucheva, D. 1964. Subclass Sarcopterygii. Pp. 420–98 in Orlov, I.A. (ed), *Fundamentals of palaeontology: vol. 11. Agnatha and Pisces.* I.T.P.P., Jerusalem.

Vorobyeva, E., and Obrucheva, H. D. 1977. Rhizodont crossopterygian fishes (family Rhizodontidae) from the Middle Palaeozoic deposits of the Asiatic part of the USSR. Pp. 89–97, 162–63 in Menner, V.V. (ed), *Ocherki po filogenii i sistematike iskopaemykh ryb i beschelyustnick.* Akademia Nauka, Moscow.

Young, G.C., Long, J.A., and Ritchie, A. 1992. Crossopterygian fishes from the Devonian of Antarctica: Systematics, relationships and biogeographic significance. *Records of the Australian Museum Supplement* 14: 1–77.

Chapter 13. The Greatest Step in Evolution

Ahlberg, P.E. 1991. Tetrapod or near tetrapod fossils from the Upper Devonian of Scotland. *Nature* 354: 298–301.

Ahlberg, P.E. 1995. *Elginerpeton pancheni* and the earliest tetrapod clade. *Nature* 373: 420–25.

Ahlberg. P.E. 2008. *Ventastega curonica* and the origin of tetrapod morphology. *Nature* 453: 1199–1204.

Ahlberg, P.E, Clack, J.A., and Luksevics, E. 1996. Rapid braincase evolution between *Panderichthys* and the earliest tetrapods. *Nature* 381: 61–64.

Ahlberg, P.E., Luksevics, E., and Lededev, O. 1994. The first tetrapod finds from the Devonian (Upper Famennian) of Latvia. *Philosophical Transactions of the Royal Society of London* B343: 303–28.

Ahlberg, P., and Milner, A.R. 1994. The origin and early diversification of tetrapods. *Nature* 368: 507–12.

Boisvert, C.A. 2005. The pelvic fin and girdle of *Panderichthys* and the origin of tetrapod locomotion. *Nature* 438: 1145–47.

Boisvert, C., and Ahlberg, P.E. 2008. The pectoral fin of *Panderichthys* and the origin of digits. *Nature* 456: 633–36.

Brazeau, M., and Ahlberg, P.E. 2006. Tetrapod-like middle ear architecture in a Devonian fish. *Nature* 439: 318–21.

Callier, V., Clack, J.A., and Ahlberg, P.E. 2009. Contrasting developmental trajectories in the earliest known tetrapod limbs. *Science* 324: 364–67.

Campbell, K.S.W., and Bell, M.W. 1977. A primitive amphibian from the Late Devonian of New South Wales. *Alcheringa* 1: 369–81.

Clack, J.A. 1988. New material of the early tetrapod *Acanthostega* from the Upper Devonian of East Greenland. *Palaeontology* 31: 699–724.

Clack, J.A. 1989. Discovery of the earliest tetrapod stapes. *Nature* 342: 425–30.

Clack, J.A. 2001. *Eucritta melanolimnetes* from the Early Carboniferous of Scotland, a stem tetrapod showing a mosaic of characteristics. *Transactions of the Royal Society of Edinburgh, Earth Sciences* 92: 72–95.

Clack, J.A. 2002. *Gaining ground—The origin and evolution of tetrapods.* Indiana University Press, Bloomington.

Clack, J.A. 2006. The emergence of early tetrapods. *Palaeogeography, Palaoeclimatology, Paleoecology,* 232: 167–89.

Clack, J.A., and Finney, S.M. 2005. *Pederpes finneyae,* an articulated tetrapod from the Tournaisian of western Scotland. *Journal of Systematic Palaeontology* 2: 311–46.

Coates, M.I. 1996. The Devonian tetrapod *Acanthostega gunnari* Jarvik: postcranial anatomy, basal tetrapod interrelationships and patterns of skeletal development. *Transactions of the Royal Society of Edinburgh, Earth Sciences* 87: 363–421.

Coates, M.I., and Clack, J.A. 1990. Polydactyly in the earliest known tetrapod limbs. *Nature* 347: 66–69.

Coates, M.I., and Clack, J.A. 1991. Fish-like gills and breathing in the earliest known tetrapod. *Nature* 352: 234–36.

Coates, M.I., Ruta, M., and Friedman, M. 2008. Ever since Owen: changing perspectives on the early evolution of tetrapods. *Annual Reviews of Ecology, Evolution and Systematics* 39: 571–92.

Daeschler, E.B., Shubin, N.H., Thomson, K.S., and Amaral, W.W. 1994. A Devonian tetrapod from North America. *Science* 265: 639–42.

Daeschler, E.B., Shubin, N.H., and Jenkins, F.A. 2006. A Devonian tetrapod-like fish and the evolution of the tetrapod body plan. *Nature* 440: 757–63.

Jarvik, E. 1996. The Devonian tetrapod *Ichthyostega. Fossils and Strata* 40: 1–213.

Lebedev, O.A., and Coates, M.I. 1995. The postcranial skeleton of the Devonian tetrapod *Tulerpeton curtum* Lebedev. *Zoological Journal of the Linnean Society* 114: 307–48.

Long, J.A. 1990. Heterochrony and the origin of tetrapods. *Lethaia* 23: 157–66.

Long, J.A., and Gordon, M. 2004.The greatest step in vertebrate history: a paleobiological review of the fish-tetrapod transition. *Physiological and Biochemical Zoology* 77: 700–19.

Long. J.A., and Holland, T. 2008. A possible elpistostegalid fish from the Middle Devonian of Victoria. *Proceedings of the Royal Society of Victoria* 120: 186–93.

Niedźwiedzki, G., Szrek, P., Narkiewicz, K., Narkiewicz, M., and Ahlberg, P.E. 2010. Tetrapod pathways from the early Middle Devonian period in Poland. *Nature* 463: 43–48.

Panchen, A.L. 1967. The nostrils of choanate fishes and early tetrapods. *Biological Reviews of the Cambridge Philosophical Society* 42: 374–420.

Panchen, A.L. 1977. Geographical and ecological distribution of the earliest tetrapods. Pp 723–728 in Hecht, M.K. (ed), *Major patterns in vertebrate evolution.* Plenum Press, New York.

Panchen, A.L., and Smithson, T.R. 1987. Character diagnosis, fossils and the origin of the tetrapods. *Biological Review* 62: 341–438.

Schultze, H.-P., and Arsenault, M. 1985. The panderichthyid fish *Elpistostege:* a close relative of tetrapods? *Palaeontology* 28: 293–310.

Shubin, N.H., Daeschler, E.B., and Jenkins, F.A., Jr. 2006. The pectoral fin of *Tiktaalik* and the origin of the tetrapod limb. *Nature* 440: 764–71.

Thulborn, T., Warren, A., Turner, S., and Hamley, T. 1996. Early Carboniferous tetrapods in Australia. *Nature* 381: 777–80.

Warren, A., Jupp, R., and Bolton, B. 1986. Earliest tetrapod trackway. *Alcheringa* 10: 183–86.

Warren, A., and Turner, S. 2004. The first stem tetrapod from the Early Carboniferous of Gondwana. *Palaeontology* 47: 151–84.

Warren, J.W., and Wakefield, N. 1972. Trackways of tetrapod vertebrates from the Upper Devonian of Victoria, Australia. *Nature* 228: 469–70.

Index

Italic numerals indicate an illustration.